T0306027

BASIC AERODYNAMICS

In the rapidly advancing field of flight aerodynamics, it is important for students to completely master the fundamentals. This textbook, written by renowned experts, clearly presents the basic concepts underlying aerodynamic prediction methodology. These concepts are closely linked to physical principles so that they may be more readily retained and their limits of applicability fully appreciated. The ultimate goal is to provide the student with the necessary tools to confidently approach and solve practical flight-vehicle design problems of current and future interest. The text is designed for use in a course in aerodynamics at the advanced undergraduate or graduate level. A comprehensive set of exercise problems is included at the end of each chapter.

Gary A. Flandro is the Boling Chair (Emeritus) of Mechanical and Aerospace Engineering at the University of Tennessee Space Institute. He is also Vice President and Chief Engineer of Gloyer-Taylor Laboratories, LLC. He is a Fellow of the American Institute of Aeronautics and Astronautics. His research interests include acoustics, aerodynamics, rocket propulsion, flight mechanics and performance, hypersonic aerodynamics, propulsion, and vehicle design. Dr. Flandro received the National Aeronautics and Space Administration Exceptional Achievement Medal (1998) and the Fédération Aéronautique Internationale Diamond Soaring Badge (1979) for his work.

Dr. Howard McMahon was a Professor Emeritus of Aerospace Engineering at the Georgia Institute of Technology. Following his graduate work at Santa Clara and doctoral research at the California Institute of Technology, he worked for CARDE (the Canadian Armament and Research Development Establishment) near Quebec City, with top rocketry researchers in Canada, including Gerald Bull. His desire to teach and expand his research work led him to the Georgia Institute of Technology, where he guided undergraduate and graduate students for 26 years. His particular area of focus was in aerodynamics, and he spent many years guiding research in the university wind tunnel with projects involving rotary aviation, compressible flow, and fluid mechanics. Following his retirement from the university in 1990, he continued to collaborate with both teaching and research faculty until his death in 2008.

Robert L. Roach currently teaches mathematics and science at the Kfar Hayarok School in Ramat Hasharon, Israel. He was formerly an Assistant Professor of Aerospace Engineering at the University of Tennessee Space Institute. He also taught aerodynamics, rocket propulsion, and mathematics at the Georgia Institute of Technology and at the U. S. Air Force Institute of Technology. He is an Associate Fellow of the American Institute of Aeronautics and Astronautics. His research interests include numerical solutions of the canonical equations of engineering, propulsion, and many aspects of solar energy.

Cambridge Aerospace Series

Editors: Wei Shyy and Michael J. Rycroft

Basic Aerodynamics

Incompressible Flow

Gary A. Flandro
University of Tennessee Space Institute

Howard M. McMahon
Georgia Institute of Technology

Robert L. Roach
formerly University of Tennessee Space Institute

CAMBRIDGE
UNIVERSITY PRESS

University Printing House, Cambridge CB2 8BS, United Kingdom

One Liberty Plaza, 20th Floor, New York, NY 10006, USA

477 Williamstown Road, Port Melbourne, VIC 3207, Australia

314-321, 3rd Floor, Plot 3, Splendor Forum, Jasola District Centre, New Delhi - 110025, India

79 Anson Road, #06-04/06, Singapore 079906

Cambridge University Press is part of the University of Cambridge.

It furthers the University's mission by disseminating knowledge in the pursuit of education, learning and research at the highest international levels of excellence.

www.cambridge.org
Information on this title: www.cambridge.org/9780521805827

© Cambridge University Press 2012

First published 2012

A catalogue record for this publication is available from the British Library

Library of Congress Cataloging in Publication data
Flandro, G. A. (Gary A.), 1934–
Basic aerodynamics : incompressible flow / Gary A. Flandro, Howard M. McMahon, Robert L. Roach.
    p.   cm. – (Cambridge Aerospace Series)
Includes bibliographical references and index.
ISBN 978-0-521-80582-7 (hardback)
1. Air flow – Textbooks.   2. Aerodynamics – Textbooks.   I. McMahon, Howard M. (Howard Martin), 1927–   II. Roach, Robert L.   III. Title.
IV. Series.
TL574.F5F53   2011
629.132'3–dc23   2011022458

ISBN 978-0-521-80582-7 Hardback

For Howard

Dr. Howard McMahon spent 26 years of his career as a professor and researcher at the Georgia Institute of Technology in Atlanta. Having been born just a few weeks after Charles Lindbergh's flight across the Atlantic, he moved through the age of biplanes to the modern era of jets and rockets, working first as a researcher and followed by his years as a teaching professor. His retirement from education allowed him to focus his energy on the completion of this textbook in collaboration with Dr. Gary Flandro and former student Dr. Bob Roach. Though he did not live to see the benefits of this book for aerospace students, we, his wife and children, are happy and proud that his work will be recognized.

Dr. McMahon's tenure at Georgia Tech occurred during a time of rapid change within the university. His work in the classroom and as a departmental colleague and leader helped to define the Aerospace Engineering department's identity and solidify the school's recognition for excellence and high standards. Dr. McMahon's strengths were his attention to meticulous detail and understanding of people. He directed the undergraduate fluid dynamics laboratories and guided students through cooperative work-study programs, placing them in research and laboratory settings throughout the country.

Whether at the university or at home, Dr. McMahon's office was an obvious destination for anyone who needed advice on how to handle a difficult problem. He blended a knowledge and love of people with an ability to always make the complex seem simple. He shared his talents with much energy and passion during his tenure at the university, having taught more than 4000 aerospace engineers. We know that through the gift of this textbook many more people will learn from his wisdom and insights. As family members, we were fortunate to have that experience throughout our own lives.

With loving memory,

The McMahon Family
June 2011

Contents

Preface

This textbook presents the fundamentals of aerodynamic analysis. Major emphasis is on inviscid flows whenever this simplification is appropriate, but viscous effects also are discussed in more detail than is usually found in a textbook at this level. There is continual attention to practical applications of the material. For example, the concluding chapter demonstrates how aerodynamic analysis can be used to predict and improve the performance of flight vehicles. The material is suitable for a semester course on aerodynamics or fluid mechanics at the junior/senior undergraduate level and for first-year graduate students. It is assumed that the student has a sound background in calculus, vector analysis, mechanics, and basic thermodynamics and physics. Access to a digital computer is required and an understanding of computer programming is desirable but not necessary. Computational methods are introduced as required to solve complex problems that cannot be handled analytically.

The objective of this textbook is to present in a clear and orderly manner the basic concepts underlying aerodynamic-prediction methodology. The ultimate goal is for the student to be able to use confidently various solution methods in the analysis of practical problems of current and future interest. Today, it is important for the student to master the basics because technology is advancing at such a rate that a more directed or specific approach often is rapidly outdated. In this book, the basic concepts are linked closely to physical principles so that they may be understood and retained and the limits of applicability of the concepts can be appreciated. Numerous example problems stress solution methods and the order of magnitude of key parameters. A comprehensive set of problems for home study is included at the end of each chapter.

Physical insights are developed primarily by constructing analytical solutions to important aerodynamics problems. In doing this, we follow the example set by Theodore von Kàrmàn and we subscribe to Dr. Küchemann's concept of "ingenious abstractions and approximations." However, after graduation, the student in the workplace will encounter many numerical-analysis techniques and solutions. Thus, the textbook introduces the fundamentals of modern numerical methods (as they are used in aerodynamics and fluid mechanics) as well. Physical understanding plays a valuable role in computational analysis because it provides an important check on the expected ranges of magnitude of numerical solutions that are generated by these techniques.

A feature of this textbook is a companion Web site (www.cambridge.org/flandro) that contains numerical-analysis codes of three types: (1) codes for performing routine algebraic calculations for evaluating atmospheric properties or compressible flow properties, which are often found only in tables or charts; (2) menu-driven codes that allow the student to observe the effects of parametric variations on solutions that are developed in the text; and (3) numerical-analysis codes for complex flow problems. The latter codes arise when solving linear problems using panel methods or nonlinear problems using finite-difference methods. Sample applications of these codes are presented as needed to illustrate their use in addressing realistic aerodynamics problems.

The authors express their gratitude to members of the aerodynamics faculties at Georgia Institute of Technology (GIT) and the University of Tennessee Space Institute (UTSI) for many helpful discussions during the writing of this textbook. In addition, Professors Jagoda (GIT) and Collins (UTSI) kindly used draft copies of certain chapters in their classes to provide valuable feedback. We are indebted to Professors Harper and Hubbartt of GIT for allowing the use of materials developed in their classroom notes. Finally, the first two authors thank their teachers and research advisors for insight into the inner workings of fluid mechanics and aerodynamics attained during their graduate studies in aeronautics at the California Institute of Technology. Giants such as Clark Millikan, Hans Liepmann, Lester Lees, Frank Marble, and Anatol Roshko deserve special mention for their influence on our understanding of this subject.

The three authors of this book represent more than 90 years of teaching and practical experience in aeronautics and associated disciplines. We wish you success in your study of aerodynamics and hope that it is as fulfilling to you as it has been to us.

Gary A. Flandro
Howard M. McMahon
Robert L. Roach

1 Basic Aerodynamics

This was the last first in aviation, we had always said, a milestone, and that made it unique. Would we do it again? No one can do it again. And that is the best thing about it.

Jeana Yeager and Dick Rutan
"Voyager" 1987

1.1 Introduction

Aerodynamics is the study of the flow of air around and within a moving object. Its main objective is understanding the creation of forces by the interaction of the gas motion with the surfaces of an object. Aerodynamics is closely related to hydrodynamics and gasdynamics, which represent the motion of liquid and compressible-gas flows, respectively.

Aerodynamics is the essence of flight and has been the focus of intensive research for about a century. Although this might seem to be a rather long period of development, it is really quite short considering the time span usually required for the formulation and full solution of basic scientific problems. In this relatively short time, mankind has advanced from the first gliding and primitive-powered airplane flights to interplanetary spaceflight.

Perhaps the most important motivation for this rapid development is the challenge to the human spirit represented by manned flight. However, practical needs also have strongly affected these endeavors, and we often find periods of rapid growth in aerodynamic knowledge associated with the solution of compelling problems in transportation, military applications, industry, and even sporting competitions. Figure 1.1 illustrates this growth in terms of the maximum speed attained by manned aircraft. Speed is a key measure of performance in almost every aspect of flight. It is of obvious vital importance in commercial flight as well as in military operations.

Even a casual study of the history of aviation yields considerable insight into the pressures that have motivated periods of almost explosive growth in the technology of flight. Much of the increase in speed during the 1940s and several subsequent decades was motivated strongly by military considerations. However, notice that two radical departures (shown as dashed lines in Fig. 1.1) from the curve for conventional airplanes occur in the 1920s and in the two decades between 1940 and 1960.

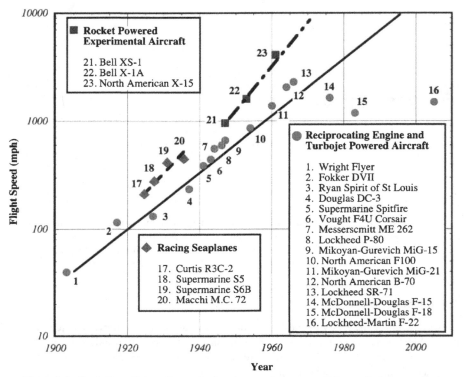

Figure 1.1. Evolution of aircraft speeds showing nearly exponential growth. Jet propelled aircraft speeds level off in the 1980's as turbojet performance Mach number limits are exceeded.

The phenomenal growth in speed from the 100 miles per hour (mph) range to over 400 mph that occurred during the period 1920–1932 was spurred by international competition in the guise of the Schneider Trophy Seaplane Races. Rapid progress not only in high-speed aerodynamics but also in aeropropulsion took place during that period. Similar growth occurred during the 1950s in supersonic flight. More recently, the international competitive spaceflight activities brought about rapid growth in propulsion, electronics, and materials, if not much in aerodynamics. There are signs that a new international competition is underway in the area of hypersonic aerodynamics and related technologies as policy decisions are made regarding the need for low-cost single-stage-to-orbit (SSTO) space vehicles.

Several of the key historical aircraft identified in Fig. 1.1 are illustrated in Figs. 1.2–1.12. Progressive improvements in aerodynamic configuration are apparent. In this textbook, we focus on the physical laws and related analytical and computational methods used to arrive at the aerodynamic-problem solutions implied in the evolving vehicle shapes depicted in the aircraft illustrated in this chapter.

Figure 1.2. Beginnings: The 1903 Wright Flyer (Smithsonian Institute).

Figure 1.3. Early Schneider Trophy seaplane: Curtis R3C-2 (1923).

Figure 1.4. Outright winner of the Schneider Trophy: Supermarine S6B (1931).

Figure 1.5. Supermarine Spitfire Mk II – an outcome of the Schneider Trophy racing seaplane research.

Figure 1.6. Messerschmitt ME 262. The first operational jet-propelled aircraft, 1944.

Figure 1.7. Bell XS-1: First supersonic airplane.

Figure 1.8. X-15: First hypersonic airplane.

Figure 1.9. Lockheed-Martin F-22 Raptor supersonic fighter.

Figure 1.10. Boeing 787 Dreamliner transonic jet transport during first flight test.

Figure 1.11. Concorde supersonic transport.

Figure 1.12. NASA X-43 hypersonic test vehicle using Scramjet propulsion.

Although it is not required to gain an understanding of the text material, the student is encouraged to supplement the text coverage with a parallel study of the history of aeronautics. Those interested in engaging in creative work as their careers in aeronautics develop will benefit greatly from this extra effort. References at the end of each chapter and frequent historical notes (usually provided as footnotes) serve as a guide for such an in-depth study. Much useful material is now available on the Internet and other large-scale computer networks.

In solving the problems of aerodynamics, those involved have been required to create basic technology along with the associated mathematical and experimental methodology. It is vital that the student understand the framework of this technology in detail and learn not only the application of the tools but also the deeper physical meaning they represent. This textbook is designed to promote this type of critical study of the subject. A carefully paced discussion of the traditional tools, such as mathematical analysis and experiment along with modern computational methods, is used to provide the student a broad understanding of both the physical meaning and the modern implementation of a wide variety of techniques and problem solutions. It is significant that the book outline follows closely the historical outline in terms of the need for each successive new idea and problem solution.

1.2 The Fundamental Problem of Aerodynamics

The basic problem addressed in this book is the accurate determination of the aero-dynamic force system acting on a body moving through the air. Figure 1.13 shows a typical force system on an airplane in flight; only the resultant forces are shown. An important task is to relate these resultant effects to the distributions of pressure and frictional forces that create them. We focus only on the part of the force system related to the interaction of the airflow with the vehicle—that is, on the creation of lift (**L**) and drag (**D**), the aerodynamic forces normal and parallel to the vehicle velo-city vector, as defined in the figure. However, much aerodynamic influence also is implied in the creation of thrust, as depicted by the vector **T** in Fig. 1.13. Interaction of the vehicle flow field with propellers, engine exhaust stream, cooling air inlets and outlets, and airbreathing propulsion system ducting must be accounted for. These important related aspects of applied aerodynamics are used at several points in the book to emphasize their strong dependence on knowledge of the vehicle flow field.

Aerodynamic Force System

Figure 1.13 illustrates several features of the force system on a moving body that is the focus of considerable attention throughout this book. The aerodynamic forces represent the integrated effect of a continuous distribution of pressure and shear forces acting on all of the exposed surfaces of the vehicle. The shape of the vehicle plays a crucial role in determining both the magnitude of the forces and their line

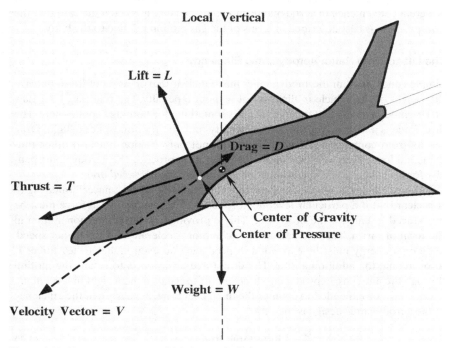

Figure 1.13. Force system on a vehicle in steady flight.

of action. In particular, notice that the net aerodynamic forces act at a point defined in the figure as the *center of pressure*. Clearly, to predict the vehicle-flight character-istics, it is necessary to know the precise location of this point relative to, say, the center of mass through which the vehicle weight acts. In level equilibrium flight, a component of the lift force is called on to balance the weight, **W**, of the airplane, expressed in vector form as **W** = m**g**, where m is the mass of the vehicle and **g** is the gravitational acceleration vector. Recalling the fundamental concepts of statics, it is clear that the balance of moments about the mass center also must be considered. Again, the latter considerations form the subject of stability and control; we concentrate here on the generation of the forces and moments and their relationship to properties of the airflow and vehicle shape.

Galilean Transformation

In Fig. 1.13, the point of view is assumed to be that of an observer fixed with respect to the atmosphere with the vehicle moving past with speed $V = |$**V**$|$ in the direction of its velocity vector **V**. It often is simpler for an observer to be moving with the vehicle. Then, the airflow relative to the vehicle is in the direction -**V** at a sufficiently great distance upstream of the body—that is, far enough upstream that the effect of the presence of the vehicle has not yet affected the relative flow of the air particles. This change in point of view is useful and often referred to as a *Galilean Transformation*. As long as there are no acceleration effects present (i.e., no vehicle acceleration relative to the airmass or angular motion about the mass center), then the force system on the body can be taken to be independent of the choice of coordinate frame. This is a great convenience in aerodynamic modeling because it is often the case that the flow problem is best described in terms of the gas motion relative to the body.

The Lift-to-Drag Ratio: Aerodynamic Efficiency

Simple concepts from thermodynamics make it clear that creation of lift to balance the weight of the vehicle in flight is not without a penalty. The rule (i.e., First Law of Thermodynamics) that "you cannot get something for nothing" applies here. The drag force is a measure of the cost or penalty function for atmospheric flight. Drag results from complex interactions involving not only friction but also other fundamental loss effects involved in the lift-generation process. The aerodynamic penalty for lift generation is the production of what is called *induced drag*.

Figure 1.14(a) shows the force balance for a vehicle in level, unaccelerated flight. It is clear that if a particular level flight speed is to be maintained, a force must be introduced to balance the drag force. This is provided by the propulsion system in the form of thrust, **T**. If the vehicle is to climb or accelerate to a yet higher speed, additional energy must be expended in producing an even higher thrust force, **T**, to overcome the additional drag. The drag force always acts to retard the motion through the air. Thus, in producing sufficient lift to balance the weight in level flight, energy must be expended to counter the drag. Therefore, a measure of the efficiency of the aerodynamic design is the ratio:

$$\frac{L}{D} = \text{Lift to Drag Ratio} \tag{1.1}$$

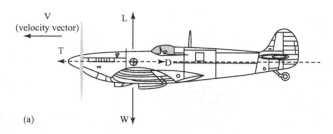

Figure 1.14. Force balance for (a) level, unaccelerated flight, and (b) gliding flight.

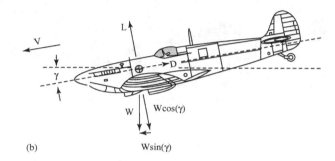

It is obvious that we strive to make this value as large as possible within a set of design constraints and the mission requirements for the vehicle. The range, speed, rate of climb, and many other performance factors depend on achievement of high values for this ratio.

Key performance elements depend on how well a designer addresses the many competing requirements. For example, it is shown readily by computing the power required for level flight and determining the rate at which fuel must be consumed to produce this power that the range of a propeller-driven airplane is:

$$R = \text{Range} = \left(\frac{\eta}{c}\right)\left(\frac{L}{D}\right)\ln\left(\frac{W_{\text{initial}}}{W_{\text{final}}}\right), \tag{1.2}$$

the famous Breguet range equation. In this formula, η is the efficiency of the propeller, c is the specific fuel consumption (i.e., rate at which fuel is used to produce the motor power output), and $W_{\text{initial}}/W_{\text{final}}$ is the ratio of the initial weight of the airplane to the weight after the fuel has been expended. This result clearly indicates the importance of aerodynamic efficiency in attaining long-range flight.

In the special field of sailplane (i.e., motorless airplane or glider; cf. Figs. 1.14(b) and 1.15) competition, recent research in drag reduction has led to the achievement of lift-to-drag (L/D) ratios approaching 60. The motivations for a high L/D ratio in this application are obvious. In Fig. 1.14(b), notice that in equilibrium, unpowered flight, the airplane must fly *downward* relative to the airmass at the glide angle γ. Therefore, the lift must balance only the cosine component of the weight. The other weight component (proportional to sin γ) is required to counteract the drag. A measure of the sailplane performance is the "flatness" of the glide (i.e., smallness of angle γ). By equating the balancing-force components, we see that:

Figure 1.15. Typical high-performance racing sailplane.

$$\frac{L}{D} = \frac{W\cos\gamma}{W\sin\gamma} = \frac{1}{\tan\gamma} \qquad (1.3)$$

Therefore, a high L/D ratio means a very flat glide. A sailplane with L/D = 60 can cover a distance of 60 miles while losing only 1 mile of altitude. The glide-slope angle γ is less than 1 degree! Thus, the L/D ratio often is referred to as the *glide ratio*. Much of the work needed to achieve this phenomenal performance was done using mathematical tools and concepts of the type studied in detail in this book, as well as careful wind-tunnel testing of aerodynamic refinements.

Similar attention to drag reduction has led to more economical commercial flight in the transonic flight regime and to higher cruising speeds in the supersonic regime. More complex dependence on lift and drag is involved in those cases.

1.3 Plan for Study of Aerodynamics

In the chapters that follow, we present the theory of aerodynamics mainly by means of a set of increasingly more realistic models of key flow phenomena. Chapter 2 discusses basic flow behavior and indicates strategies to represent the physical problems in mathematical form. Each chapter contains many examples that illustrate the techniques for solving key problems. An important aid to learning the material is a set of computer programs that demonstrate application of the solutions and numerical approaches required in the solution of more complex situations. The problem set at the end of each chapter provides students the opportunity to develop their skill with the material.

Organization of Text Material

It is demonstrated in Chapter 2 that a convenient way to study aerodynamics is to organize it into a set of *flow regimes*, each of which focuses on an important special

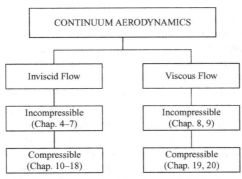

Figure 1.16. Organization of textbook material into flow regimes.

case governed by special features of the fluid motion. Chapter 3 is a review of the fundamental physics and modeling strategies needed to describe aerodynamic flows. These concepts are developed in the mathematical language needed to formulate and solve the key problems of aerodynamics.

Chapter 3 emphasizes the need for several methods of approach in problem formulation and solution. A major goal is to instruct the student regarding the correct choice of the most effective strategies for problem solution. Features of real-life aeronautical situations calling for either analytical, numerical, or experimental approaches (or combinations of these fundamental problem-solving methods) are introduced throughout the book by using examples and the problem sets based on them.

Figure 1.16 presents the book layout in terms of the flow regimes, which are organized so that each part forms the foundation for those that follow. Thus, we begin in Chapter 4 to study problems that can be described in terms of a flow field that is both frictionless and incompressible. This leads to the simplest type of mathematical formulation, which can be treated using many of the engineering mathematics tools acquired in other courses. For instance, methods of linear differential equation theory, vector analysis, and integral theorems are applied in an elegant solution method usually referred to as *potential-flow theory*. This forms the basis for the solution of many problems of major importance in aeronautics. A key example is the treatment of flow around an airfoil, which leads to an understanding of the generation of lift—a central theme of aerodynamics.

The fluid-mechanics solutions are progressively extended to include problems that are affected strongly by compressibility and viscous effects. The former study leads us into the important flow regimes of transonic, supersonic, and hypersonic flight. Inclusion of viscous forces leads finally to a comprehensive model in which all important features of the flow about an object at any speed or any altitude can be accommodated. Numerical techniques are introduced to solve these problems when the complexity precludes an analytical approach.

In addition to the categories shown in Figure 1.16, we could add further unsteady and low-density or free-molecular flows in each regime illustrated. In this book, however, these latter types of problems do not have a central role.

PROBLEMS

1.1. Referring to the free-body diagram in Fig. 1.13, write the three equations for static equilibrium. Show that the center of pressure must coincide with the center of mass so that the system is in equilibrium flight.

1.2. Draw the free-body diagram of a glider in steady flight (i.e., constant velocity) in still air. Note that the velocity vector must lie at a downward angle, γ, to the local horizon so that the force system is in equilibrium. By examining the components of the lift and drag, show that the tangent of this angle is proportional to the inverse of the L/D ratio. Also show that the ratio of the distance covered along the ground to the altitude lost is proportional to the L/D ratio.

1.3. Compare the gliding distance in still air of the following three airplanes if power is shut off at an altitude of 10,000 feet:

(a) Curtis Jenny with a glide ratio of L/D = 8
(b) Boeing 767 with a glide ratio of L/D = 14
(c) high-performance sailplane with a glide ratio of L/D = 55

1.4. If the measured range of a certain airplane in its aerodynamically unrefined form is 800 miles, what percentage reduction in the drag, **D** is needed if the range must be increased to 1,200 miles?

1.5. A small airplane is to be designed to fly nonstop around the earth. Describe some of the design features that you would incorporate. Then, compare your design to the Rutan Voyager that actually accomplished this task. Why does it look like a sailplane?

REFERENCES AND SUGGESTED READING

Anderson, John D., Fundamental Aerodynamics, McGraw-Hill Book Company, New York, 1984.

Crouch, Tom D., A Dream of Wings, Americans and the Airplane, 1875–1905, W. W. Norton & Co., New York, 1981.

Gabrielli, G., and von Kàrmàn, T, "What Price Speed?," Mechanical Engineering, Vol. 72, 1950, pp. 775–781.

Hallion, Richard P., Supersonic Flight: Breaking the Sound Barrier and Beyond, The Smithsonian Institution, 1972.

Hirsch, Robert S., Schneider Trophy Racers, Motorbooks International, 1993.

Kinnert, Reed, Racing Planes and Air Races, Aero Publishers, 1967.

Yeager, J., and Rutan, D., Voyager, Alfred A. Knopf, Inc., 1987.

2 Physics of Fluids

2.1 Aerodynamic Forces

Because the objective of aerodynamics is the determination of forces acting on a flying object, it is necessary that we clearly identify their source. Lift and drag forces, for example, are the result of interactions between the airflow and vehicle surfaces. Part of the force must be a result of pressure variations from point to point along the surface; another part must be related to friction of gas particles as they scrub the surface. Clearly, the key to understanding these forces is found in details of the fluid motions. The application of simple molecular concepts provides considerable insight into these motions.

Modeling of Gas Motion

As a branch of fluid mechanics, aerodynamics is concerned with the motion of a continuously deformable medium. That is, when acted on by a constant shear force, a body of liquid or gas changes shape continuously until the force is removed. This is unlike a solid body, which only deforms until internal stresses come into equilibrium with the applied force; that is, a solid does not deform continuously.

To understand the motion of a fluid, it is necessary to apply a set of basic physical laws, which consist of some or all of the following:

- conservation of mass (the continuity equation)
- Newton's Second Law of Motion (the momentum equation)
- First Law of Thermodynamics (the energy equation)
- Second Law of Thermodynamics (the entropy equation)

One skill that the student must develop is the effective application of approximation and simplification methods. A proper set of approximations may make unnecessary the use of some laws to produce a practical yet accurate solution for a given problem. This approach is possible only if a clear understanding of the physics of a fluid motion is attained. It is, of course, possible to construct a mathematically correct solution to an incorrectly formulated problem or a solution that is based on an inappropriate set of assumptions. In such cases, the results can be confusing or misleading or can even lead to costly mistakes. There is no substitute in aerodynamics for a sound understanding of the fundamental physical laws on which fluid mechanics is based.

It also may be necessary to introduce additional mathematical models or relationships to supplement those in the preceding list. For example, it often is necessary to use equations that characterize a working fluid. These *equations of state*, or *constitutive equations*, describe the physical attributes of a fluid. For example, it often is the case in practical problems that the fluid is an *ideal* or *perfect gas*, for which the equation of state is:

$$p = \rho RT, \tag{2.1}$$

which is a special relationship among the thermodynamic-state variables, pressure p, density ρ, and temperature T, needed to describe the behavior of a gas. R is the gas constant—a constant of proportionality—that is determined by the molecular configuration of the gas. This and other equations are discussed in considerable detail as needed throughout this chapter.

With regard to details of the molecular structure of a working fluid, we can choose to approach problems from the standpoint of the molecular motion, or a *continuum model* can be applied that does not attempt to address directly the actual small-scale particle motion. The former is called *statistical mechanics*, or the *kinetic theory of gases*. These are fascinating disciplines; however, we do not need a full description of molecular motion to predict forces on an aerodynamic vehicle unless it is flying at an extremely high altitude. In this book, we concentrate almost exclusively on a continuum model, which takes advantage of simplifications that result from the relatively large geometrical scale in realistic engineering situations when compared with an atomic or molecular-length scale.

In contrast, it is important to be aware of the molecular origin of physical quantities. In an ideal gas, the size of individual molecules is small when compared with the average distance between them. On this basis, the change in kinetic energy due to mutual attraction is negligible. Also, collisions between molecules can be considered as perfectly elastic so that there is no loss of energy due to permanent deformation of the molecular configuration. The average distance traveled between collisions is the mean free path, λ, which is important in the kinetic theory of gases. The mean random velocity of molecules is represented herein by the symbol c.

Continuum

Throughout this book, a continuum assumption is implicit, which means that no matter how small the fluid particle being considered, it is assumed to contain many molecules of the fluid or the gas such that the behavior of individual molecules is not important. Another way to say this is that the mean free path of the molecules (i.e., the distance between molecular collisions) is assumed to be small when compared with any length scale considered, such as a body length or diameter or an average airfoil-chord length—even if the length scale is infinitesimal.

Size of a Particle in the Continuum Model

When applying the continuum idea, we often refer to the motion of a *fluid particle*, which represents an element of the medium that is small compared with the basic scale of the problem. For example, it could represent a particle of air moving over the surface of a wing. In that case, we assume that the particle is small compared with the distance traveled from the leading to the trailing edge—that is, the *chord length*. It is important to form an accurate picture of just how small the particle is. If it were the size of a molecule, then the concept of a continuum would fail because it would be necessary to account for individual molecular motion, as in the previous example. Figure 2.1 illustrates a "thought experiment" to select the correct size for a particle. Consider a particle in the form of a cubical box of finite length L on each side and imagine measuring the variation of a property of the fluid across the element. The pressure is a convenient property to use in this thought experiment. We want to choose a particle size such that the property in question exhibits a constant value throughout the element. Thus, if L were the chord length of the wing, it would not work because the pressure changes radically in moving from the leading to the trailing edge. As Fig. 2.1 shows, if L is too large, then there are large variations in the value of the property across the particle. Conversely, if the particle is too small (e.g., the size of a few molecules), then the concept of pressure is lost. There would be huge oscillations in the pressure, depending on how many molecules happened to be inside the test volume at a given instant. On this basis, it appears that a practical lower limit for the size of a valid particle is a fairly large number of mean free paths, λ.

We must ensure that there is a sufficient number of molecules in the particle so that the influence of individual molecular motion is "smoothed out." Therefore, the practical answer to the question, "How large is a fluid particle?," is "Very, very small, but not too small."

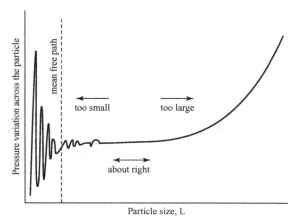

Figure 2.1. How large is a fluid particle?

Resolution of Aerodynamic Forces in Normal and Parallel Components

In addition to the pressure, other fluid properties are needed to model aerodynamic forces. In particular, the effects of energy dissipation or friction must be considered. In the simplest picture, forces normal to a surface are created by elastic molecular collisions. This simplified model suffices in some situations; however, the effects of forces parallel to the surface arising from viscous interactions can be of crucial importance in many aerovehicle-design problems.

Figure 2.2 illustrates the basic concept for resolving the interaction-force increment at a point into components normal and parallel to the surface. Consider the force $\mathbf{dF}$ acting on a small area element, dA, at any location on the body shown. The normal component $\mathbf{dF}_n$ most often is due to the action of pressure on an element of the surface:

$$\mathbf{dF}_n = -(\text{pdA})\mathbf{n}. \tag{2.2}$$

The negative sign indicates that the local pressure force acts *toward* the surface—that is, in a direction opposite to the outward-pointing unit normal vector, $\mathbf{n}$. The pressure can be due to momentum exchange from both the random molecular motion (i.e., the thermodynamic pressure) and the directed motion of the particles due to flow of the gas in the continuum sense. The latter pressure component often is referred to as the *dynamic pressure* and is an important element in subsequent chapters, which discuss control of dynamic pressure by the shape and orientation of the body. This is crucially important in the generation and control of lift (e.g., on a wing surface in external flows) and in the production of thrust (e.g., in a ducted propulsion system with internal flow).

Molecular Origin of the Thermodynamic Pressure

As an example of the molecular basis of important properties, we first examine the origin of the thermodyamic pressure. Consider a uniform gas contained in a box.

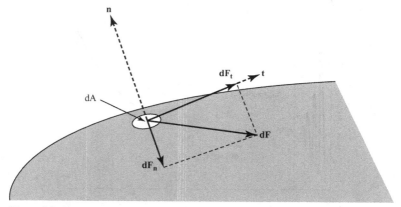

Figure 2.2. Resolution of surface forces into parallel and normal components.

From kinetic theory, we know that the pressure is the reaction force generated by the interactions of the molecules with the box surfaces. *Pressure* is defined as the force per unit area arising from these collisions. Because the interaction is (in the first approximation) an elastic collisions, then the net reaction force vector is perpendicular to the surface at any point. If one-third of the molecules are traveling in each of three mutually perpendicular directions (e.g., axes parallel to the edges of the box) at a given time, and half of those going in any particular direction are traveling away from the surface on average, then one-sixth of the molecules in a layer of thickness dx strikes the surface in a time $dt = dx/c$, where c is the average velocity, as already defined. If m is the mass of each molecule and the momentum is reversed in the elastic collision with the surface, then the reaction force in each collision is $dF = 2mc$. Thus, if n is the number of molecules per unit volume, the pressure is:

$$p = \frac{n(2mc)dx}{6} = \frac{1}{3}nmc^2 = \frac{2}{3}\left(\frac{1}{2}\rho c^2\right) \tag{2.3}$$

because $\rho = nm$ is the density, or mass per unit volume. This shows that the pressure is proportional to the kinetic energy per unit mass due to random molecular motion.

Molecular Origin of Viscous Forces

The part of the interaction force parallel to the surface, $\mathbf{dF_t}$, often is described as the *shear force* or *viscous force*. As the latter term suggests, it is the result of frictional effects and is a major source of drag force on aerodynamic vehicles.

As in the case of the pressure, the simple ideas of kinetic theory based on random molecular motion provide a useful physical picture of the origin of shearing stresses in a fluid or a gas. The random motion tends to eliminate discontinuities that may form in a gas-velocity distribution. For example, consider the gas flow over a surface, as shown in Fig. 2.2. The parallel velocity distribution is expected to look somewhat like that shown in Fig. 2.3.

The molecules in the immediate vicinity of the surface must be brought to rest. The result is known as the *no-slip condition* and is satisfied in almost all practical situations.* Gas particles lose momentum in the tangential direction at the surface due to collisions with the irregular-surface molecular lattice, as depicted in the blowup of a small region near the surface in Fig. 2.3. Clearly, if the mean free path is short, as it almost always is, there will be molecular collisions between the crystal-lattice structures of the surface (which, on the microscopic scale, represent a rough surface texture) involving lateral-momentum transfer. Gas particles must be brought to rest in the tangential direction at the surface.

Consider this simple thought experiment based on events leading up to the steady-state situation illustrated in Fig. 2.3. The surface is suddenly introduced into uniform flow parallel to the surface moving at velocity U. This initially creates a discontinuity in the velocity distribution because the no-slip condition applies only

---

* This may not be a correct picture of the motion in the case of *free molecular flow* at extremely high altitudes, where the mean free path can be very long.

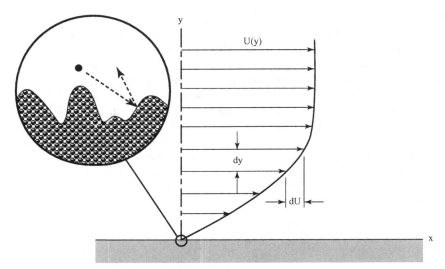

Figure 2.3. Viscous shearing stress in a boundary layer.

to gas particles close to the surface. This discontinuity must be adjusted by the molecular motion. Gas molecules at a distance have yet to be affected by the presence of the surface. Molecules moving outward from the surface vicinity have zero momentum parallel to the surface. The mixing of these two types of molecules must finally produce a velocity distribution similar to that shown in Fig. 2.3. (It is the task of subsequent chapters to determine the exact form of this distribution.) The effect can be visualized as the dragging-along of slower fluid particles located closer to the surface by those at a greater distance, and a similar retardation of molecules farther from the surface by those that have been affected already by its presence. A careful measurement of the force on the surface would show that there is a shearing stress, τ, that tends to move the plate in the original direction of the fluid. There is, of course, an equal and opposite retarding stress on the fluid.

If two layers of fluid separated by a vertical distance, dy, are considered, a change of velocity, dU, is evident between the layers, which can be interpreted as the net result of molecular mixing at that particular location. That is, it is a measure of the local shearing stress in the fluid. The force per unit area parallel to the surface, the shear stress, can be written as[†]

$$\tau = \mu \frac{dU}{dy}, \qquad (2.4)$$

where the constant of proportionality is the coefficient of viscosity, μ. The region in which the adjustment of the velocity is made by molecular diffusion or mixing is called a *boundary layer* and is studied in a subsequent chapter. The value of the shearing stress, τ, at the surface has the maximum value because the velocity gradient

---

[†] This equation often is referred to as Newton's Law of Viscosity. Gases or liquids that behave as indicated are called *Newtonian fluids*; those that do not are called *non-Newtonian fluids* and are rarely encountered in aeronautics.

is largest there. The viscosity coefficient, μ, mainly depends on the temperature of gases; it also may be sensitive to pressure in other media.

Effects of Compressibility

All gases and even solids are compressible. The speed of sound, a, represents the speed of propagation of weak disturbances due to compressibility of the medium. These weak waves often are referred to as *acoustic waves*. For example, sound waves in water propagate at the acoustic speed—or sound speed—characteristic of that liquid.

Other variables become important when additional physical effects must be accounted for, such as compressibility. For instance, as the speed of a vehicle approaches a significant percentage of the speed of sound, several changes take place in the flow-field characteristics and in the accompanying forces on the vehicle. If the representative aircraft shown in Figs. 1.2–1.12 in Chapter 1 are studied, it becomes apparent that drastic changes in design accompany the increases in speed. Compare the pictures of the high-speed seaplanes used in the Schneider Cup races with the supersonic F-22 and the Concorde transport; quite different shapes are required when the speed is higher than the speed of sound. Similarly, there are important design differences between low-speed aircraft and those that must operate in the transonic range, which begins when the speed is about 80% of the speed of sound. Other families of shapes are demanded when a vehicle must operate at hypersonic speeds; that is, when the velocity is five or more times higher than the acoustic speed.

The dimensionless ratio of the flow speed to the speed of sound represents a measure of the relative importance of compressibility. The thermodynamics and gas-dynamics of this situation are carefully worked out as needed in subsequent chapters. It suffices at this point to introduce the speed of sound for isentropic wave propagation in air:

$$a = \sqrt{\gamma \frac{p}{\rho}}, \tag{2.5}$$

where γ is the ratio of specific heats. For air, this parameter has a value of $\gamma = 1.4$. For an ideal gas, it is easy to see that by substituting Eq. 2.1, the speed of sound depends on only the temperature, T. We find:

$$a = \sqrt{\gamma R T}. \tag{2.6}$$

For sea-level air, the speed of sound is approximately $a = 1,116$ ft/sec $= 340$ m/s.

An important parameter is the ratio of vehicle speed to speed of sound—that is, the Mach number:

$$M \equiv \frac{V}{a}, \tag{2.7}$$

which is vital in determining the forces on a vehicle. This dimensionless quantity is an example of a *similarity parameter*, one of several parameters that determine the flow-field behavior and, thus, the force system. How this happens is discussed in the following section.

Table 2.1. *Effects of compressibility*

M	Flow Regime	Characteristics of Flow Field
0	Incompressible	No dependence on compressibility; density is constant everywhere.
<0.85	Subsonic	Compressibility effects are sometimes negligible.
0.85 to 1.05	Transonic	Compressibility greatly affects flow. There is a major increase in drag. Shock waves may appear at points on the body where flow is locally higher than sonic speed.
1.05 to 5	Supersonic	The flow field is dominated by shock waves. Major drag increases appear that are related to the shock waves.
>5	Hypersonic	Shock waves may lie close to body surfaces and may interact with viscous boundary layers.

Whether M is zero or less than or greater than unity determines which of several flow "regimes" within which a vehicle operates. Each regime is characterized by distinct flow-field features, which are summarized in Table 2.1. The ramifications of several flow regimes are examined later in great detail because they are important in determining aerodynamic performance and in solving practical vehicle-design problems.

2.2 Aerodynamic Variables

In Section 2.1, we encountered two fundamental forces: (1) the force perpendicular to a surface due to normal stress, or pressure; and (2) the tangential force due to viscous-shearing stress. These two forces are clearly related to the lift and drag on a body. That is, integration of the differential force contributions over the wetted surface of a body moving through air leads to the force system discussed in Chapter 1. The integrated force components can be resolved into lift and drag and moments about the center of gravity, as previously defined. The emphasis in Section 2.1 was on associating these forces with processes occurring at the molecular level. The benefits of using a macroscopic point of view based on a continuum model were also described. This section explores this concept in more detail. Rather than concentrating on microscopic interactions, we attempt to discover how much can be learned by applying the simple ideas of dimensional analysis. In this process, we identify important properties, both physical and geometrical, that influence the generation of the force system on a body moving through a fluid.

Many benefits accrue from the process. For instance, we discover key scaling parameters, or similarity parameters. Understanding the physical content of these parameters is a major step in classifying types of aerodynamic problems in ways that simplify the modeling process and lead to approximate formulations of practical flow problems that are both accurate and easy to use. The results also have an acritical application in guiding the design of experimental procedures and in the correct interpretation and correlation of experimental measurements in aeronautics as well as other fields.

Dependence of Forces on Flow Parameters and Geometry

When designing a flight vehicle, it is necessary to clearly understand how aerodynamic forces are related to the following:

- the shape, size, and orientation of a body
- the properties of the airstream in which a body moves

Flow-field properties may include density, flight speed, and pressure. Common experience tells us that the force on a body moving through a fluid depends in some way on variables of this type.

Conduct the following simple aeronautical experiment: Hold your hand out the window of a moving automobile and observe that a force can be perceived by the need to exert muscular forces to oppose those created by the airflow. Notice that if your hand is held nearly perpendicular to the air motion, there is a strong drag effect. If you hold it at a shallow angle, you feel both lift and drag. There is an obvious dependence on the orientation of your hand as well as the speed of the car, and so on. If you were to guess at the set of contributing physical variables involved based on your observations, you might choose to represent the aerodynamic force F as a mathematical function such as:

$$F = F \ (speed, density, pressure, size, shape). \tag{2.8}$$

The first three variables describe the airflow; the last two indicate that the size and orientation of your hand (or another object) with respect to airflow probably affect the forces.

The basic problem of aerodynamics is defined in Chapter 1 as the determination of mathematical details of physical relationships of this type. To accomplish this, it is necessary to understand *all* of the important interactions of *all* of the participating variables and their influence on the force system. The initial difficulty is determining which of myriad possible variables matter. In the following subsections, we explore practical means to make this crucial determination.

Systems of Units

The first step is to establish precise definitions for all of the variables likely to be important. It must be possible to express each of these and their interactions in mathematical form. Of vital importance are the dimensions and associated systems of units needed to describe the variables. Such considerations likely have already had a major role in a student's technical training, therefore, only a brief review is necessary.

A basic set of physical variables significant in aerodynamics is defined in Table 2.2. The dimensions of each property are given in terms of familiar mass-based (M, L, T) and force-based (F, L, T) systems of units. M represents mass (in kg and slugs in Système International [SI] and English systems of units, respectively).* For example, F represents the magnitude of the force (N or lbf) and t is the time in seconds.

---

* Standard SI and English-system symbols and units are used throughout this book. For example, N represents force in newtons, lbf is force in pounds, and kg is mass in kilograms; a *slug* is the basic unit of mass in the traditional English engineering system. Length in English units is in *ft* (feet) or *in* (inches); m is the metric unit of length in meters. Time in seconds is s or *sec*.

Table 2.2. *Important variables and fluid properties*

Quantity	Symbol	Definition	Dimensions (mass-based units[*])	Dimensions (force-based units[†])
Flow speed	V	$V = |\mathbf{V}|$, where $\mathbf{V}$ is a velocity vector of gas motion at a point in the flow	$\left[\dfrac{L}{T}\right]$	$\left[\dfrac{L}{T}\right]$
Sound speed	a	Speed of propagation of weak disturbances; a measure of compressibility effects	$\left[\dfrac{L}{T}\right]$	$\left[\dfrac{L}{T}\right]$
Acceleration	$\dfrac{dV}{dt}$	Local rate of change of speed of gas particles; proportional to forces acting on particles	$\left[\dfrac{L}{T^2}\right]$	$\left[\dfrac{L}{T^2}\right]$
Force	F	Force due to interaction of a gas flow with an object	$\left[\dfrac{ML}{T^2}\right]$	$[F]$
Pressure (or normal stress)	P	$p = \lim_{dA \to 0}\left(\dfrac{dF}{dA}\right),$ where dF is the force on the surface element dA; force per unit area normal to a reference surface	$\left[\dfrac{M}{LT^2}\right]$	$\left[\dfrac{F}{L^2}\right]$
Shear stress	τ	Force per unit area parallel to a reference surface	$\left[\dfrac{M}{LT^2}\right]$	$\left[\dfrac{F}{L^2}\right]$
Density	ρ	$\rho = \lim_{dv \to 0}\left(\dfrac{dm}{dv}\right),$ where dm is mass contained within the volume element, dv	$\left[\dfrac{M}{L^3}\right]$	$\left[\dfrac{FT^2}{L^4}\right]$

[*] Metric SI units are based on length, L; time, T; and mass, M. Force is a derived unit.
[†] Traditional English engineering units are based on length, L; time, T; and force, F. Mass is a derived unit.

For the most part, these are familiar thermodynamic variables; therefore, only a brief review is necessary. Other variables may be needed to describe the fundamental characteristics of a flowing liquid or a gas. They are defined by basic physical relationships; they are sometimes empirical in nature, requiring knowledge of experimentally determined parameters. For example, in the case of an ideal gas, the thermodynamic variables are related by the familiar gas law, sometimes called the equation of state or constitutive equation, expressed in Eq. 2.1. The gas constant, $R \equiv \Re/M$, that appears in this equation is known from extensive experimental studies and from statistical mechanics and the kinetic theory of gases. $\Re$ is the universal gas constant and M is the molecular weight.

Similarly, forces and acceleration of gas particles are related by Newton's Second Law of Motion, so that (from the point of view of physical dimensions) for a gas particle of mass m, we write:

$$F = m\frac{dV}{dt}. \tag{2.9}$$

If we must account for viscous stresses, τ, we need to know that a relationship of the type deduced in Eq. 2.4 might be involved. Table 2.2 illustrates how the correct dimensions for possibly unfamiliar physical variables such as the coefficient of viscosity, μ, in Eq. 2.4 can be determined by means of the *principle of dimensional homogeneity*. For Eqs. 2.4 and 2.9 to be correct, the dimensions of quantities represented on the left and right sides of the equation must be identical.

EXAMPLE 2.1 *Required*: Verify the dimensions of force in a mass-based system of units as listed in Table 2.2. Determine the corresponding standard metric SI and English engineering units for force.

Approach: Use the principle of dimensional homogeneity and Newton's Second Law (Eq. 2.9).

Solution: The defining equation and the corresponding dimensions are:

$$F = m\frac{dV}{dt}.$$

$$\therefore \ [F] = \left[m\frac{dV}{dt}\right] = M\frac{L/T}{T} = \frac{ML}{T^2},$$

where the brackets denote dimensions of the enclosed variable.

Discussion: Often, the symbols denoting basic units are the same as those appearing in the defining equation. Here, the symbols for generic mass, m, and time, t, are the same as the corresponding quantities in Newton's Second Law.

From this, it is clear that 1 unit of force in the SI system is expressed (in mass-based units) as:

$$1\frac{\text{kg } m}{\text{s}^2} = 1 \text{ newton}$$

because mass is measured in kilograms, length in meters, and time in seconds. Similarly, 1 unit of force in the English engineering system is:

$$1\frac{\text{slug ft}}{\text{sec}^2} \equiv 1 \text{ lbf}$$

because the basic unit of mass is the slug in English units. The most common abbreviations for various quantities are used.

EXAMPLE 2.2 *Required*: Find the weight of an object with a mass of 1 slug at the earth's surface.

Approach: Use Newton's Second Law in English engineering units.

Solution: Because the acceleration due to gravity at the earth's surface is approximately 32.17 ft/sec², the gravitational force (i.e., the weight) is:

$$\mathbf{W} = 1 \text{ slug} \times 32.17 \text{ ft/sec}^2 = 32.17 \text{ lbf}.$$

Discussion: It is useful to use the abbreviation *lbf* to denote forces to distinguish between two common uses for the pound (*lb*). The pound-mass (*lbm*) often is used in some technical fields: 1 *lbm* is the amount of mass that would weigh 1 pound at the earth's surface. In compatible units, this is equivalent to 1/32.17 ≈ 0.031 slug. At the earth's surface, the weight of an object with a mass of 1 slug is about 32.17 lbf.

Notice that English units are a force-based system using the pound as the basic unit. However, Newton's Second Law (Eq. 2.9) indicates that 1 *lbf* can be written as:

$$1 \text{ lbf} = (1 \text{ slug})\left(1\frac{\text{ft}}{\text{sec}^2}\right).$$

Notice that this defines the mass unit that is compatible with the chosen units for acceleration. It is incorrect to replace slug with lbm because, by definition, 1 lbm is the mass that weighs 1 lbf at the earth's surface.

Of course, equations can be written that contain the conversion factors needed to adjust the units. In some fields of study (e.g., thermodynamics and heat transfer), it is traditional to use *lbm* as the mass unit when working in English engineering units. It is then necessary to insert the factor $g_o = 32.17$ lbm/slug in any numerical calculations.

It is better to use compatible units so that serious numerical errors can be avoided. It was once common practice in engineering textbooks to display constants such as g_o or J (i.e., the number of ft-lb/sec per BTU) in the fundamental equations to remind users to adjust the units. This is not done in this book! Dimensional numerical constants should not appear in the fundamental equations. It is necessary to check dimensional homogeneity only to be sure that the correct units are applied to represent each variable involved.

EXAMPLE 2.3 *Required*: Find the dimensions of the coefficient of viscosity using Newton's Viscous Force Law (Eq. 2.4). Then, determine the corresponding units in standard SI and English units.

Approach: Use the principle of dimensional homogeneity.

Solution: From Table 2.2, the shearing stress (i.e., force per unit area) in mass-based units has the following dimensions:

$$[\tau] = [ML/T^2]/[L^2] = [M/LT^2],$$

and the dimensions of the velocity gradient are:

$$\left[\frac{\partial u}{\partial y}\right] \rightarrow \left[\left(\frac{L}{T}\right)\frac{1}{L}\right] \rightarrow \left[\frac{1}{T}\right].$$

Then, for dimensional homogeneity, the correct dimensions for the viscosity coefficient must be:

$$[\mu] \rightarrow \left[\frac{M}{LT}\right].$$

Thus, in the SI system, the viscosity coefficient must have kg/m-sec units. In the force-based engineering system, the units are lbf sec/ft^2.

The application of Newton's Second Law shows that viscosity also can be written in terms of the units slug/ft sec. By similar reasoning, it is correct to write the viscosity coefficient with N-s/m^2 units.

Dimensional Analysis

With the simple tools now at our disposal, we can gain considerable insight into the nature of aerodynamic forces and related flow-field effects. A powerful tool is the requirement for dimensional homogeneity. This makes it possible to uncover relationships between parameters without complete information about the actual physics of any underlying phenomena, which may not be known at the outset. The method we now construct has many applications, not the least important of which is the planning and implementation of experimental studies of complex aerodynamic phenomena.

We return now to the ideas expressed in Eq. 2.8. We expect that on the basis of our observations, the drag force is a function of several flow parameters and on the geometrical configuration of the object immersed in the gas flow. This is true for any aerodynamic force component, F, so let us generalize by writing:

$$F = F(V, \rho, \mu, a, d), \tag{2.10}$$

where we now have chosen a set of parameters that reflect our expectations that speed, V; gas density, ρ; and size may affect the force. We also include the speed of sound, a, and the coefficient of viscosity to account for possible compressibility and shear-force effects. The pressure is excluded because the force is clearly the result of pressure distribution over the body; to include it simply would double the dependence on the same quantity. The influence of the size of the object is reflected by inclusion of a length d, which could be any one of the dimensions of the object (e.g., length or width). A parameter describing the shape is not incorporated because it is not possible to represent this by a mathematical parameter. Instead, we employ the concept of *geometric similarity*. The relationship to be constructed holds for any family of objects, possibly of different sizes but with the same shape (as illustrated in Fig. 2.4)—that is, a family of *similar* shapes.

We now examine Eq. 2.10 from the standpoint of dimensional reasoning. The method to be used often is generalized in an elegant form known as the Buckingham Pi Theorem. The full machinery is not needed at this point in the discussion; a simpler intuitive approach is more instructive as an introduction.

The first step is to choose an appropriate mathematical form for the relationship implied in Eq. 2.10. To accomplish this, we simply note that in the majority of such situations, physical relationships involving several variables are in the form of products of powers of the several variables. There is no guarantee that this will work here; however, subject to subsequent careful analysis, we ask what such a relationship would yield in the current problem.

With this in mind, assume that the aerodynamic force F is a function of products of the several variables to unknown powers. This choice is based on the observation that mathematical descriptions of natural phenomena often take this form. Write:

$$F = C V^{c_1} \rho^{c_2} \mu^{c_3} a^{c_4} d^{c_5}, \tag{2.11}$$

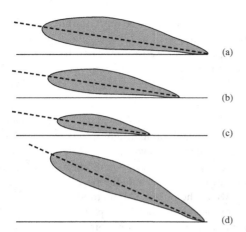

Figure 2.4. **Geometric similarity**. All shapes shown
are identical, although sizes are different. Because
(a)–(c) also are oriented at the same angle to
the reference line, they are geometrically similar.
Shape (d) is identical to (a), but it is oriented
differently with respect to the reference line.
Therefore, shape (d) is *not* geometrically similar to
the other shapes.

where C is an arbitrary (i.e., dimensionless) constant and c_n ($n = 1$ to 5) is a set of
initially undetermined exponents. Thus, in terms of the dimensions of each of the six
factors, it is necessary for dimensional homogeneity that:

$$
\left[\frac{mL}{t^2}\right] = \left[\frac{L}{t}\right]^{c_1} \left[\frac{m}{L^3}\right]^{c_2} \left[\frac{m}{Lt}\right]^{c_3} \left[\frac{L}{t}\right]^{c_4} [L]^{c_5}
$$
$$
= [m]^{(c_2+c_3)}[L]^{(c_1-3c_2-c_3+c_5)}[t]^{(-c_1-c_3-c_4)}, \tag{2.12}
$$

where the mass-based set defined in Table 2.2 is used. The same results would be
forthcoming if the force-based dimensions were used instead. Now, for Eq. 2.11 to
be dimensionally correct, it is necessary that each of the three basic units enters the
problem with the same exponent on each side of the equation. Therefore, for dimensional homogeneity, it is necessary that

$$
\begin{cases}
1 = c_2 + c_3 \\
1 = c_1 - 3c_2 - c_3 + c_4 + c_5 \\
-2 = -c_1 - c_3 - c_4.
\end{cases} \tag{2.13}
$$

Unfortunately, there are five unknown exponents but only three equations. Therefore, the best that can be done is to write two of the exponents in terms of three that
we choose corresponding to an appropriate set of reference variables. Because it is
likely that the velocity, air density, and size of the object are of prime importance in
controlling the magnitude of the aerodynamic force, we choose them as the principal
variables. That is, we choose to write the exponents c_1, c_2 and c_5 in terms of c_3 and c_4.
This is readily accomplished by solving Eq. 2.13 simultaneously. We find that:

$$
\begin{cases}
c_1 = 2 - c_3 - c_4 \\
c_2 = 1 - c_3 \\
c_5 = 2 - c_3.
\end{cases}
$$

Therefore, the aerodynamic force can be written as:

$$F = CV^{(2-c_3-c_4)}\rho^{(1-c_3)}\mu^{c_3}a^{c_4}d^{(2-c_3)}.$$

Collecting the terms with the same unknown exponent, we find:

$$F = C\rho V^2 d^2 \left(\frac{\mu}{\rho V d}\right)^{c_3} \left(\frac{a}{V}\right)^{c_4}. \tag{2.14}$$

It is not possible to determine the exponents c_3 and c_4 using dimensional homogeneity alone. However, the expression we found contains vital information regarding the dependence of the aerodynamic forces on the parameters that we identified as those most likely to matter. Notice that the expressions raised to the powers c_3 and c_4 are dimensionless groups of the variables. Because the exponents are unknown, it would be correct to rewrite Eq. 2.14 as:

$$F = C\rho V^2 d^2 \left(\frac{\rho V d}{\mu}\right)^{d_3} \left(\frac{V}{a}\right)^{d_4}, \tag{2.15}$$

where two new exponents (d_3 and d_4, which are obviously the negative values of the original two) are used. The two dimensionless groups clearly have physical significance in the system behavior. They are called similarity parameters and are the focus of considerable attention throughout this book. One of the parameters, arising in a natural way in this simple analysis, is the Mach number V/a that was encountered in the introduction to compressibility effects previously in this chapter. The second, called the Reynolds number, clearly is related to effects of viscosity.

Similarity Parameters

Any physical problem can be analyzed in the manner just illustrated. It always happens that dimensional reasoning alone cannot show in detail how each parameter (e.g., the viscosity or speed of sound) enters the mathematical expression governing the problem. However, the dimensionless groupings that appear have enormous significance. Hundreds of these groups have been identified as important in engineering applications; they usually are assigned the name of the individual who first recognized its importance. For example, in the aerodynamic force expression derived in Eq. 2.14, the two groups that appear are as follows:

$$\text{Mach number} \qquad M = \frac{V}{a}, \tag{2.16}$$

$$\text{Reynolds number} \qquad R_e = \frac{\rho V d}{\mu}. \tag{2.17}$$

The first group was identified as a key parameter in describing the effects of compressibility in aerodynamic flows. It was introduced by the Austrian physicist Ernst Mach in the 1870s. The second is named for Osborne Reynolds, who showed in the 1880s that this group governs the role of viscous forces in the flow of liquids or gases. In particular, it controls the transition between laminar and turbulent flow.

Table 2.3. *Basic fluid forces*

Inertia force	$\rho V^2 L^2$	~ Mass × acceleration
Pressure force	$\Delta p L^2$	~ Pressure increment × area
Viscous force	$\mu V L$	~ Shear stress × area
Gravity force	$\rho g L^3$	~ Mass × gravitational acceleration
Surface-tension force	σL	~ Surface tension × length
Compressibility force	$\rho \dfrac{dp}{d\rho} L^2$	~ Pressure required to change density

Each dimensionless group can be identified as the ratio of sets of forces governing the fluid flow. Clearly, the ratio of any two such forces is a dimensionless number like those we found. Some key forces are:

$$\text{Inertia force} \approx m\frac{dV}{dt} \rightarrow [\rho L^3]\left[\frac{V^2}{L}\right] = [\rho V^2 L^2] \tag{2.18}$$

$$\text{Viscous force} \approx \tau A = \mu \frac{dV}{dy} A \rightarrow \left[\mu \frac{V}{L} L^2\right] = [\mu V L], \tag{2.19}$$

where the dimensional form is expressed by means of the notation introduced in Table 2.2. Thus, the ratio:

$$\frac{\text{inertia force}}{\text{viscous force}} = \frac{[\rho V^2 L^2]}{[\mu V L]} = \left[\frac{\rho V L}{\mu}\right] = R_e \tag{2.20}$$

is the Reynolds number. All dimensionless groups can be expressed similarly as ratios of the forces governing the fluid motion. Table 2.3 lists a set of forces and their dimensions that may occur in fluid-dynamics problems. Of these, only the viscous force, pressure force, and force related to compressibility are likely to be encountered in aerodynamics problems.

It is important to demonstrate the power of the concept of similarity parameters. Using the Reynolds number to illustrate, consider what the difference would be in a flow with a high R_e compared to one with a low R_e value. For example, the flow over an airplane wing in low-speed flight typically has a R_e number (based on the flight speed and average width of the wing) of two million or three million. From Eq. 2.20, it is clear that this means that the inertia forces—those due to acceleration or deceleration of gas particles moving over the wing—are enormous compared to the viscous forces due to the shearing stresses along the surface. For a second example, consider the flow of ketchup from an upturned bottle. Here, the R_e is small, indicating the overwhelming importance of the viscous forces in controlling the flow.

Aerodynamic-Force Coefficients

Using these ideas, a helpful way to write the expression for the aerodynamic force is:

$$F = C(M, R_e)\rho V^2 d^2, \tag{2.21}$$

where the effects of the dimensionless groups now have been incorporated into a coefficient of proportionality, C. For example, what is usually written in aerodynamics problems involving, for example, the lift and drag on an airplane wing is:

$$\left\{ \begin{array}{l} \text{Lift} = L = \frac{1}{2}\rho V^2 S C_L \qquad\qquad\qquad\qquad\qquad (2.22) \\[2mm] \text{Drag} = D = \frac{1}{2}\rho V^2 S C_D, \qquad\qquad\qquad\qquad (2.23) \end{array} \right.$$

where the coefficients $C_L(M, R_e)$ and $C_D(M, R_e)$ are functions of the appropriate similarity parameters. These are the well-known lift and drag coefficients, respectively. The factor of 1/2 is introduced here to take advantage of the fact that the combination

$$\text{dynamic pressure} \equiv q = \frac{1}{2}\rho V^2, \qquad\qquad\qquad\qquad (2.24)$$

which has the dimensions of pressure, appears throughout the book as an important aerodynamic parameter. The area, S, is introduced as a convenient substitute for the squared characteristic length, d, which clearly has units of area. Any convenient reference area can be used.

In the study of lift on three-dimensional wings, the reference area, S, usually is chosen as the projected area of the wing surface. Other choices can be made. For instance, in describing the drag of bodies of revolution, we often use the projected frontal area of the body as the reference area, S.

Notice that, by using a combination of physical and dimensional reasoning, we now have reduced the aerodynamic-force problem to a convenient and practical form, which can be applied readily in airplane design, for instance.

Consider the application of Eq. 2.22 for determining, for example, the lift on an airplane wing. Three main elements are involved. One factor, the dynamic pressure, describes the environment—that is, the flight speed and altitude as represented by air density (which, as we shall see, is strongly dependent on altitude). An important geometrical dependence is seen in the reference area, S. Finally, a coefficient, C_L, implies a dependence on compressibility and viscous effects because, as demonstrated, its value is a function of the Mach and Reynolds numbers characterizing the flow field. It must be understood that aerodynamic coefficients also are dependent on geometry.

The situation would be simplified if the coefficients were thought of as constants. Then, it would be necessary to compute or measure them only once for a given shape to provide design information for using that shape in a practical application. However, our analysis shows that they vary with special parameters—in this case, the Reynolds and the Mach numbers.

To use Eqs. 2.22–2.23, for example, in a wing-design problem, we must have values for all three factors. The first two are known from the design operating conditions and the geometry. However, how can we determine the lift coefficient? Several approaches naturally come to mind:

1. Develop a theory for the flow and resulting pressure distributions on aerodynamic surfaces. Use integrated pressure forces to determine the force coefficients.
2. Carry out a numerical simulation of the flow field over the wing.
3. Build a model of the wing and measure the forces in a wind tunnel.

Why are there different approaches and how do we determine which is the correct one to use in a given situation? These are important practical questions, and we endeavor to provide the answers in the following discussion. A major objective of this book is to introduce methods for accomplishing Method 1. An introduction to Method 2 also is discussed in considerable detail at appropriate points. The numerical approach is rapidly gaining in popularity as large-scale computing costs drop and workstation and desktop computer capabilities improve. Method 3, experimentation, is the traditional approach and still is used extensively for reasons that will become clear. It is also important to observe the relationship between Methods 2 and 3. In many ways, Method 2 is another method of experimentally determining the required information. It represents "virtual" experimentation, which can take the place of a usually more costly physical experiment.

It is of the greatest importance that the student realizes that experimentation, either numerical or physical, cannot be used as a substitute for a thorough theoretical understanding of the problem to be solved. The flow-similarity ideas discussed in this chapter show how important this physical understanding can be. Even lacking an in-depth theory of gas flow over a body, using simple dimensional reasoning, we have identified the correct relationship between the key variables in Eqs. 2.22 and 2.23 for determining aerodynamic forces. Basically, it reduces the problem to one of finding coefficients such as C_L and C_D.

To illustrate how important such theoretical results can be, consider the following example: Suppose it is your job to design the wing of a new airplane. You consider yourself an experimentalist and do not have much use for theoreticians and their incessant equations. However, you fully understand the thinking that led to Eq. 2.10. You know that there are five key variables. You also understand geometrical similarity, so you proceed to build a scale model of the airplane to be tested in the wind tunnel. Now, if you are not armed with the right theoretical understanding, you probably will think that it is necessary to vary all five of the variables to obtain a full set of data describing how the forces on the wing vary with speed, density, viscosity, size, and so on. If it takes a minimum of five values of each variable over their expected range to determine how they affect the aerodynamic forces, then it is easy to see that it will require $5^5 = 3,125$ separate test runs to acquire the needed information. Keeping in mind the high cost of laboratory space, instrumentation, and technicians, this would turn out to be an expensive and time-consuming exercise; you would most probably lose your job.

Conversely, if you base your experiments on your understanding of the theory, you will see that there are only two parameters to keep track of on a model shaped and oriented like the real prototype: the Reynolds number and the Mach number. The number of experiments needed can be reduced drastically.

Dynamic Similarity Principle

We already discussed geometrical similarity. In conducting experiments using scale models to determine force coefficients, it is clear that both the shape and orientation relative to airflow are important. However, suppose that we accurately measure the lift coefficient under conditions of geometric similarity on a model, for example, one-tenth the size of the actual vehicle. Is this enough? Can we now use this lift coefficient to estimate the lift on the prototype? The results of Eq. 2.22 show that this might not work. We must conduct the test so that the Reynolds and the Mach numbers on the model match those on the prototype. If this is done, we will achieve dynamic similarity and the predictions most likely will work if we correctly identified all of the key variables.

This is a concept of enormous value in the planning of experiments. Notice that even a complete theory or a complete numerical study often must be verified by experiments. It is therefore crucial to do this in a cost-effective and efficient manner. The principle of dynamic similarity is the key. In the wing-design example in the previous subsection, it is clear that if we had known the flight Reynolds and Mach numbers for the prototype, we would be able to design a test matrix that held these quantities fixed; the number of required test conditions could be reduced drastically.

It may not always be easy to match both the test Reynolds and Mach numbers to the prototype values. For instance, in testing a model of a hypersonic airplane, to fly at a Mach number of, for example, $M = 11$, configuring the test apparatus using a small model (no large wind tunnels can operate continuously in this Mach-number range) probably would lead to a R_e that is too small to match that experienced on the full-sized prototype. However, it often may be the case that one or more of the similarity parameters is of lesser importance. In the hypersonic airplane example, it may be that dependence on viscous effects is secondary to compressibility effects, which are important in high Mach-number flight. Then, useful data still can be obtained without a matching R_e. What is required here is an intimate knowledge of the sensitivity of the results to the parameters involved. This knowledge comes with experience. We attempt throughout this book to aid the student in acquiring at least some of the needed experience.

EXAMPLE 2.4 *Situation*: A 1:10 scale model sailplane wing is tested in a wind tunnel to determine the aerodynamic characteristics. The span of the prototype is 15 m and the average chord (width) of the wing is 0.8 m. Sea-level air is the working fluid for both the model and full-sized wings. Data from the test are shown in the table.

Speed (m/s)	Drag (N)
15.00	0.31
18.20	0.44
21.70	0.61
25.10	0.80
29.20	1.08
32.10	1.30
34.10	1.46
36.80	1.70
38.90	1.90

Required: (1) Sensitivity of the drag coefficient to the R_e based on the average chord length; and (2) drag estimate for the prototype at a speed of 135 km/hr.

Approach: Use the dynamic similarity principle. If the drag coefficient can be shown to be insensitive to the R_e, then the model data can be applied directly in estimating the drag.

Solution: The drag coefficient can be found from:

$$C_{D_m} = \frac{D_m}{\frac{1}{2}\rho V_m^2 S_m},$$

where the subscript, m, refers to the model. Standard sea-level air has a density of 1.23 kg/m^3, and the area of the model wing is $S_m = \frac{0.8}{10} \times \frac{15}{10} = 0.120$ m^2.

Thus, the drag coefficient is $C_{D_m} = \dfrac{13.55\, D_m}{V_m^2}$,

which is dimensionally correct if the drag measurements are inserted in newtons (N) and the velocity is measured in m/s.

Because the kinematic viscosity for sea-level air is:

$$v = \frac{\mu}{\rho} = 1.45 \times 10^{-5}\, \frac{m^2}{s},$$

the R_e for the model based on the average chord is found from:

$$R_m = \frac{\rho V_m c_m}{\mu} = \frac{0.08}{1.45 \times 10^{-5}} V_m = 0.552 \times 10^4\, V_m.$$

If the data are inserted and plotted, we find:

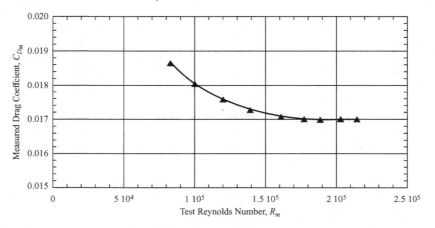

This shows that above a R_e of about 1.5×10^5, the drag coefficient is insensitive to the R_e (this may break down at yet higher speeds). This corresponds to a speed of about 27 m/s. Therefore, above this value, the drag coefficient has a nearly constant value of about $C_D = 0.017$. If the prototype speed is:

$$V_p = 135\, \frac{km}{hr} = 135\, \frac{km}{hr} \times \frac{1}{3,600} \frac{hr}{sec} \times 1,000\, \frac{m}{km} = 37.50\, \frac{m}{s},$$

then the corresponding R_e is:

$$R_p = \frac{\rho V_p c_p}{\mu} = \frac{0.8}{1.45 \times 10^{-5}} V_p = 5.52 \times 10^4 \times 37.50 = 2.07 \times 10^5.$$

Because dynamic similarity is achieved, $C_{D_p} = C_{D_m} = 0.017$.

The projected wing area is $S = 15.0 \times 0.8 = 12.0\, m^2$. Therefore, the full-scale drag at 135 km/hr is:

$$D_p = \tfrac{1}{2}\rho V_p^2 S C_{D_p} = 0.5 \times 1.23\, \frac{kg}{m^3} \times \left(37.50\, \frac{m}{s}\right)^2 \times 12.0\, m^2 \times 0.017 = 176.43\, N.$$

Appraisal: This example illustrates several useful features of the dynamic-similarity approach. It often happens that the aerodynamic force is insensitive to one or more of the similarity parameters. In this case, the Reynolds-number dependence was weak in the speed range of interest. Therefore, the drag coefficient measured in the wind tunnel could be used directly. Notice that there was no mention of possible Mach-number dependency; this is because the speed range is so low compared to the speed of sound that compressiblity effects are unimportant.

2.3 Mathematical Description of Fluid Flows

Several concepts of great value in later derivations of the defining equations are reviewed in this section. It is important that students are familiar with these ideas, especially the mental images that they require. Students probably are already familiar with some or all of these ideas from their fundamental courses in mechanics or thermodynamics.

Lagrangian versus Eulerian Mathematical Description

In studying the motion of particles and rigid bodies in mechanics courses, students learned the Lagrangian approach in which each element of the system is represented by a detailed model of its absolute motion and its motion relative to other parts of the system. In aerodynamics (or fluid mechanics), two different mathematical viewpoints may be adopted. Following the Lagrangian methods used in mechanics, physical laws are applied to individual particles in a flow. Equations then are derived that predict the location and status of individual fluid particles as a function of time as they move through space. The equations of dynamics typically use the Lagrangian viewpoint, time being the only independent variable. In these situations, it is necessary to keep track of the positions and velocities of all elements that comprise the system. However, for fluid-flow problems, so many particles are involved (virtually, an infinite number) in the field that this modeling stratagem usually is not practical.

Aerodynamics is concerned with a continually streaming fluid and the task is to determine distributions of flow velocities and fluid properties within a flow-field region rather than tracking the motion of specific particles passing through the region. For example, pressures at the surface of an airfoil or at points of interest in the flow are the main concern, not the behavior of the individual particles that exert these pressures. For this reason, a *field*, or *Eulerian, representation* is preferred in most aerodynamics problems and this viewpoint is taken herein. In this representation, the fluid properties and velocities throughout the flow are expressed in terms of position and time. Thus, flow variables such as pressure and velocity are described in terms of the flow-field coordinates and time, so that for an unsteady flow in 3-dimensional Cartesian coordinates (x,y,z), the pressure $p = p\,(x,y,z,t)$ and the velocity vector $\mathbf{V} = \mathbf{V}(x,y,z,t)$.

System

The physical laws evoked in derivation of the defining equations all apply to a system (e.g., recall that the First Law of Thermodynamics refers to a system). A *system* is

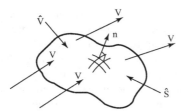

Figure 2.5. Fixed finite control volume.

a collection of fluid particles of *fixed identity*. Visualize a sealed plastic bag filled with liquid or gas that is proceeding downstream in a flow. The contents of the bag constitute a system. Note that if the flow is compressible, the density of the material inside the sealed bag may change. That is, the volume enclosed by the plastic bag may change but the mass of material contained within the bag must remain constant. A system may be finite in size or infinitesimal. An infinitesimal mass of fluid of fixed identity is termed a *fluid particle*.

Control Volume

The fundamental physical laws or principles expressing conservation of mass, momentum, and energy are expressed with respect to a system, as previously mentioned. For example, the First Law of Thermodynamics states that the heat added to a system minus the work done by the system equals the change in internal energy of that system. In aerodynamics, however, it is more convenient to derive these conservation equations by applying the physical laws to a control volume rather than to a system.

A *control volume* is a volume *fixed* in *space*. Flow may pass into or out of the control volume through the control surface that surrounds the control volume. Consider a surface that encloses a volume of arbitrary shape. Imagine that this surface is made of wire screen and that it is fixed in space in a flowing medium. Fluid may pass in or out through the porous screen, as illustrated in Fig. 2.5. The wire screen constitutes a control surface enclosing a control volume. This control volume may be finite or infinitesimal in size.

Review the definition of a system; it is different than a control volume, and the two concepts must be clearly understood. Both have important applications because the mathematical tools needed for describing a flowing medium are derived in subsequent chapters.

Flow Lines

One characteristic of fluid flow that makes it challenging to study is that in most cases, unlike solid mechanics, the medium is invisible. Experimental fluid mechanics makes use of smoke or dye to obtain visual insight into the physics of a flow. Likewise, it is helpful in analysis to obtain an analogous mathematical "visual" representation of a flow field whenever possible. The following three types of flow lines are of great value in flow visualization:

1. *Streamlines.* These lines are everywhere tangent to the velocity vector. They are most useful in describing a steady flow because in that case, each stream-line is the path that a fluid particle traces as it traverses the flow field. In an unsteady flow, the velocity vector changes in both direction and magnitude with time at any given point; therefore, only instantaneous streamlines are useful in visual interpretation of the field. That is, they describe how fast and in what direction the fluid particles are moving at a given instant. Because streamlines are always parallel to the velocity vector, flow cannot pass across a streamline.

2. *Streaklines.* These lines would be traced out by a marker fluid that is con-tinuosly injected into the flow stream at a given point. This is a commonly used experimental flow-visualization technique. For example, smoke often is injected into a flow to make the details visible. If the flow is steady, the fil-aments of smoke particles trace out the streamlines that pass through the injection points.

3. *Pathlines.* These are lines traced out in time by a given particle as its motion would be explained in a Lagrangian mathematical description.

Streamlines

Of the visualization lines described previously, the streamlines have the most value as a flow-visualization device throughout this book. This is because for the most part, we concentrate on steady-flow applications. Because streamlines are such a useful tool, we now must learn how to describe them mathematically.

Suppose that we have carried out an analysis for a flow problem that results in a solution for the velocity vector as a function of position in an appropriate coordi-nate system. That is, we found the velocity vector throughout the domain in Eulerian form:

$$\mathbf{V} = \mathbf{V}(\mathbf{r},t),$$

where $\mathbf{r}$ is the position vector for any point in the domain. This might be a compli-cated algebraic function that is difficult to interpret physically. Mathematical flow visualization is necessary to bring out the details. To illustrate the method, consider a case in which $\mathbf{V}$ is described in terms of a Cartesian coordinate system in which a point in the flow is identified by its position vector, $\mathbf{r} = x\mathbf{i} + y\mathbf{j} + z\mathbf{k}$, where $\mathbf{i}, \mathbf{j}$, and $\mathbf{k}$ are the unit vectors along the three coordinate axes. Then, in component form,

$$\mathbf{V} = u(x,y,z,t)\mathbf{i} + v(x,y,z,t)\mathbf{j} + w(x,y,z,t)\mathbf{k},$$

where u, v, and w are the velocity components as a function of position in the field. How can we find the streamline for this velocity vector that passes through a given point? By definition, the streamline must be parallel to the velocity vector. There-fore, the slope of the streamline projected on each of the three planes normal to the coordinate directions must be equal to the slope of the projection of the velocity vector. Thus, we must have:

$$\frac{dy}{dx} = \frac{v}{u}, \quad \frac{dz}{dx} = \frac{w}{u}, \quad \frac{dz}{dy} = \frac{w}{v}$$

as the condition of tangency at any given time. Because these can be written in the form:

$$\frac{dy}{v} = \frac{dx}{u}, \quad \frac{dz}{w} = \frac{dx}{u}, \quad \frac{dz}{w} = \frac{dy}{v},$$

it is clear that this information can be conveyed best in the simpler form:

$$\frac{dx}{u} = \frac{dy}{v} = \frac{dz}{w}. \tag{2.25}$$

To find the equation of the streamline at an instant of time, it is necessary to integrate these equations to define, for example, $z = z(x,y)$, which is an equation for any streamline. To find a particular streamline, which can be plotted for flow visualization, we simply evaluate the constants of integration by inserting values of x, y, and z corresponding to a point through which that streamline passes.

This is a simple process for two-dimensional problems because we then must integrate only the simple equation:

$$\frac{dy}{dx} = \frac{v}{u} \tag{2.26}$$

to find $y = y(x)$, which describes the required streamline.

The following example illustrates the technique used in finding a mathematical description of streamlines in a given flow. Notice that the presence of a function of time is of no concern because only the *instantaneous* streamline is meaningful.

EXAMPLE 2.5 *Required*: Find the equation for streamlines in a flow field described by the unsteady velocity vector:

$$\mathbf{V} = x(1 + at^2)\mathbf{i} + ye^{bt}\mathbf{j}.$$

Find the specific streamline passing through the point $(3, -2)$ at time $t = 0$.

Approach: Integrate Eq. 2.26.

Solution: Substituting the velocity components into Eq. 2.26 yields:

$$\frac{dy}{dx} = \frac{v}{u} = \frac{ye^{bt}}{x(1+at^2)}.$$

The equation is separable so, for a given time, we can write:

$$\frac{dy}{y} = \left(\frac{e^{bt}}{1+at^2}\right)\frac{dx}{x},$$

which immediately can be integrated to give:

$$y = Cx^{\left(\frac{e^{bt}}{1+at^2}\right)},$$

where C is the constant of integration. Notice that the streamline is time-dependent. Because we want the streamline passing through point $(3, -2)$ at time $t = 0$, C is readily found to be $C = -2/3$, and the required streamline equation is:

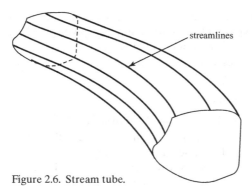

Figure 2.6. Stream tube.

$$y = -\frac{2}{3}x,$$

which is obviously a straight line at $t = 0$. The shape of the streamline changes as time progresses unless constants a and b are both zero.

Stream Tubes

In addition to visualization, there are other useful applications of streamlines. Consider, for instance, a bundle of streamlines that pass through a reference volume of arbitrary shape and size. The outermost set of streamlines form the walls of a *stream tube* and define the surface. Figure 2.6 shows a length of a typical stream tube. If the cross-sectional area of a stream tube is small, it is often referred to as a *stream filament*. The "streamlines" in an experiment using smoke for flow visualization in reality are stream filaments because the smoke must be injected from orifices of small but finite size. Obviously, all the streamlines passing through neighboring points must be nearly parallel, although they may diverge or become even more closely packed farther downstream.

Because there is no flow across any streamline, a stream filament behaves as a miniature pipe. We show in the next chapter that the average velocity at any point along a stream tube or stream filament is related to its cross-sectional area. It is almost obvious that if the flow is incompressible, the speed must increase when the area decreases and vice versa.

2.4 Behavior of Gases at Rest: Fluid Statics

In the previous sections, introductions to the problem of modeling fluid flows in aerodynamics and some calculations for aerodynamic forces are presented. At several points, information about properties of the atmosphere are required in completing the calculations. For example, it is frequently the case that we need to know the value of the atmospheric temperature, density, or pressure when estimating the magnitude of aerodynamic forces or evaluating propulsion-system performance. This need provides the opportunity to gain further practice in handling fluid-mechanics modeling and, in the process, to review methods for representing atmospheric effects. Because only the simplest type of physical relationship is involved, we can accomplish this now without using the more sophisticated modeling needed in subsequent problems. Only the ideas of statics and the ideal gas equation are required.

Modeling a Fluid at Rest

Because there is no motion, the ideas of statics yield the information that we need. That is, simple force and moment balances provide all of the necessary relationships among variables. In a fluid such as a gas, the forces acting on a fluid element are related to pressure or possibly to surface tension. Viscous forces do not appear because they depend on the presence of a velocity gradient.

Consider the fluid element shown in Fig. 2.7.

This rectangular element has cross-sectional area dA normal to the vertical axis, z. Only the forces in the z-direction are shown because all forces in the lateral direction are mutually balanced. As the figure suggests, the weight of the element, dW, is balanced by the pressure forces acting on the faces of the element. This provides a way to determine how the pressure changes in a body of fluid such as a water tank or the earth's atmosphere. We use the latter application to illustrate the principles involved.

For static equilbrium in the vertical directions, it is necessary that $\sum F_z = 0$, so that:

$$pdA - (p + dp)dA - dW = 0. \tag{2.27}$$

Because the mass of the element is equal to the volume times the density of the material inside, we can write the differential weight as $dW = dmg = (\rho dV)g$, where g is the local acceleration due to gravity. If the height of the element is dz, then the volume is $dV = dAdz$, and the force balance yields $dp + \rho gdz = 0$, which can be expressed in differential-equation form as:

$$\frac{dp}{dz} = -\rho g. \tag{2.28}$$

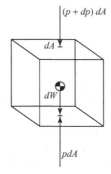

Figure 2.7. Fluid element in static equilibrium.

This result is often called the *hydrostatics equation* and it shows how pressure and density are interrelated when there is no motion of the fluid. If the density is known as a function of pressure and the gravitational acceleration is known as a function of z, this equation can be integrated to yield information about the pressure as a function of z. In our application, we interpret $z = h$ to be measured from the earth's surface; it is referred to as the *geometric altitude*. There are other definitions for *altitude* that may be useful on occasion. For example, we refer to the *pressure altitude* as the altitude corresponding to a standard atmospheric value for the static pressure measured outside the vehicle. Equation 2.28 indicates that if the density and gravity were constant, the pressure would decrease linearly with altitude. However, this does not account for observable features of the atmosphere; effects of temperature and density changes also must be considered.

Standard-Atmosphere Models

Equation 2.28 is the basis of a set of atmospheric representations called *standard atmospheres*. In fact, there is no accepted standard. Tables of atmospheric properties displayed in many aerodynamics textbooks usually are based on the Air Force ARDC = Air Research and Development Command model atmosphere published in 1959. This model, briefly described herein, is based on radiosonde balloon observations of the temperature distribution in the atmosphere.

Understand that the values predicted by this approximate model represent only an average set of properties. No diurnal or seasonal changes and no geographical effects are reflected in this model. The definition of the model is shown in Fig. 2.8. The temperature distribution through the atmosphere is represented by a set of straight lines. Vertical lines represent "isothermal" layers such as the stratosphere in which temperature changes are negligible. Other layers, such as in the *troposphere*, are characterized by the rate at which the temperature changes with altitude in an assumed linear manner. The rate of change in a given "gradient" layer is called the *lapse rate*, denoted by symbols a_n in the figure. What is significant is that this set of lines defines the temperature, $T = T(h)$, from which all further information can be found by means of the ideal gas equation (Eq. 2.1) and the hydrostatics equation (Eq. 2.28).

Atmospheric Property Variations in Gradient Layers

The thermodynamic properties for the various layers can be estimated by solving the appropriate equations. In a gradient layer, such as the troposphere, the temperature is assumed to change linearly in such a way that:

$$a_n = \frac{dT}{dh} \tag{2.29}$$

is a constant. Therefore, by integration for a given gradient layer,

$$T(h) = T_i + a_n(h - h_i), \tag{2.30}$$

where T_i is the temperature at some altitude h_i in the gradient layer, n ($n = t,s,i$, and e stand for troposphere, stratosphere, ionosphere, and so on). Then, from the ideal-gas

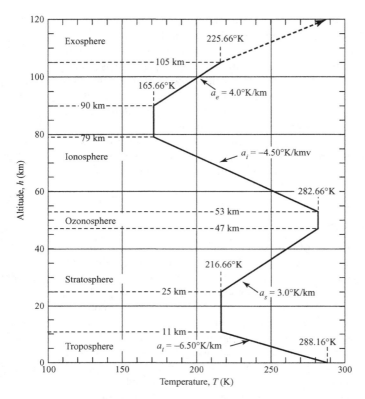

Figure 2.8. Typical "standard" atmosphere.

equation of state (Eq. 2.1), we can find a relationship among the pressure, altitude, and density:

$$\rho = \frac{g}{R(T_i + a_n(h - h_i))} p. \tag{2.31}$$

This then can be inserted into Eq. 2.28 to replace the density so that only the pressure and altitude variables remain. Thus,

$$\frac{dp}{p} = -\left(\frac{g}{R}\right)\frac{dh}{(T_i + a_n h_i + a_n h)}, \tag{2.32}$$

where the variables were separated. This simple differential equation can be solved easily by integration, with the result:

$$ln\frac{p}{C_1} = -\left(\frac{g}{a_n R}\right)ln(T_i + a_n h_i + a_n h),$$

where C_1 is the constant of integration. Rearranging,

$$p = C_1[T_i + a_n(h - h_i)]^{-\left(\frac{g}{a_n R}\right)} = C_1 T^{-\left(\frac{g}{a_n R}\right)} \tag{2.33}$$

can be evaluated by using known values of p and T at a given altitude. For example, in the troposphere, we can evaluate the constant of integration by using the fact that the sea-level pressure and temperature are $p_{SL} = 1.01325 \times 10^5$ N/m^2 and $T_{SL} = 288.16°$K. Then, with $a_n = a_t = -6.5°$K/km in the troposphere, and letting the reference point be at sea level ($h_i = 0$, $T_i = 288.16°$K), the constant of integration is:

$$C_1 = p_{s1} T_{sl}^{(g/a_n R)}.$$

Now, all the basic properties are available at any altitude in the troposphere gradient layer: Eq. 2.30 gives the temperature, Eq. 2.33 gives the pressure, and Eq. 2.31 gives the density.

Other properties such as the speed of sound and coefficient of viscosity that are mainly functions of temperature also can be determined. For example, the speed of sound, from Eq. 2.6, is:

$$a = \sqrt{\gamma R T},$$

and using Sutherland's Law, the viscosity coefficient is

$$\mu = \mu_0 \left(\frac{T}{T_0}\right)^{\frac{3}{2}} \frac{T_0 + S_1}{T + S_1}, \tag{2.34}$$

where for air, $S_1 = 111°$K, and $\mu_0 = 1.81 \times 10^{-5}$ kg/m s and $T_0 = 273.2°$K are reference values. The kinematic viscosity, ν, often is needed. It is simply the ratio:

$$\nu = \text{kinematic viscosity coefficient} = \frac{\mu}{\rho} \tag{2.35}$$

of quantities already known at a given altitude.

Notice that armed with these results, the atmospheric properties can be evaluated at any altitude within the layer and at the upper edge of the layer. The latter information then is used to evaluate conditions in the next layer.

Atmospheric Property Variations in Isothermal Layers

In an isothermal layer, the temperature is constant; therefore, the equation of state reduces to a relationship between pressure and density:

$$\rho = \left(\frac{1}{RT}\right) p.$$

Thus, substituting for the density, the hydrostatics equation becomes:

$$\frac{dp}{p} = \left(\frac{g}{RT}\right) dh,$$

where the combination in parentheses can be considered constant. This simple differential equation can be solved by direct integration, with the result:

$$p = C_2 \exp\left[-\left(\frac{g}{RT}\right) h\right], \tag{2.36}$$

where C_2 is a constant of integration that can be evaluated by inserting the initial conditions for a particular isothermal layer from the information in Fig. 2.8 and from results of the previous gradient layer evaluated at the initial altitude of the isothermal layer. Knowledge of the temperature allows evaluation of the speed of sound and the viscosity coefficients.

The Web site that accompanies this book includes a program labeled **STDATM** that evaluates the equations described in the preceding paragraphs to give atmospheric properties at any altitude. Running the program, the user is asked to choose the system of units (i.e., either English or SI) and to insert the altitude either in meters or feet. The program then returns the corresponding temperature, pressure, density, speed of sound, dynamic-viscosity, and kinematic-viscosity values. Also, the program can be called from within other program modules (described later) to determine automatically the required atmospheric data.

2.5 Summary

This chapter is a review of mathematical concepts for modeling fluid flows such as those experienced in most aerodynamics applications. Important ideas from continuum mechanics and thermodynamics are reviewed, and modeling of the earth's atmosphere is described using a simple application of fluid statics.

The most important part of this chapter is the introduction of the powerful method of dimensional analysis and the associated ideas of similarity. Application of these ideas reduces the basic problem of aerodynamics—namely, the estimation of the principal aerodynamic forces and moments—to its simplest form. We found that the key to solving this problem is the estimation of dimensionless coefficients such as lift and drag coefficients. An important discovery is that these coefficients are dependent on basic similarity parameters that can be identified as ratios of the fundamental set of forces that characterize the gas motion. In aerodynamics, by far the most important similarity parameters are the Mach and Reynolds numbers, which represent effects of compressibility and viscosity, respectively. Other forces frequently may enter an aerodynamics problem. By applying the examples given in this chapter, students can address such situations when they arise. The following problem set provides the opportunity for students to test their level of understanding.

Another important feature of the similarity approach is its great utility in planning efficient experiments. Understanding how the many variables interact through the dimensionless similarity parameters enables an experimenter to drastically reduce the number of tests needed to determine the significant interactions. It also provides a useful tool for classifying and interpreting various experimental results.

Finally, the stage is now set so that we can explore in more detail, both analytically and numerically, the fascinating problem of flow over aerodynamic bodies and the attendant production of forces that can be used to design efficient flight vehicles.

PROBLEMS

2.1 A well-known demonstration of aerodynamic forces consists of a vertical jet of air. A light ball (e.g., a ping-pong ball) can be supported in equilibrium in this jet. The equilibrium position, distance h from the jet, is found to depend on the ball radius, R; the jet radius, r; the ball weight, W; the jet velocity, V; and the density and viscosity of the airflow. Find the dimensionless similiarity parameters that govern this problem by using the ideas of dimensional analysis.

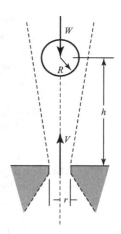

2.2 A racing car is to travel through standard sea-level air at a speed of 200 km/hr. An experiment is devised to determine the pressure distribution over the body using a one-sixth scale model in a water-tunnel test. Which water speed should be used? Which factors must be considered to achieve dynamic similarity?

2.3 The drag of an airfoil depends on the density, viscosity, and flow speed as well as chord length, c. A one-fifth scale model is tested in a wind tunnel at a test R_e of 4.1×10^5, based on chord length, c. Test conditions are $T = 15°C$ and $p = 10$ atmospheres. The prototype airfoil has a chord length of 1.75 m and is to be operated in standard sea-level air. Find the following:
(a) velocity at which the wind-tunnel model was tested
(b) velocity at which the prototype is to operate

2.4 An airship is to operate at 20 m/s in standard sea-level air. A one-twenty scale model is tested in a wind tunnel at the same air temperature to determine the drag.
(a) Which criteria assure dynamic similarity?
(b) If the model is tested at 20 m/s, which pressure should be used in the wind tunnel?
(c) If the model drag is 300 N, what is the drag of the prototype in lbf?

2.5 A model test is performed to determine the flight characteristics of a Frisbee. Dependent parameters are the drag force, D, and the lift force, L. The independent parameters are the diameter, d; speed, V; and angular velocity, w; as well as density and viscosity. Determine suitable dimensionless parameters and express the functional dependence among them.

2.6 Consider the water flow around a circular cylinder of diameter, D, and length, L. In addition to geometry, the drag force is known to depend on liquid speed, V; density, ρ; and viscosity, μ. Express the drag force in dimensionless form as a function of all relevant variables. The static-pressure distribution on a cylinder measured in a laboratory can be expressed in terms of the dimensionless pressure coefficient:

$$C_p = \frac{p - p_\infty}{\frac{1}{2}\rho V^2}.$$

The lowest value of the pressure coefficient is found to be $C_p = -2.4$ at the location of the minimum static pressure on the cylinder surface. Estimate the maximum speed at which the cylinder could be towed in water at atmospheric pressure.

2.7 A Boeing 747 with a gross weight of 650,000 lbf is flying at 35,000 ft at a speed of 465 knots (1 knot = 1nm/hr = 1.152 statute miles/hr = 1.69 ft/sec). This airplane has a wing area $S = 5,500$ ft^2. The atmospheric density at this altitude is

0.0007382 slug/ft^3. Estimate the value of the lift coefficient produced by the wing under these conditions.

2.8 If the Boeing 747 in Problem 2.6 has a L/D ratio of 16.5, what is the drag coefficient under the conditions described?

2.9 A Piper Super Cub is cruising at 115 mph at 7,000 ft ($\rho = 0.001927$ slug/ft^3). The Cub has a wing area of 157 ft^2 and a weight of 1,500 lbf. It is flying at a drag coefficient of $C_D = 0.031$. Find:
   (a) drag force
   (b) lift coefficient
   (c) L/D ratio

2.10 The drag force exerted by air friction on a moving automobile must be overcome by power produced by the engine. Assuming that this power is a function of the velocity, frontal area, drag force, air density, and viscosity, use dimensional analysis to find an expression for the power.

2.11 The efficiency of a pump is assumed to be a function of the flow rate Q (cubic ft/sec), pressure increase Δp, pipe diameter d, and density and viscosity of the fluid. Use dimensional analysis to find an expression for the pump efficiency.

2.12 You are asked to determine the drag force exerted on a low-drag automobile design at a speed of 150 mph and an air temperature of 62°F. A one-fifth scale model is to be tested in a water tunnel at 70°F. Determine the required water velocity for dynamic similarity between the prototype car and the model. If the drag force on the model is measured at 10 lbf, determine the expected drag on the automobile.

2.13 Derive an expression for the terminal velocity of a parachutist falling through air if the velocity V depends on the parachute diameter d, air density, viscosity, acceleration due to gravity, and mass of the parachutist.

2.14 A jet-propelled boat takes in water at the bow, passes it through pumps, and discharges it through ducts at the stern. The propulsive force T is a function of the velocity of the boat, V; water velocity at the duct exit, u_e; exit pressure, p_e; pump power, P_p; and water density. Use dimensional analysis to find an expression for the propulsive thrust.

2.15 In the design of an all-terrain vehicle, it is decided to not use a propeller to move the vehicle through water. Instead, the rotation of the wheels is expected to provide the necessary thrust. The propulsive force T is determined to be a function of rotational speed of the wheels, w; diameter of the tires, d; width of the tires, w; and the density and viscosity of the water. Determine the expression for the propulsive force by means of dimensional analysis.

2.16 A rocket expels high-velocity gases as it travels through the atmosphere. The thrust, T, depends on the exit pressure of the gases, ambient pressure, and exit gas velocity and nozzle exit area. Use dimensional analysis to derive an expression for the rocket thrust.

2.17 An airplane wing of chord length 0.8 m travels through sea-level air. The wing velocity is 100 m/s. A one-tenth scale model of the wing is tested in a wind tunnel at 25°C and 101 kPa. What must the tunnel velocity be for dynamic similarity?

2.18 A jet engine takes in air at atmospheric pressure, compresses it, and uses it in a combustion process to generate high-temperature gases. These gases then are passed through a turbine to provide power for compressing the inlet air. After passing through the turbine, the gases are exhausted to the atmosphere through a nozzle to produce thrust. The thrust T developed is a function of inlet air density and pressure, ρ_i and p_i; pressure after the compressor, p_c; energy content of the fuel, E; exhaust gas pressure, p_e; and exit velocity, v_e. Use dimesional analysis to find an expression for the thrust.

2.19 The drag characteristics of a blimp traveling at 4 m/s are to be studied by experiments in a water tunnel. The prototype is 20 m in diameter and 110 m long. The model is one-twentieth scale. What velocity must the model have for dynamic similarity? If the drag of the model is 15 N, what thrust must the blimp propulsion system generate to fly at constant velocity?

2. 20 Liquid flows horizontally through a bed of sand. The grains are all spherical and have the same diameter. Use dimensional analysis to obtain an expression for the pressure drop through the sand bed.

2.21 A common drinking glass is 7 cm in diameter and filled to a depth of 12 cm with water. Calculate the pressure difference between the top and bottom of the glass sides.

2.22 At an elevation of 12 km, atmospheric pressure is 19 kN/m^2. Determine the pressure at an eleveation of 20 km if the temperature is $-57°C$.

2.23 A two-dimensional steady flow is described by the velocity components:

$$u = 2x; \qquad v = -6x - 2y.$$

Find an equation for a streamline passing through point (1,1). Sketch this streamline and its neighbors.

2.24 A two-dimensional velocity field is described by the vector:

$$\mathbf{V} = (2xy^2)\mathbf{i} + (2x^2y)\mathbf{j}.$$

Find the equation for a streamline that passes through point (1,7).

2.25 The absolute value of the velocity and the equation of the streamlines in a velocity field are:

$$|\mathbf{V}| = V = \sqrt{x^2 + 2xy + 2y^2}; \quad y^2 + 2xy = \text{constant, respectively.}$$

Find expressions for the two Cartesian velocity components, u and v.

REFERENCES AND SUGGESTED READING

Anderson, John D., *Fundamental Aerodynamics*, McGraw-Hill Book Company, New York, 1984.

Kuethe, A. M., and Schetzer, J. D., *Foundations of Aerodynamics*, 2nd ed., John Wiley & Sons, New York, 1961.

Liepmann, H. W., and Roshko, A., *Elements of Gasdynamics*, Dover Publications, Inc., New York, 1985.

Prandtl, L., and Tietjens, O. G., *Fundamentals of Hydro- and Aeromechanics*, Dover Publications, Inc., New York, 1957.

Shevell, Richard S., *Fundamentals of Flight*, Prentice-Hall, Inc., NJ, 1989.

von Kàrmàn, Theodore, *Aerodynamics, Selected Topics in the Light of Their Historical Development*, Dover Publications, Inc., New York, 2004.

3 Equations of Aerodynamics

3.1 Introduction

To solve the fundamental problems of aerodynamics defined in Chapter 1, it is necessary to formulate a mathematical representation of the underlying fluid dynamics. The appropriate mathematical expressions or sets of equations may be algebraic, integral, or differential in character but will always represent basic physical laws or principles. In this chapter, the fundamental equations necessary for the solution of aerodynamics problems are derived directly from the basic laws of nature. The resulting mathematical formulations represent a large class of *fluid mechanics* problems within which aerodynamics is an important subclass.

Some problems in aerodynamics require solutions for all of the variables needed to describe a moving stream of gas—namely, velocity, pressure, temperature, and density. Because velocity is a vector quantity (i.e., with magnitude and direction), in a general case there are three scalar velocity components. Thus, in many cases of interest, there is a total of six unknowns: three velocity components and the scalar thermodynamic quantities of pressure, temperature, and density.[*] This requires six independent equations to be written to solve for the six unknowns. The physical laws of *conservation of mass, momentum,* and *energy* supply five such equations (i.e., the momentum equation is a vector equation; therefore, conservation of momentum leads in general to three component equations). For all of the subject matter in this book, the assumption of an ideal gas is physically realistic. Thus, the perfect gas law (i.e., equation of state) $p = \rho RT$, which relates pressure, density, and temperature, supplies the final equation needed to solve for the six unknowns.

3.2 Approach

The equations for mathematical representation of a moving fluid are derived here for the most general case of an unsteady, compressible, viscous flow. It is most often the case that full generality is not required in a particular application. It may not be

---

[*] In some cases, it may be necessary to include additional dynamic variables such as angular momentum in rotating fluids and, in other cases, additional thermodynamic properties such as the entropy, which is used to account for the reversibility of flow processes.

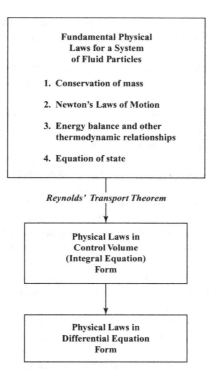

Figure 3.1. Procedure for development of equations of motion.

necessary to account for compressibility effects in the flow over an airplane if the speed of flight is sufficiently low. Similarly, viscous forces may be negligible in some applications. An important skill emphasized in this chapter is the decision-making process used to reduce the general formulation to a simpler one that accounts for only the needed physical effects. In the examples mentioned, the density can be held constant if the flow is incompressible or the viscous terms can be dropped when frictional effects are not crucial. Such considerations may result in vast simplification of the mathematical problem.

The approach to be followed is illustrated in Fig. 3.1. We start by briefly reviewing the basic physical laws pertaining to conservation of mass, momentum, and energy. These are naturally expressed in familiar form by representing the behavior of a well-defined system of fluid particles. In this Lagrangian form, the equations are of limited use in handling the problems we address. What usually is needed is the effect of the motion of these particles on the forces of interaction on a body immersed in the flow. It is not required to determine a detailed history of the motion of the original group of fluid particles (i.e., the system) as it moves downstream. The Eulerian control-volume approach as introduced in Chapter 2 provides a powerful tool for determining the properties of the fluid flow at any time and any location in the domain of interest. Review the definitions in Chapter 2 to understand the distinction between a Lagrangian and a Eulerian approach.

As indicated in Fig. 3.1, the first step is to translate the physical laws into control-volume form. The Reynolds' Transport Theorem provides the mechanism for passing from the Lagrangian-system model to the more convenient Eulerian control-volume representation. The result is a set of integral equations that provide a useful way to describe the effects of fluid motion without knowledge of the details at each point in the region of interest. In other words, it gives an average or overall picture of the motion. There is a significant benefit here because the control volume can be defined in a variety of useful ways to fit the geometry of any particular problem. The shape and location of the boundaries of the control volume can be chosen to accommodate the required size and shape of the problem domain. A control volume can translate, accelerate, or rotate as necessary to fit the definition of the application. Any number of control volumes can be applied in a given problem to provide information about specific parts of the flow field.

Although in most applications we apply directly the differential equations of fluid motion in solving aerodynamics problems in later chapters, there are many useful and practical benefits of the intermediate control-volume integral formulation. The control-volume point of view often is a valuable tool in gaining physical understanding in a given situation before undertaking to solve the applicable differential equations. For example, if we need to determine only the *total* force acting on an object immersed in a flow, then this information frequently can be found without establishing all the details of the local-gas motion. If, however, we need to know how the fluid-interaction force elements that lead to this net force are distributed over the body, then a "sharper" representation is needed in which the local behavior of the fluid is determined. In this example, it may be necessary to find the actual distribution of pressure and viscous-shear forces over the body surface. This can be done only if we pass from the control-volume *integral equation* form to *differential equation* form so that we can study how the fluid properties change locally from point to point. Again, Fig. 3.1 illustrates the "plan of attack." We use the integral control-volume equations to determine the underlying differential equations. As these analytical steps are carried out in this chapter, it is important to follow the procedure focusing mainly on the reduction from the general form to specific applications. The final result is a set of mathematical tools that enables the eventual solution of any fluid-dynamics problem. The remainder of the book is devoted to the application of these tools to analyze a wide variety of aerodynamics problems.

3.3 Physical Laws for Motion of a System

Throughout the discussions in this chapter, we use the simplifications available by application of the modeling ideas discussed in Chapter 2. Thus, the medium is taken to be composed of a continuum of contiguous fluid particles rather than individual molecules. Because we are interested in the dynamics of fluid flow, it is natural to express each physical law describing the motion of these particles as a rate of change of a physical variable required to characterize the material. As examples, we may be interested in the rate at which mass flows across a cross section of a duct, how fluid particles are accelerated by external forces, and the rate at which work is done on the fluid by pressure forces.

All of the physical effects needed in representing the fluid motion can be given in rate form for a system of particles. To aid in visualization of what we are modeling, consider a flow of fluid in which we identify at some arbitrary initial time a distinct "glob" of air consisting of many individual particles. Imagine that the glob can be marked by coloring it with dye. Then, as it moves downstream, it is stretched and deformed by its interaction with the surroundings and by forces acting among the constituent particles. The identity is preserved although its shape, size, and other characteristics may be altered by the motion, by forces applied at its boundaries, or by *body forces* (e.g., gravity or magnetic force) that act on the entire system, or by energy flowing into or out of the surroundings. Henceforth, we refer to this marked glob as the *system*.

Each of the required physical laws is expressed in system form. These should be already familiar to students from the study of basic physics and thermodynamics.

Mass Conservation

If the mass of the system is m at the initial instant of time, then its mass is preserved for all time afterward unless there is a process involving this group of particles that changes part of its mass into energy. This could happen if, for instance, nuclear processes occur in which mass is converted to energy. We do not consider such processes in this book; so, for all cases of interest, we can state with confidence that the mass remains constant. Thus, in rate form, we write:

$$\frac{dm}{dt} = 0 \tag{3.1}$$

for the specified system of fluid particles. The mass can be expressed conveniently as an integral over the volume containing only the identified particles:

$$m = \int_{\substack{\text{system} \\ \text{mass}}} dm = \iiint_{\substack{\text{system} \\ \text{volume}}} \rho d\hat{V}, \tag{3.2}$$

where *system mass* is always meant to identify the mass of the particular glob of particles constituting the defined system; and *system volume*, $\hat{V}$, is that volume of space enclosing the system glob at any given instant of time. Notice that the system volume might be required to change with time (although the mass remains constant) if the density of the glob, ρ, changes from point to point in the flow field. If Eq. 3.1 is satisfied, we ensured that system mass is not created or destroyed; in other words, *mass is conserved*. All other required laws of motion are expressed in similar rate form in the following subsections.

Newton's Laws of Motion

These important laws of nature specify how the system of particles moves in response to external forces or moments. For the system (described in an inertial frame of reference), we write:

$$\begin{cases} \dfrac{d\mathbf{P}}{dt} = \mathbf{F} \tag{3.3} \\[2mm] \dfrac{d\mathbf{H}}{dt} = \mathbf{M}, \tag{3.4} \end{cases}$$

where **P** and **H** are the linear and angular momenta of the system, respectively, and **F** and **M** are the externally applied sums of forces and moments. Because we are dealing with a system of individual particles, the force **F** must include the interaction forces between the particles that constitute the system. These occur in equal and opposite pairs, so that it is usually the case that only the forces acting on the boundary of the system must be considered. It may be helpful to review the treatment of systems of particles in textbooks on dynamics.

Remember that the coordinate system in which the momentum vectors, **P** and **H**, are expressed must be a *Newtonian* or *inertial* coordinate frame; that is, it cannot be accelerating or rotating. If it is necessary or convenient to use noninertial coordinates in describing the system motion, then it is necessary to introduce corrective terms (e.g., centripetal and Coriolis effects) into Eqs. 3.3–3.4. This normally is not required in the applications considered in this book because the emphasis is on steady flow without spin.

The linear and angular momentum for the system can be written as integrals over the mass or volume of the system as follows:

$$
\mathbf{P} = \int\limits_{\substack{\text{system}\\\text{mass}}} \mathbf{V}\, dm = \iiint\limits_{\substack{\text{system}\\\text{volume}}} \rho \mathbf{V}\, d\hat{V} \tag{3.5}
$$

$$
\mathbf{H} = \int\limits_{\substack{\text{system}\\\text{mass}}} \mathbf{r} \times \mathbf{V}\, dm = \iiint\limits_{\substack{\text{system}\\\text{volume}}} \rho(\mathbf{r} \times \mathbf{V}) d\hat{V}, \tag{3.6}
$$

where **V** is the velocity vector at a point in the system glob and **r** locates the corresponding point in the inertial coordinate system.

Although both linear motion of the system as expressed by the linear momentum **P** and rotational motion expressed by the angular momentum **H** are defined here for completeness, only the former appears in most aerodynamics applications. Thus, we carry out further manipulation of Newton's laws in detail only for the linear momentum changes of the system. The student should notice that completely analogous developments can be made for the angular momentum if needed in a particular application of interest, where angular motion or spin is an important feature.

The First Law of Thermodynamics

A fundamental law describes changes of the energy of a system by interaction with its surroundings. The First Law of Thermodynamics is often written in differential form in the notation of thermodynamics as:

$$
dE = \delta Q - \delta W, \tag{3.7}
$$

where dE is the change in system energy and δQ and δW represent the differential flow of heat energy to and differential work done by the system. In some textbooks, the work is defined as that done *on* the system *by* the surroundings. Then, the sign on δW is reversed. We use the more common engineering form expressed in Eq. 3.7. In rate form, the First Law of Thermodynamics may be written as:

$$\frac{dE}{dt} = \dot{Q} - \dot{W}, \tag{3.8}$$

where the system energy is:

$$E = \int\limits_{\substack{\text{system} \\ \text{mass}}} e'dm = \iiint\limits_{\substack{\text{system} \\ \text{volume}}} \rho e'd\hat{V}. \tag{3.9}$$

The energy per unit mass is:

$$e' = e + \frac{V^2}{2} + \text{potential energy}, \tag{3.10}$$

where e is the internal energy due to random molecular motion at any point within the system and $V^2/2$ is the kinetic energy per unit mass. An additional term is often added to this expression to account for the potential energy due to position in a gravity field. This normally is not needed in the gas flows of interest in this book. It could be important in flows of liquid involving large elevation changes.

The Second Law of Thermodynamics

It may be that compressible processes must be tested for reversibility by means of the Second Law of Thermodynamics. For completeness, we show this here in the form of the entropy-rate inequality. In differential form, the Second Law of Thermodynamics is:

$$dS \geq \frac{\delta Q}{T}, \tag{3.11}$$

where S is the system entropy and T is the (absolute) temperature. S can be written in terms of the corresponding intensive property, s (entropy per unit mass) as:

$$S = \int\limits_{\substack{\text{system} \\ \text{mass}}} s\, dm = \iiint\limits_{\substack{\text{system} \\ \text{volume}}} \rho s d\hat{V}. \tag{3.12}$$

Then, the required rate equation form for the Second Law of Thermodynamics is:

$$\frac{dS}{dt} \geq \frac{\dot{Q}}{T}. \tag{3.13}$$

Additional Physical Laws

The "laws" describing the physics of fluid motion in rate form usually must be supplemented by additional information so that a complete mathematical representation is achieved. Important supplementary physical relationships may come from thermodynamics, chemistry, heat transfer, or additional mechanical models to fully describe the fluid medium. For example, in Chapter 2, supplementary physical relationships are introduced in the form of the equation of state (see Eq. 2.1) and Newton's Law of Viscosity (Eq. 2.4). These laws are incorporated as needed to fully describe any given situation.

3.4 Physical Laws in Control-Volume Form

Throughout the analyses in this chapter, we use the simplifications afforded by describing the fluid motion in Eulerian control-volume form rather than as a Lagrangian system of fixed identity. It is therefore necessary to transform the physical laws expressed in system form into a more convenient control-volume form. To do this, we change our focus from an identifiable glob of fluid to a region of space in which we are interested. There is really no need to follow the glob (i.e., the system) downstream. All we need to establish is the effect that its motion has on the local flow-field characteristics in the region, the control volume, that encompasses the part of the flow to be studied. Then, we must account for the passage of mass, momentum, energy, or other fluid properties through the boundaries of the control volume. The mathematical rule that facilitates the change from the system to the control-volume form often is called the *Reynolds' Transport Theorem*.

The Reynolds' Transport Theorem

Examine each physical law described in Section 3.3. Notice that each law involves an *extensive* and a related *intensive* property of the fluid. For example, in Eq. 3.9, the extensive property E (i.e., the total energy of the entire system) is given by the integral over the system volume in which the differential element of energy is $e'\rho d\hat{V}$ (i.e., e' the intensive variable, energy per unit mass, multiplied by a differential mass element $\rho d\hat{V}$). Table 3.1 shows each extensive variable involved in the rate equations and the related intensive variable. One of the intensive variables is a constant, 1, because the differential mass element is $1\,\rho d\hat{V}$.

Two of the variables are scalars (e' and s) and two are vectors ($\mathbf{V}$ and $\mathbf{r} \times \mathbf{V}$). What is needed is a way to represent the rate of change of each extensive variable as a volume integral (we also need surface integrals) over a fixed control volume instead of the moving system volume. Actually, it is not necessary that the control volume be fixed, but the analytical method is demonstrated here for the simplest case of a control volume fixed in space, as shown in Fig. 3.2.

The region illustrated in Fig. 3.2 is filled with a fluid moving relative to the control volume (dashed outline). There is no restriction on the size, shape, or location in the flow field of the control volume. How these geometrical features are chosen to analyze a given problem is shown by example later in this chapter. For now, we choose the control volume so that at some arbitrary initial time, t, it coincides with our moving system glob, as illustrated in Fig. 3.2(a) with a solid outline. At a later

Table 3.1. *Extensive and intensive variables*

Property	Extensive Variable	Associated Intensive Variable
Mass	m	1
Momentum	$\mathbf{P}$	$\mathbf{V}$
Angular momentum	$\mathbf{H}$	$\mathbf{r} \times \mathbf{V}$
Energy	E	e'
Entropy	$\mathbf{S}$	s

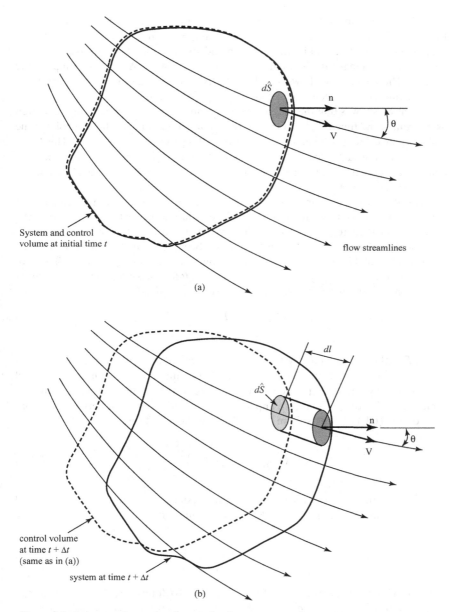

Figure 3.2. System and associated control volume.

instant of time $t + \Delta t$—for example,—the system has moved with the stream to a new location slightly displaced from the original one in Fig. 3.2(b). From the standpoint of the control volume, depicted by the dashed boundary, some of the system mass and its associated properties including momentum and energy flowed out through the surface, and new material similarly passed in to take the place of the part that moved downstream. Notice that because a small but finite time is required for this to happen, it may be that there are corresponding small changes in both the physical properties of the fluid and in the direction and speed of the flow. We use the small-ness of these changes in the limit of a small time change, $\Delta t \to 0$, to simplify the analysis. The result shows how to replace the time derivative of the system extensive variable into an equivalent control-volume representation.

What we show is that the rate of change of any intensive property can be written as the sum of the time rate of change of the volume integral over the control volume and a surface integral over the surface bounding the control volume. The latter accounts for the fact that the rate of change is due in part to the passage of mass and the associated physical properties through the surface of the control volume, as previously described.

The derivation is simply an extension of the ideas of differential calculus. It is useful to carry out the analysis for a representative property—for example, the energy—and then to generalize the final result for use with any other property. The energy balance is a good sample property because most students have worked extensively with it in basic thermodynamics courses. To use the First Law of Thermo-dynamics (Eq. 3.8), it is necessary to evaluate dE/dt for the system and then convert it into control-volume form. From the definition of the derivative:

$$\frac{dE}{dt} = \lim_{\Delta t \to 0} \left(\frac{E_{t+\Delta t} - E_t}{\Delta t} \right), \tag{3.14}$$

where E_t is the energy of the system of fluid particles at the initial time, t, and $E_{t+\Delta t}$ is the system energy at $t + \Delta t$, a short time later. At the initial time, the energy contained in the control volume is equal to that in the system because they occupy exactly the same space. For the moment, we append subscripts to distinguish between the con-trol volume (CV) and the system (SYS). Thus, at the initial time:

$$E_t = (E_{SYS})_t = (E_{CV})_t = \left(\iiint_{CV} e' \rho d\hat{V} \right)_t. \tag{3.15}$$

At $t + \Delta t$, we must account for the fact that the system is displaced from the control volume, as illustrated in Figure 3.2(b). Thus, the integral:

$$(E_{CV})_{t+\Delta t} = \left(\iiint_{CV} e' \rho d\hat{V} \right)_{t+\Delta t} \tag{3.16}$$

is *not* equal to the system energy at the later time. To determine the system energy, $E_{t+\Delta t}$, for use in the limiting process in Eq. 3.14, we must correct the control-volume integral (Eq. 3.16) in two ways. We must account for new energy that has passed into the control volume with the flow to make up for the void left by the system. This energy then must be *subtracted* from the energy given by Eq. 3.16. We also must *add* the energy in the part of the system that already has left the region of the control volume (i.e., that part of the system shown in Figure 3.2(b) to the right of the dashed boundary of the control volume), properly updated to the later time.

The two corrections can be handled simultaneously and simply by working with a differential volume such as that illustrated in Figure 3.2(b). Consider a differential surface element, $d\hat{S}$, at any point on the surface of the control volume at time t (see Figure 3.2[a]). The outward pointing unit vector, $\mathbf{n}$, is perpendicular to the control surface at that point. Notice that the velocity vector $\mathbf{V}$ at that point in the flow may be in any direction at an angle (e.g., θ) to the normal direction. Note, however, that at locations on the control surface where $\theta < 90°$, the flow is *out* of the control volume and that at locations where $\theta > 90°$, the flow is *into* the control volume. When $\theta = 90°$, there is no flow across the boundary because the streamlines are then parallel to it. At the later time shown in Fig. 3.2(b), the part of the system bounded by the surface element $d\hat{S}$ moved with the flow to a new location at distance:

$$\Delta\ell = |v|\Delta t = V\Delta t \tag{3.17}$$

downstream. It is easy to see that in the limit of $\Delta t \to 0$, the surface element sweeps out a volume:

$$\Delta\hat{V} = \cos\theta \; d\hat{S}\Delta\ell = d\hat{S}\frac{\mathbf{n}\cdot\mathbf{V}}{V}\Delta\ell = \mathbf{n}\cdot\mathbf{V}\Delta t \; d\hat{S}, \tag{3.18}$$

where $\cos\theta \; d\hat{S}$ is the cross-sectional area of the stream tube and $\cos\theta$ is equal to the dot product between the normal $\mathbf{n}$ and the unit vector parallel to the local velocity vector $(\mathbf{V}/V)$. The mass contained in that volume element is equal to the density times the volume; hence, the amount of energy it contains is:

$$\Delta E = e\rho\mathbf{n}\cdot\mathbf{V}\Delta t \; d\hat{S}. \tag{3.19}$$

To account for this energy flux, it is necessary to sum only the differential areas $d\hat{S}$ over the entire *surface* of the control volume, which is abbreviated as CS. That is, CS is the area of the bounding surface containing the volume CV. Notice that both of the required corrections to the energy in the system are provided because $\mathbf{n}\cdot\mathbf{V}$ is *positive* for the part going out of the control volume that must be added; $\mathbf{n}\cdot\mathbf{V}$ is *negative* for the flow of new energy into the control volume from upstream that must deducted. Therefore, the *system* energy at $t + \Delta t$ is:

$$E_{t+\Delta t} = (E_{SYS})_{t+\Delta t} = (E_{CV})_{t+\Delta t} + \left(\iint_{CS} e\rho\mathbf{n}\cdot\mathbf{V}\Delta t \; d\hat{S}\right)_{t+\Delta t} =$$
$$= \left(\iiint_{CV} e\rho d\hat{V} + \Delta t\iint_{CS} e\rho\mathbf{n}\cdot\mathbf{V} \; d\hat{S}\right)_{t+\Delta t}, \tag{3.20}$$

where the Δt in the surface integral is independent of the geometry and can be moved out of the integral as shown in the last term. Inserting Eqs. 3.15 and 3.20 into the definition for the rate of energy change results in:

$$\frac{dE}{dt} = \lim_{\Delta t \to 0}\frac{1}{\Delta t}\left(\left(\iiint_{CV} e\rho d\hat{V} + \Delta t\iint_{CS} e\rho\mathbf{n}\cdot\mathbf{V} \; d\hat{S}\right)_{t+\Delta t} - \left(\iiint_{CV} e\rho d\hat{V}\right)_{t}\right). \tag{3.21}$$

The surface term can be removed from the limit process because the Δt divides out. Then, using the fundamental definition for a derivative, we find that:

$$
\frac{dE}{dt} = \lim_{\Delta t \to 0} \frac{1}{\Delta t} \left(\left(\iiint_{CV} e' \rho d\hat{V} \right)_{t+\Delta t} - \left(\iiint_{CV} e' \rho d\hat{V} \right)_{t} \right) + \iint_{CS} e' \rho \mathbf{n} \cdot \mathbf{V} d\hat{S} =
$$

$$
= \frac{\partial}{\partial t} \iiint_{CV} e' \rho d\hat{V} + \iint_{CS} e' \mathbf{n} \cdot \mathbf{V} \rho d\hat{S},
$$

(3.22)

where partial-differentiation notation is used because the quantities inside the integrals may vary spatially as well as temporally. Expressed in words, Eq. 3.22 indicates that the rate of change of the energy in a system of particles is the sum of the rate of change of the energy residing at a given time within an arbitrarily chosen control volume that encompasses the system, plus the *flux* of energy through the surface bounding the control volume.

Notice that the integrands are composed of the intensive property (in this case, the energy per unit mass, e') multiplied by other factors that describe the local flow field and fluid properties. In the volume integral, the factor is simply the mass of a volume element so that the integration gives the energy contained within the control volume. In the surface integral, e' is multiplied by the factor $\mathbf{n} \cdot \mathbf{V} \rho d\hat{S} = dm/dt = \dot{m}$, which can be interpreted as the mass flow rate through the differential area element at any location on the surface. Then, multiplying this by the energy per unit mass gives the flux of energy per unit area through the surface. Integration yields the net flux of energy from the control volume. The combination of the volume and the surface integrals then represents the time rate of change of the extensive property E, whose corresponding intensive property is e'.

It is easy to see that we can replace E and e' by any other pair of extensive variables (e.g., the pairs identified in Table 3.1). Then, we can use this tool to transform the fundamental laws of nature into the useful control-volume form for solving fluid-mechanics or aerodynamics problems. Writing Eq. 3.22 in general form, for any extensive property (e.g., J and its corresponding intensive property j), we have the Reynolds' Transport Theorem:

$$
\underbrace{\frac{dJ}{dt}}_{\substack{\text{rate of} \\ \text{change} \\ \text{of variable}}} = \underbrace{\frac{\partial}{\partial t} \iiint_{CV} j \rho d\hat{V}}_{\substack{\text{rate of change of} \\ \text{variable residing within} \\ \text{control volume}}} + \underbrace{\iint_{CS} j\, \mathbf{n} \cdot \mathbf{V} \rho d\hat{S}}_{\substack{\text{flux of variable} \\ \text{through surface of} \\ \text{control volume}}}
$$

(3.23)

It is now used to write each of the required natural laws in control-volume form in the manner described previously for the energy. In the discussions in the following subsections, the student will gain full confidence in this representation of fluid motion. However, the student will see that there are inherent limitations that characterize this approach. Later in the chapter, we pass to the differential-equation formulation of fluid dynamics to overcome these limitations. Nevertheless, the control-volume formulation provides useful insight into many basic fluid-flow problems of practical interest. In what follows, each law is recast in control-volume form, and the application of the results is demonstrated in a series of examples.

Table 3.2. *Continuity equation in control-volume form*

General form (Eq. 3.24)	$\dfrac{\partial}{\partial t}\iiint\limits_{V}\rho\,d\hat{V} + \iint\limits_{S}\mathbf{n}\cdot\mathbf{V}\rho\,d\hat{S} = 0$	
Steady compressible flow	$\iint\limits_{S}\mathbf{n}\cdot\mathbf{V}\,\rho\,d\hat{S} = 0$	(3.25)
Incompressible flow	$\iint\limits_{S}\mathbf{n}\cdot\mathbf{V}\,d\hat{S} = 0$	(3.26)

Conservation of Mass: The Continuity Equation

For the mass of the system of particles to stay constant, Eq. 3.1 must be satisfied. Then, in control-volume form, using the Reynolds' Transport Theorem and noting that the extensive property is simply unity as shown in Table 3.1, it is necessary that:

$$\frac{\partial}{\partial t}\iiint\limits_{CV}\rho\,d\hat{V} + \iint\limits_{CS}\mathbf{n}\cdot\mathbf{V}\,\rho\,d\hat{S} = 0. \tag{3.24}$$

This is the general control-volume statement of the continuity equation. It remains to apply this equation in a variety of situations to illustrate its utility in problem solving. Before discussing actual problems, it is useful to define several special cases that frequently arise. For example, in many problems, the flow is *steady*—that is, the properties at a given location in the flow field may not change with time. Then, the volume integral does not change with time and its derivative vanishes. In other situations, the flow may be *incompressible*, in which case the density ρ is constant and can be removed from the integrals. Table 3.2 summarizes several special cases of interest.

Notice that the equation for incompressible flow holds for both steady and unsteady flows because the volume of the control volume does not change with time. The constant density can be factored out of the equation and the derivative term vanishes whether or not the flow is time-dependent. The notations CV and CS indicating the volume and surface of integration often are replaced with $\hat{V}$ and $\hat{S}$, as shown in the table; there is no difference in meaning.

Mass Flow Rate and Volume Flow Rate

The integral:

$$\dot{m} = \iint\limits_{A}\mathbf{n}\cdot\mathbf{V}\,\rho\,d\hat{S} \tag{3.27}$$

over a selected part of a control surface of area A through which fluid passes often is referred to as the *mass flow rate, m*`, across that surface area. If English units are used, the velocity is in [ft/sec], the compatible unit of density is [slug/ft^3], and the area is in [ft^2]. Then, the mass flow rate has units of mass units per unit time, or [slug/sec]. A common practice in thermodynamics is to express the flow rate in [lbm/sec], but this is not a truly compatible unit (see Examples 2.1 and 2.2). It is more correctly called the *weight flow rate* because an factor of g was applied. The usual cautions of mixing force and mass units apply. In metric units, the flow rate often is expressed in [kg/sec].

When the density is constant, as in an incompressible flow, mass flow rate still can be used as defined. In some problems, there is no dependence on the density; therefore, the rate of flow can be described in terms of the volume flow rate:

$$Q = \iint_A \mathbf{n} \cdot \mathbf{V}\, d\hat{S}. \qquad (3.28)$$

Typical units for volume-flow rate are [ft³/sec] and [m³/sec]. The following examples show how these measures of the rate of fluid flow enter problems naturally.

Applications of the Continuity Equation in Control-Volume Form

The integral form of the continuity equation is now applied in typical situations. Students should attempt to work through each example to check their understanding of the principles involved and then compare results to the detailed solutions. It is imperative that students develop a thorough understanding of the continuity result. Mathematical representations of flow fields that obey the continuity rules are said to be *physically realizable* flow patterns. We see later that it is possible to find solutions of certain problems that do not conform to reality because they are not continuous in the sense of satisfying continuity requirements.

EXAMPLE 3.1 *Given*: Water flows steadily through a pipe at moderate pressure. The cross-sectional areas at Stations 1 and 2 are 2 m², and 0.5 m², respectively. The velocity at Station 1 is measured as 3 m/s. The flow is assumed to be one-dimensional; that is, variations in velocity across any cross-sectional area of the pipe are small compared with velocity variations along the length of the pipe. Thus, the velocity in the pipe at any cross-sectional area may be considered constant across the area.

Required: Predict the velocity at Station 2.

Approach: The flow is steady and a liquid is essentially incompressible. No details are needed between the two stations. Thus, apply the continuity Eq. 3.26.

Solution: First, the control surface, $\hat{S}$, must be selected for this problem. Although any control surface may be chosen and gives the same answer, some thought usually suggests a choice that makes the calculation easiest. In this case, a cylindrical control surface (i.e., the dotted line) along the inner surface of the pipe with ends perpendicular to the flow velocities at the two ends is the best choice. Because there is no flow through the sides of the pipe, the integrand is zero everywhere except at the two ends. At the ends of the control volume, the velocity and the unit outward normal are collinear so that $(\mathbf{V} \cdot \mathbf{n})$ becomes simply the magnitude of the velocity, with signs positive for outflow and negative for inflow. Thus,

$$\iint_S (\mathbf{V} \cdot \mathbf{n}) \, d\hat{S} = \iint_{[1]} (\mathbf{V} \cdot \mathbf{n}) d\hat{S} + \iint_{[2]} (\mathbf{V} \cdot \mathbf{n}) d\hat{S}$$

$$= -V_1 \iint_{[1]} d\hat{S} + V_2 \iint_{[2]} d\hat{S} = -V_1 A_1 + V_2 A_2 = 0$$

$$\text{So, } V_2 = V_1 \frac{A_1}{A_2} = 3 \frac{(2)}{(0.5)} = 12 \text{ m/s}$$

Appraisal: The pipe area decreased from Station 1 to Station 2 and the velocity at Station 2 is found to be larger than at Station 1. This confirms experience from observations of water flow.

The point was made in Example 3.1 that *any* control-surface choice leads to the same result. Suppose, for example, that the control surface at Station 2 was chosen so as to be at an angle δ with respect to flow velocity at that point, as follows:

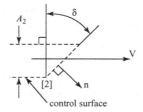

Then,

$$\iint_{[2]} (\mathbf{V} \cdot \mathbf{n}) \, d\hat{S} = \iint_{[2]} (V_2 \cos \delta) d\hat{S} = V_2 \cos \delta \iint_{[2]} d\hat{S} = V_2 \cos \delta \frac{A_2}{\cos \delta} = V_2 A_2,$$

as before.

EXAMPLE 3.2 *Given:* A liquid (density 2.0 slugs/ft³) flows steadily through the pipe connection shown here. There is a mass flux into Station 1 with a magnitude $\dot{m} = 20$ slugs/s. The mass flux out at Station 3 is $\dot{m} = 8$ slugs/s. The flow through the pipe at the three stations may be assumed to be one-dimensional:

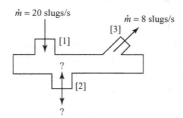

Required: Predict the volume flow and the flow direction (i.e., in or out?) at Station 2.

Approach: Choose a control volume that coincides with the inside surface of the pipe and is perpendicular to the inflow/outflow vectors at Stations 1, 2, and 3. Because the fluid is a liquid, assume incompressible flow. However, use Eq. 3.25 because the integrand corresponds to mass flux, which is the given for the problem even though the density, ρ, is constant. If the mass flux at Station 2 can be found, then the volume flow rate follows because the density is known. There must be three terms in the equation, corresponding to inflow or outflow at the three stations.

Solution: Using Eq. 3.25,

$$\iint_{\hat{S}} \rho(\mathbf{V} \cdot \mathbf{n})d\hat{S} = -\dot{m}_1 + \dot{m}_3 \pm \dot{m}_2 = 0,$$

where the signs at Stations 1 and 3 correspond to inflow ($\mathbf{V} \cdot \mathbf{n} < 0$) and outflow ($\mathbf{V} \cdot \mathbf{n} > 0$), respectively, and the sign of the term representing the mass flux at Station 2 is, for the moment, unknown. Substituting

$$-20 + 8 \pm \dot{m}_2 = 0,$$

it follows that $\dot{m}_2$ must be positive, so that the flow at Station 2 must be outward and $\dot{m}_2 = 12$. Now, $\dot{m}_2 = 12 = \rho VA = 2.0 \, VA$, where A is the area of the pipe at Station 2. Solving, VA = volume flow rate = 6 ft^3/s.

Appraisal: The physical fact is that for this steady flow, the mass flux into the connector must equal the mass flux out; there is no change in the mass of liquid within the connector. Because the density is constant, this states further that the volume flow rate out must be equal to the volume flow rate coming in.

The Momentum Equation in Control-Volume Form

The Conservation of Momentum principle for a fluid system corresponds to the word statement of Newton's Second Law of Motion that the net force acting on a system is equal to the time rate of change of momentum of the system. Because both force $\mathbf{F}$ and momentum $\mathbf{P}$ are vector quantities (i.e., magnitude and direction), the resulting equation is a *vector* equation. For brevity, we discuss only the details of the linear momentum balance in this subsection, which can be applied to the angular momentum $\mathbf{H}$ if the torque, $\mathbf{M}$, is accounted for and applied to the control volume. The results to be derived for the linear momentum can be extended easily to handle problems in which angular momentum may be important. This is accomplished in the example applications presented herein.

To apply the Reynolds' Transport Theorem to Eq. 3.3, we must first define the force vector, $\mathbf{F}$, acting on the fluid in the control volume. That is, the rate of change of the system momentum is determined by this force field. Although the momentum equation often is referred to as "conservation of momentum," momentum is conserved, strictly speaking, only if $\mathbf{F}$ is zero.

It is useful to distinguish between two types of forces that may act on the fluid. The first type acts on all of the fluid particles in the control volume—for example, the gravity force or the magnetic force acting on a conducting fluid. The differential body force on each element of volume enclosed by the control surface is given by $\rho \mathbf{b} d\hat{V}$ (units are [slug/ft^3][lbf/slug][ft^3] = [lbf]), where $\mathbf{b}$ is the vector body force per unit mass. In general, **b** can change from point to point and also may be a function of time. Summing over the entire control volume:

$$\text{body force} = \mathbf{F}_b = \iiint_{\hat{V}} \rho \mathbf{b} d\hat{V}. \tag{3.29}$$

The second type of force acts on the surface surrounding the control volume. In general, there is a force on each element of the surface of the control volume due to viscous and pressure stresses arising from the presence of surrounding parts of the flow field. Therefore, this force conveniently can be decomposed (see Fig. 2.2 and the related discussion) into components normal and tangential to the surface. Thus,

$$\text{surface force} = \mathbf{F}_S = \iint_{\hat{S}} \boldsymbol{\tau} d\hat{S} = \iint_{\hat{S}} \tau_n \mathbf{n} d\hat{S} + \iint_{\hat{S}} \tau_t \mathbf{t} d\hat{S}, \tag{3.30}$$

where $\boldsymbol{\tau}$ is the surface force per unit area (i.e., surface stress); τ_n and τ_t are the normal stress and tangential stress components, respectively. Thus, the total force acting on the system is:

$$\mathbf{F} = \mathbf{F}_b + \mathbf{F}_s. \tag{3.31}$$

F also may involve mechanical forces applied directly to the control volume. To use this result, it is necessary to *model* the force system to represent the particular problem of interest. For example, the normal stress in Eq. 3.31 may be represented by the pressure; the tangential stress may be due to viscous shear of the type that is well represented by Newton's Law of Viscous Force (see Eq. 2.4 and the related discussion). The body force may be due to a gravitational field, $\mathbf{b} = \mathbf{g}$, where the magnitude of vector **g** is the gravitational acceleration and its direction is along the gravitational field lines.

In applying the Reynolds' Transport Theorem (Eq. 3.23), we see that the intensive property corresponding the system momentum, **P**, is now the velocity vector, **V**, as indicated in Table 3.1. Note that $\rho\mathbf{V}$ is the momentum per unit volume. For a fixed control volume, Newton's Second Law can be stated in words as "The net force acting on the fluid within the control volume is equal to the time rate of change of momentum within the control volume (the unsteady contribution) plus the net momentum flux (outflow–inflow) through the surface of the control volume." Thus, we must have:

$$\mathbf{F} = \iiint_{\hat{V}} \rho \mathbf{b} \, d\hat{V} + \iint_{\hat{S}} \boldsymbol{\tau} d\hat{S} = \frac{\partial}{\partial t} \iiint_{\hat{V}} \rho \mathbf{V} d\hat{V} + \iint_{\hat{S}} \rho \mathbf{V} (\mathbf{V} \cdot \mathbf{n}) d\hat{S} \tag{3.32}$$

The first term on the right side represents the rate of change of the linear momentum contained within the control volume. Because it is no longer necessary to distinguish between the system and the control volume, the more common notation for the volume and surfaces integrated over are used in Eq. 3.32. The net momentum flux across the surface of the control volume is equal to the vector momentum per unit mass, **V** multiplied by $\rho(\mathbf{V} \cdot \mathbf{n})d\hat{S}$, the mass flow rate through the area $d\hat{S}$.

Equation 3.32 is the most general form of the momentum equation. It may not be necessary to retain all of this generality in a particular application in aerodynamics. Again, the student should acquire the necessary physical understanding to confidently drop unneeded terms. Consider the following simplifying assumptions.

It often is the case that flow may be assumed realistically to be steady. Then, the first term in Eq. 3.32 can be dropped. Most of the aerodynamic models in this book

Table 3.3. *Momentum equation in control-volume form*

General form Eq. 3.32	$\dfrac{\partial}{\partial t}\iiint\limits_{V}\rho\mathbf{V}d\hat{V}+\iint\limits_{S}\rho\mathbf{V}(\mathbf{V}\cdot\mathbf{n})d\hat{S}=\iiint\limits_{V}\rho\mathbf{b}d\hat{V}+\iint\limits_{S}\tau d\hat{S}$
Steady flow	$\iint\limits_{S}\rho\mathbf{V}(\mathbf{V}\cdot\mathbf{n})d\hat{S}=\iiint\limits_{V}\rho\mathbf{b}d\hat{V}+\iint\limits_{S}\tau d\hat{S}$ (3.33)
Incompressible flow	$\dfrac{\partial}{\partial t}\iiint\limits_{V}\mathbf{V}d\hat{V}+\iint\limits_{S}\mathbf{V}(\mathbf{V}\cdot\mathbf{n})d\hat{S}=\iiint\limits_{V}\mathbf{b}d\hat{V}+\dfrac{1}{\rho}\iint\limits_{S}\tau d\hat{S}$ (3.34)
Flow without body forces	$\dfrac{\partial}{\partial t}\iiint\limits_{V}\rho\mathbf{V}d\hat{V}+\iint\limits_{S}\rho\mathbf{V}(\mathbf{V}\cdot\mathbf{n})d\hat{S}=\iint\limits_{S}\tau d\hat{S}$ (3.35)
Inviscid flow without body forces	$\dfrac{\partial}{\partial t}\iiint\limits_{V}\rho\mathbf{V}d\hat{V}+\iint\limits_{S}\rho\mathbf{V}(\mathbf{V}\cdot\mathbf{n})d\hat{S}=-\iint\limits_{S}\mathbf{p}\mathbf{n}d\hat{S}$ (3.36)
Steady, incompressible, inviscid flow	$\iint\limits_{S}\rho\mathbf{V}(\mathbf{V}\cdot\mathbf{n})d\hat{S}=-\iint\limits_{S}\mathbf{p}\mathbf{n}d\hat{S}$ (3.37)

use this approximation. Flow over an airplane wing flying at constant speed clearly is steady; however, aerodynamic modeling of a dragonfly wing, for example, requires full application of the unsteady formulation.

In this book, we concentrate on aerodynamics of incompressible flows. Then, the density can be treated as a constant and divided out of all but the last term in Eq. 3.32.

A commonly used simplification is to neglect the body force by setting **b** to zero. Body forces are negligible compared to other forces in most aerodynamic applications. For example, the effect of gravitational forces on the motion of air around a body is important only if significant changes in elevation are involved. Review the model of the earth's atmosphere presented in Chapter 2 to see that elevation changes like those involved in the flow over an airplane wing or fuselage are likely to involve only minor changes in fluid properties due to the gravitational acceleration. A possible exception is in airflow over a very large structure (e.g., a dirigible or a tall building). Even in these cases, the body-force effects often are negligible compared to pressure and viscous forces.

Other simplifications that often are appropriate include representing the surface tractions only by the pressure forces. Then, we set $\tau = \tau_n = \mathbf{p}$, where **p** is the (vector) normal stress due to pressure. This amounts to the assumption that the flow is *inviscid*—that is, that viscous effects are negligible. Of course, no liquid or gas is truly inviscid but, in many cases, the viscous forces are not decisively large compared with other forces that dominate the problem. The assumption greatly simplifies the problem while still leading to useful and practical results, and we use it in several crucial analyses.

Special cases of the momentum equation are displayed in Table 3.3.

In the last two cases shown in the table, the surface force is represented only by the pressure. Then, the negative sign on the force term indicates that a positive pressure exerts a force normally inward at the surface. That is, the positive pressure force is in the opposite direction to the surface normal, **n**. The final special case, Eq. 3.37, has many applications and often is referred to as the *Momentum Theorem*. Expressed in words, it states that if the flow is steady, inviscid, and incompressible, then the momentum flux through the surface is controlled entirely by the pressure forces acting at the control surface. This is a useful form of the integral-momentum equation because, for steady flow, the volume-integral term drops out and the remaining terms require information only on the surface of the control volume, not within it. Eq. 3.37 is a vector equation and, in the most general case, it stands for three component equations in the coordinate directions.

Of course, there are other special cases of the momentum equation. Inspection of the forms shown in Table 3.3 reveals that it is a simple matter to adjust for any situation. It is only necessary to thoroughly understand and apply the meanings of the words *steady, incompressible*, and *inviscid*.

Problem Solving with the Momentum Equation in Control-Volume Form

In applying the integral-momentum equation to a given problem, there are two important things to remember:

1. In choosing a suitable control surface, some choices may lead to simpler solutions than others.
2. *All* of the surface- and body-forces and momentum-flux terms must be accounted for with special attention given to the sign of each term.

Again, recall that we are working with a vector equation, and it may be necessary to account for all three components in some situations.

Notice the analogy between the control volume as used here for fluid-mechanics problems and the *free-body diagram* learned in mechanics problems. Whenever we work with forces and moments, it is necessary to use the methods of dynamics in relating the force system to the motion of the system of particles. When we define a control volume that fits the geometry of a given problem, it also is useful to visualize it as a free-body diagram. The forces on the boundaries of the control volume represent interaction with the outside world. It is important that *all* forces of interaction are represented. In Table 3.3, only the forces due to fluid-dynamic interactions are displayed. It is possible that other forces may need to be accounted for; for example, suppose that the control volume is attached to a beam, strut, spring, or other structure. Such directly applied forces also must appear in the force balance. Incorporation of such mechanical constraints should be second nature from earlier studies of statics and dynamics. We review the freebody-diagram method as extended to fluid dynamics by means of the example problems in this section.

So as not to miss a term or a sign in assessing the control volume for a given problem, a four-step procedure is suggested for working problems using the integral-momentum equation, as follows:

1. Survey the control surface and note the magnitude and coordinate direction of any pressure (or shear) force acting on the control surface. If the force is not

along a coordinate direction, it must be resolved into components in the coordinate system chosen.

2. Identify all forces, including those due to mechanical constraints; some may have unknown magnitude and direction. If the direction of a force component is known, apply the correct sign depending on whether it is in the positive or negative coordinate direction.

3. Identify the parts of the control surface across which there is a flow of momentum. Recognize that the momentum-flux integrand contains two signed terms. The mass flow part, $\rho(\mathbf{V} \cdot \mathbf{n})d\hat{S}$, is a *scalar* and therefore has no dependence on coordinate direction; the sign depends only on whether the flux is *out* of the control volume (positive sign) or *into* the surface of the control volume (negative sign). The $\mathbf{V}$ part of the integrand is the momentum vector (per unit mass) and thus carries a sign depending on how the velocity vector is pointing with respect to the coordinate axes. This sign must be consistent with the positive coordinate directions chosen for the force terms. If the momentum vector is not aligned with a coordinate direction, then a component in that direction must be included.

4. It may be necessary to use the continuity equation as well as the momentum equation to solve for all of the unknowns in the problem. It is generally easiest to do the *mass balance* (i.e., application of the continuity equation) first. If flow-channel areas are initially unknown, they usually can be determined from continuity. Recall that unknown velocity components may be in either direction. If a positive sign is assumed initially for an undetermined velocity component, the solution indicates the proper sign. That is, if it is pointed in the direction opposite to the assumed one, the final solution for the scalar magnitude includes a minus sign indicating this result.

Before presenting example problems using the integral-momentum equation, it is necessary to recall the two reference values of pressure: gauge and absolute (see Chapter 2). A *gauge pressure* is defined as the pressure above (plus) or below (minus) the atmospheric or ambient pressure. Pressure levels with respect to zero reference (i.e., a perfect vacuum) are termed *absolute pressures* and are always positive. Thus, at standard sea level, the gauge pressure in quiescent air is 0 psig and the absolute pressure is 14.7 psia. Most thermodynamic relationships, such as the First and Second Laws of Thermodynamics and the Equation of State, require the use of absolute pressure. In the case of the integral-momentum equation, either gauge or absolute pressure can be used for a given problem and leads to the same answer if all surfaces are considered. The benefit of using gauge pressure is that on surfaces on which only quiescent fluid interacts, the pressure force is zero. Of course, it is necessary to use gauge pressure to evaluate all remaining surfaces. The use of absolute pressure may lead to a more complicated solution in some cases, as illustrated in Example 3.3.

EXAMPLE 3.3 *Given*: A two-dimensional free jet of water (ρ = 62.4 lbm/ft^3) impinges with a steady flow onto a stationary flat plate. The jet divides into two streams flowing in opposite directions along the plate. The entire flow is assumed to be frictionless (i.e., no flow losses and no frictional force along the plate surface). Because there are no flow losses, the velocity and static pressure of the two jets leaving the plate are the same as in the jet. The static pressure in the flow is everywhere ambient pressure because the flow forms a free surface with respect to the ambient air, and there can be no pressure difference across a free surface. Assume all flow segments to be one-dimensional. Certain velocity magnitudes and dimensions are known, as indicated in the following drawing:

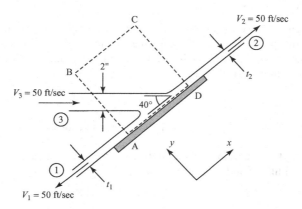

Required: Find the jet heights "t_1" and "t_2" and predict the magnitude and direction of the force exerted on the plate (per unit depth) by the impinging jet.

Approach: The flow is steady and incompressible. The momentum equation is used to predict the unknown force. Because t_1 and t_2 also are unknowns, another equation is needed—namely, the continuity equation. The control surface is arbitrary, so choose the rectangular shape A-B-C-D as indicated by the dotted line. The problem is solved in terms of gauge pressure; later, the choice of absolute pressure is examined.

Solution: From continuity Eq. 3.5:

$$\iint_{\tilde{S}} (\mathbf{V} \cdot \mathbf{n}) d\hat{S} = 0.$$

This equation states that because the density is constant, the volume flow into and out of the control surface must sum to zero. Thus,

$$V_1 A_1 + V_2 A_2 - (V_3 \cos 40°)\left(\frac{A_3}{\cos 40°}\right) = 0, \text{ where } A_3 = \frac{2}{12} \text{ft}^2.$$

Hence, $A_1 + A_2 = 2/12$ ft². Now, apply the integral-momentum equation (Momentum Theorem, Eq. 3.37) for steady flow:

$$-\iint_{\hat{S}} p\mathbf{n} d\hat{S} = \iint_{\hat{S}} \rho \mathbf{V}(\mathbf{V} \cdot \mathbf{n}) d\hat{S}.$$

With the problem expressed in terms of gauge pressure, the pressure forces on sides A-B, B-C, and C-D are zero because the pressures on all of these faces are ambient pressure, including Flow Areas 1, 2, and 3. The pressure force, F_{A-D}, on side A-D of the control volume, is the integrated pressure force of the plate acting on the control volume. A force equal and opposite to this force is the (unknown) force of the control volume (i.e., of the jet) acting on the plate. Taking the component of Eq. 3.37 in the x-direction, the net-force term on the left side of the equation is zero because there are no gauge-pressure forces present (recall that the inviscid assumption means that there are no shear forces along the plate surface A-D).

Taking the component of Eq. 3.37 in the y-direction, the only term on the left side is the pressure force, F_{A-D}, acting in the positive y-direction. There are three momentum-flux terms in the x-component equation but only one in the y-component equation. Thus,

x-direction:

$$0 = \rho(-V_1)(V_1 A_1) + \rho(V_2)(V_2 A_2) + \rho(V_3 \cos 40°)(-V_3 A_3).$$

Examine this equation and carefully consider the signs associated with each term. Simplifying and recognizing that $V_1 = V_2 = V_3 = V$,

$$V^2(A_2 - A_1) - V^2 A_3 \cos 40° = 0.$$

Eliminating V and writing A_1 in terms of A_2 from continuity, this relation becomes:

$$A_2 - (2/12 - A_2) = (2/12) \cos 40°.$$

Solving, $A_2 = 0.195$ ft^2 ($t_2 = 0.195$ ft) and $A_1 = 0.147$ ft^2 ($t_1 = 0.147$ ft).

y-direction:

$$F_{A-D} = (0) + (0) + (-V_3 \sin 40)(-V_3 A_3)$$
$$F_{A-D} = V_3^2 A_3 \sin 40° = (62.4/32.2)(50)^2(2/12)\sin 40° = 519 \text{ lbf/ft}.$$

The plus sign indicates that F_{A-D} acts in the positive y-direction. The force on the plate is equal and opposite to this—namely, a force of 519 pounds pushing the plate downward and to the right.

Note: Either of the following two control-volume choices would lead to the same result (the student should verify this). Choice (a) would make the inflow-flux calculation easier because there would be no need to resolve the velocity components:

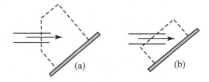

(a)   (b)

Appraisal: The predicted direction of the force on the plate agrees with physical intuition. If the problem were solved using absolute pressures, the result would be the same but the solution would not have been as straightforward. Working with absolute pressures, there would have been a pressure force on sides A-B and C-D (which still would cancel each other) and a force on B-C having a magnitude of (ambient pressure) × (length B-C). This force would have a negative sign (negative y-direction) and, when subtracted in Eq. 3.37 would have the effect of increasing the magnitude of F_{A-D} to F'_{A-D}.

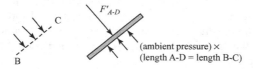

However, the force that the jet exerts on the plate is now partially resisted by the ambient pressure beneath the plate. Thus, the net force on the plate is the same as that calculated previously. The benefit of working this problem in gauge pressure is apparent.

EXAMPLE 3.4 *Given*: A steady flow of a liquid ($\rho = 2.0$ slugs/ft^3) enters and leaves the pipe coupling as shown:

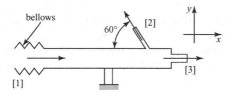

From measurement, $p_1 = 16.78$ psia and

$$A_1 = 4 \text{ ft}^2, V_1 = 40 \text{ ft/s}$$
$$A_2 = 0.5 \text{ ft}^2, V_2 = 20 \text{ ft/s}$$
$$A_3 = 1.5 \text{ ft}^2, V_3 = 20 \text{ ft/s}.$$

(Note that given any five of these areas or velocities, continuity would provide the sixth.) The flow exhausts into the atmosphere at Stations 2 and 3 as a free jet. Hence, the jet-exhaust pressure $p_2 = p_3 =$ ambient pressure (assumed to be 2,116 psfa.) Assume that the rubber bellows at Station 1 cannot withstand a horizontal force (i.e., it can support only hoop stresses).

Required: Predict the x-component, F_x, of force **F** on the support strut (i.e., magnitude and direction).

Approach: The continuity equation is not needed in this example because all of the required mass (volume) flow information is given in the problem statement. Use the integral-momentum equation to find the required unknown force. Choose a control surface that coincides with the inside of the pipe coupling. An externally applied mechanical force must be included in the momentum equation. (Note that cases displayed in Table 3.3 do include such external forces.)

Solution: For this steady flow, apply momentum Eq. 3.37. Adding the unknown external reaction force to Eq. 3.37, we must solve:

$$\iint_{S} \rho \mathbf{V}(\mathbf{V} \cdot \mathbf{n})d\hat{S} = \mathbf{F} - \iint_{S} p\mathbf{n}d\hat{S}. \tag{3.38}$$

Elect to work in gauge pressure, lbf/ft^2 (or psfa), as suggested in the discussion preceding Example 3.4. An appropriate control volume is sketched herein. Notice that this also is a free-body diagram and all forces must be displayed. The presence of the mechanical constraint due to the supporting strut is represented by the unknown reaction force, **F**, at the point where the strut connects to the pipe.

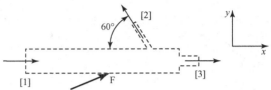

Consider the x-component of the vector-momentum equation. If gauge pressure is used, there is only one nonzero pressure-force term—namely the pressure force at Station 1 (the pressure forces are zero at Stations 2 and 3 because the pressure in the two jets has zero gauge value). There also is the horizontal component of the strut reaction, F_X that must be calculated. Then, the net force (x-component) acting on the surface of the control volume is given by:

$$p_1 A_1 + F_X = [(16.78)(144) - 2{,}116] \,(4) + F_X = 1{,}200 + F_X.$$

Because the direction of the force is unknown, it is represented as a positive unknown, F_X. If the solution is a positive magnitude, then the choice of sign was correct; that is, the reaction of the strut force, F_X, on the control surface acts to the right. If F_X is negative in the solution, then the assumed direction was incorrect and F_X actually acts to the left.

Three momentum-flux terms must be evaluated:

net momentum flux $= (V_1)(-V_1 A_1) + (-V_2 \cos 60)(V_2^2 A_2) + (V_3)(V_3 A_3)$
$= (-V_1^2 A_1 - A_2 \cos 60 + V_3^2 A_3) = -11{,}800$ lbf.

Equating the x-components of the force and flux terms as indicated in Eq. 3.38: $1{,}200 + F_X = -11{,}800$, or $F_X = -13{,}000$ pounds.

The negative sign indicates that the strut reaction force exerted *on* the control surface by the pipe is in the negative x-direction. Thus, the equal and opposite force of the control surface on the pipe (and, hence, on the strut) is to the right with a magnitude of 13,000 pounds. This is the force that the strut must resist if the pipe is to be immobilized.

Appraisal: The predicted force on the strut agrees with physical intuition.

EXAMPLE 3.5 *Given*: the drag coefficient of a two-dimensional airfoil (i.e., a section of a wing of infinite span) may be measured experimentally by installing a wing model that spans the wind-tunnel-test section. The drag of wind-tunnel models is usually measured with a force balance. However, in the two-dimensional case, an alternate method for measuring the drag is to make a velocity survey in a vertical direction through the wake suitably far downstream of the model. The model reaches from one wall to the other and, if the ends are sealed, there is no flow around the tips and the model effectively has infinite span. Application of the integral-momentum equation yields an indirect measurement of the drag. This is the example treated herein.

Consider a wing section near mid-span. There is a drag force on the airfoil section due to pressure and shear forces acting on the surface. A wake trails downstream, within which there is a velocity deficit that can be measured by suitable instrumentation. The wing section is assumed to be symmetrical and the wing is assumed to be set at zero angle of attack. Under these assumptions, the wake is symmetrical about the *x*-axis.

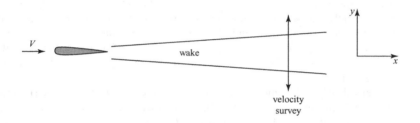

Two different choices of control volume are discussed. In both cases, it is assumed that the control volume is suitably far from the airfoil that the local pressure acting on the control surface is the undisturbed value of freestream static pressure. This requires that the wake survey be carried out several chord lengths downstream of the wind-tunnel model. The flow is assumed to be steady and incompressible. An inviscid flow in the wake is assumed, with the effects of viscosity represented by a force on the model and a resulting velocity deficit in the wake. For simplicity, the measured velocity profile is assumed to be linear. Thus, the velocity distribution within the wake is assumed to be given by $\left(V_\infty \dfrac{y}{h} \right)$. This assumption provides the correct gross wake character (i.e., velocity minimum at the center of the wake) but is not necessarily physically realistic.

Required: Predict the drag coefficient of the airfoil under test if the wake profile is linear.

Approach: Select a rectangular control surface as a simple choice. Assume that it is suitably far from the airfoil that the static pressure and velocity (except in the wake) are essentially the undisturbed tunnel-flow conditions:

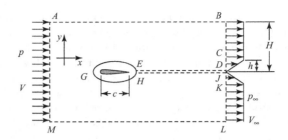

Solution: For this profile, assume that it is known that $h = 0.01c$ and arbitrarily assume H to be $H = 10h$, where c is the chord (width) of the model, and H is the half-height of the assumed control volume. Choose the rectangular control surface shown here. At "D", the control surface extends upstream to encircle the airfoil (E-G-H) and then returns on itself to "J". Because the front and rear of the control surface are far from the airfoil and at freestream static pressure, this datum is assumed to be zero gauge pressure. The pressure forces on A-M and B-L are equal and opposite, whereas the pressure forces on A-B and L-M are at right angles to the x-direction. Any pressure or viscous forces along D-E are equal and opposite to those on H-J. Thus, the only unbalanced force on the control surface in the x-direction is the force, F, of the airfoil acting on the control surface along E-G-H. This force is entered into the equations with a plus sign and the direction established later.

There clearly is a momentum flux in the left face, A-M, and out the right face, B-L. Closer inspection reveals that there also must be a mass flux out of A-B and L-M because the mass flux out through B-L is less than the mass flux in through A-M due to a mass (i.e., velocity) deficit along C-K. The detailed outflow pattern along the top and bottom is not needed but continuity requires that:

$$+\dot{m}_{AB} + \dot{m}_{LM} + \dot{m}_{BL} - \dot{m}_{AM} = 0$$

or

$$\dot{m}_{AB} + \dot{m}_{LM} = -\left[2\rho V_\infty(H-h) - 2\rho \int_0^h \frac{V_\infty y}{h} dy \right] + 2\rho V_\infty H = \rho V_\infty h.$$

This mass flux of $\rho V_\infty h$ carries with it an x-momentum given by $(\rho V_\infty h)V_\infty$. Each fluid particle exiting the top and bottom of the control surface is traveling with a velocity component in the x-direction of nearly V_∞ because the control surface is suitably far from the airfoil.

Thus, there are four flux-of-momentum terms needed for this control volume:

$$\text{flux}[A-M] + \text{flux}[(A-B)+(M+L)] + \\ + \text{flux}[(B-C)+(K-L)] + (A-B) + \text{flux}[C-K]$$

or

$$\rho V_\infty(-2V_\infty H) + \rho V_\infty^2 h + \rho V_\infty[2V_\infty(H-h)] + \\ + \rho \int_0^h \frac{V_\infty y}{h}\left(\frac{V_\infty y}{h}\right) dy = -\rho V_\infty^2\left(\frac{h}{3}\right).$$

Assembling the components of the momentum equation in the x-direction,

$$F = -\rho V_\infty^2 (h/3).$$

According to the sign, the force of the body on the control surface is to the left (i.e., upstream). Therefore, the force of the control surface on the airfoil is in the opposite direction (i.e., to the right, or downstream.) Then, the drag on the airfoil per unit span is $D' = \rho V_\infty^2 (h/3)$ and the drag coefficient for the two-dimensional airfoil is:

$$C_d \equiv \frac{D'}{\frac{1}{2}\rho V_\infty^2 c} = \frac{\rho V_\infty^2 [(0.01)(c)](/3)}{\frac{1}{2}\rho V_\infty^2 c} = 0.0067.$$

Notice that any control surface height $H > h$ gives the same result because the mass and momentum flux across side A-M through any height $H > h$ simply is canceled by the flux out across the segments B-C and K-L.

Appraisal: The drag force on the airfoil has the correct direction (i.e., downstream) and the drag coefficient is dimensionless and has a reasonable magnitude. The magnitude is not conclusive because the wake-velocity profile was idealized. Physically, the drag on the airfoil is perceived as equal to the momentum loss or defect in the wake.

Alternate Approach: As mentioned previously, there is a more natural choice of control volume that yields the same result. Consider a control volume in which the upper and lower surfaces are streamlines, as follows:

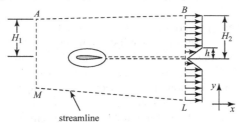

streamline

The advantage of this control-surface choice is that there is no flow across A-B and L-M because they are streamlines; hence, there is no momentum flux across these surfaces. However, because of the mass deficit in the wake, $H_2 \neq H_1$. The link between these two heights is supplied by continuity. Thus,

$$-2\int_0^{H_1} \rho V_\infty dy + 2\int_0^h \frac{\rho V_\infty y}{h} + 2\int_h^{H_2} \rho V_\infty dy = 0$$

$$H_1 = \frac{h}{2} + (H_2 - h) \Rightarrow H_2 = H_1 + \frac{h}{2}.$$

With this relationship established, the net momentum flux through the upstream and downstream faces is calculated and set equal to the force of the airfoil acting on the control surface. The wind-tunnel static-pressure forces acting normal to *A-B* and *L-M* have a component in the downstream direction, but it is balanced by the same pressure acting over the projected area $2(H_2 - H_1)$. The final result for the predicted airfoil drag is the same as that obtained by using the rectangular control surface.

The Energy Equation in Control-Volume Form

The energy equation represents the fifth of the required six equations needed for the solution of the basic aerodynamics problem defined in the chapter introduction. Specifically, it must be included in any consideration of compressible flow. The energy equation in aerodynamics is a statement of the First Law of Thermodynamics. The rate form of the First Law already was reduced to control-volume form in Eq. 3.22. Thus, for conservation of energy, we require that:

$$\frac{dE}{dt} = \frac{\partial}{\partial t} \iiint_{\hat{V}} e' \rho d\hat{V} + \iint_{\hat{S}} e' \mathbf{n} \cdot \mathbf{V} \rho d\hat{S} = \dot{Q} - \dot{W}, \tag{3.39}$$

where $\dot{Q}$ and $\dot{W}$ are the rates of heat flow and work flow from the surroundings. Only a brief discussion of the energy equation is needed here because most aspects are covered in the basic physics or thermodynamics already studied. Emphasis is on the forms needed and the notation used in compressible-flow aerodynamics.

As shown in Eq. 3.32, the rate of change of the total energy of the system of particles (expressed in control-volume form) consists of two terms: (1) the rate of change of the energy contained instantaneously within the control volume, and (2) the passage of energy through the surface of the control volume by convection with the flow.

The intensive energy variable e', shown in the integrals in Eq. 3.39, is used to represent all forms of energy that might be resident in the fluid. It consists of several parts that must be distinguished in solving compressible-flow problems. Any differential mass element in the control volume simultaneously may carry kinetic energy due to both the random motion of the molecules that comprise the element and the organized motion of the fluid. In general, there also may be potential energy due to the changing position in a body-force field (we use a uniform gravity field to illustrate). Thus, the energy, e', per unit mass often is defined in fluid-mechanics problems to consist of the following three terms:

$$e' = e + \frac{V^2}{2} + gz, \tag{3.40}$$

where e is defined as the internal kinetic energy (due to random microscopic molecular motion); $V^2/2$ is the directed kinetic energy per unit mass (with V as the magnitude of the velocity vector, a function of position and time in the field); and gz is the potential energy due to a uniform gravity field (assumed to act in the vertical direction—i.e., the z-direction in three-dimensional space). z is assumed to be measured from a convenient datum plane such as the earth's surface. This notation is chosen to conform to that used most often in textbooks and other publications describing compressible-fluid flows.

To complete the statement of the First Law of Thermodynamics, we must provide models for the heat transfer to the control volume and the rate of work done by the control volume on the surroundings as indicated by the terms to the right of Eq. 3.39.

Heat transfer may occur in the following two ways:
- internal heat release within the control volume such as by combustion, $\dot{Q}_{INT}$
- heat transfer by conduction across the surface of the control volume, represented in terms of local conditions as:

$$\dot{Q}_{COND} = \iint_{\hat{S}} \dot{q}d\hat{S}, \tag{3.41}$$

where $\dot{Q}$ is the heat-transfer rate across the surface per unit surface area and $\dot{Q}_{COND}$ is defined as the total heat-transfer rate to the control volume due to conduction from the surroundings. Heat transfer by radiation is not considered here, but it may be of crucial importance in problems in which temperatures are high (e.g., in the flow around a spacecraft reentering the earth's atmosphere). Transfer of energy by convection across the boundaries of the control volume was accounted for in the surface integral (i.e., the third term in Eq. 3.39).

It remains to provide models for the rate of work done by the control volume on the surroundings. This work rate may arise from the following three sources:

1. *Machine work (sometimes called "shaft work") done internally within the control volume but carried by mechanical linkage through the control surface (e.g., a turbine or compressor connected to an external device by a rotating shaft).* Let $\dot{W}_M$ denote the mechanical work rate done by the fluid inside the control volume, where the dot over the "W" denotes "time rate of change." $\dot{W}_M$ traditionally is assumed to be positive when it represents work done on the surroundings. Hence, it appears in the energy equation with a negative sign, as shown on the right side of Eq. 3.39.
2. *The rate of work done at the control surface due to surface stresses, which has two parts.* The part due to stresses tangent to the surface usually are due to the effects of viscosity. We denote the viscous-work-rate term by the symbol $\dot{W}_Y$, which represents the work rate done *on* the fluid within the control volume by viscous forces acting at its bounding surface. The second part is due to the normal stresses, usually related to the pressure forces. Multiplying the pressure force on the surface per unit area, $-p\mathbf{n}$, by an infinitesimal surface area, $d\hat{S}$, yields a force that may be represented as a vector everywhere perpendicular to the control surface by multiplying this force by the outward pointing unit normal vector—namely, $-p\mathbf{n}d\hat{S}$. Recall that force times distance in the same direction equals work. The work rate, then, is the force times the rate of change of distance in the same direction—that is, the force times the velocity component in the direction of the force (the convenience of using a "rate" equation is evident here). However, because $-p\mathbf{n}d\hat{S}$ is a vector, taking the dot product of $\mathbf{V}$ with this vector results in multiplying the pressure force by the component of the velocity vector in the direction of this force. Thus, $-p(\mathbf{V}\cdot\mathbf{n})d\hat{S}$ is the required work rate due to normal pressure acting at the control surface. Then, the total work rate done on the system by normal surface tractions is given by:

$$-\iint_{\hat{S}} p(\mathbf{V}\cdot\mathbf{n})d\hat{S}.$$

3. *The rate of work done on the fluid within the control volume by body forces.* This term may be represented by:

$$\iiint_V (\rho \mathbf{b} d\hat{V}) \cdot \mathbf{V},$$

where **b** is the body force per unit mass and the dot product with the velocity vector yields the rate of work.

Assembling all of these terms and dropping the potential energy term (negligible in most aerodynamic problems) gives the general energy equation, Eq. 3.42:

$$\frac{\partial}{\partial t} \iiint_V \rho \left(e + \frac{V^2}{2} \right) d\hat{V} + \iint_S \rho \left(e + \frac{V^2}{2} \right) (\mathbf{V} \cdot \mathbf{n}) d\hat{S} =$$

$$= \dot{Q}_{INT} + \dot{Q}_{COND} - \dot{W}_M + \dot{W}_V - \iint_S p(\mathbf{V} \cdot \mathbf{n}) d\hat{S} + \iiint_V \rho(\mathbf{V} \cdot \mathbf{b}) d\hat{V}. \tag{3.42}$$

Equation 3.42 may be expressed in a simpler form that is valid for a large class of aerodynamics problems by restricting the range of interest to steady flows with negligible viscous work, no combustion, negligible body-force effect, and negligible changes in potential energy. Then, the preceding equation simplifies to:

$$\iint_S \rho \left(e + \frac{V^2}{2} \right) (\mathbf{V} \cdot \mathbf{n}) d\hat{S} + \iint_S p(\mathbf{V} \cdot \mathbf{n}) d\hat{S} = \dot{Q}_{COND} - \dot{W}_M. \tag{3.43}$$

Combining the two surface integrals, this equation becomes:

$$\iint_S \rho \left(e + \frac{p}{\rho} + \frac{V^2}{2} \right) (\mathbf{V} \cdot \mathbf{n}) d\hat{S} = \dot{Q}_{COND} - \dot{W}_M.$$

The various special forms of the energy equation that we use are summarized in Table 3.4.

Table 3.4. *Energy equation in control-volume form*

	$\frac{\partial}{\partial t} \iiint_V \rho \left(e + \frac{V^2}{2} \right) d\hat{V} + \iint_S \rho \left(e + \frac{V^2}{2} \right) (\mathbf{V} \cdot \mathbf{n}) d\hat{S} =$
General form	$= \left(\begin{array}{l} \dot{Q}_{INT} + \dot{Q}_{COND} - \dot{W}_M + \dot{W}_V - \\ -\iint_S p(\mathbf{V} \cdot \mathbf{n}) d\hat{S} + \iiint_V \rho(\mathbf{V} \cdot \mathbf{b}) d\hat{V} \end{array} \right)$ (3.42)
Steady inviscid flow, zero body force, no internal heat conduction	$\iint_S \rho \left(h + \frac{V^2}{2} \right) (\mathbf{V} \cdot \mathbf{n}) d\hat{S} = \dot{Q}_{COND} - \dot{W}_M$ (3.44)
Steady inviscid flow, zero body force, no heat conduction or machine work	$\iint_S \rho \left(h + \frac{V^2}{2} \right) (\mathbf{V} \cdot \mathbf{n}) d\hat{S} = 0$ (3.46)

Now, recall that the thermodynamic state variable *enthalpy*, h, is defined by $h = e + p/\rho$, so that this equation becomes:

$$\iint\limits_{\hat{S}} \rho\left(h + \frac{V^2}{2}\right)(\mathbf{V} \cdot \mathbf{n})d\hat{S} = \dot{Q}_{COND} - \dot{W}_M. \tag{3.44}$$

Defining the *stagnation enthalpy* as $h_0 = h + V^2/2$, Eq. 3.44 may be written as:

$$\iint\limits_{\hat{S}} \rho h_0(\mathbf{V} \cdot \mathbf{n})d\hat{S} = \dot{Q}_{COND} - \dot{W}_M. \tag{3.45}$$

For an ideal gas, the internal energy is proportional to the (absolute) temperature. Therefore, we write $e = c_v T$, where c_v is the specific heat at constant volume. Also, remembering from the equation of state that the ratio $p/\rho = RT$, then h is also proportional to the temperature. Thus, it is typical to write $h = c_p T$, where c_p is the specific heat at constant pressure. In many situations, it is possible to assume that the two specific heats are constants independent of the temperature. We then speak of a *calorically perfect gas*. There exist cases in which the temperatures are so high (especially in hypersonic flows) that such assumptions may not be valid. The stagnation enthalpy, $h_0 = c_p T_0$, is useful for later discussions. The subscript "zero" denotes *stagnation conditions* (i.e., zero velocity).

EXAMPLE 3.6 *Given*: Air flows steadily through a long constant-diameter pipe at a mass-flow rate of 2.0 kg/s. The stagnation temperature has the same value (120°C) at two stations some distance apart along the pipe. Between these two stations, 2,000 Nm/s of work are added to the flowing air. Assume one-dimensional flow through the pipe. There is no combustion between the stations. Ignore work due to body forces and change in potential energy.

Required: Find the rate of heat addition (or subtraction) between these two stations by conduction through the pipe walls.

Approach: Select a convenient control volume. If the control surface coincides with the inside of the pipe, then at the walls, the rate of work due to viscosity is zero—the shear stress is large, but the velocity precisely at the wall is zero. At the two ends, the velocity is essentially constant so that the shear stresses (which are proportional to velocity derivatives) are small. Thus, viscous work may be assumed to be negligible.

Solution: Choose the control volume shown here and apply Eq. 3.45, recalling that outflow is positive and inflow is negative. Substituting in this equation yields:

$$\int\limits_{2} \rho V_2(C_p T_0)d\hat{S} - \int\limits_{1} \rho V_1(C_p T_0)d\hat{S} = \dot{Q}_{COND} - \dot{W}_M.$$

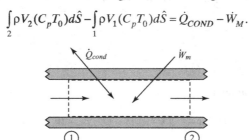

Because the flow is one-dimensional and the pipe is of constant area:

$$\rho_2 V_2 (C_p T_0)_2 A_2 - \rho_1 V_1 (C_p T_0)_2 A_1 = \dot{m} C_p (T_{02} - T_{01}).$$

Now, $T_{01} = T_{02}$ according to the given information, so that the left side of Eq. 3.25b is zero. It remains to apply the correct signs to the two terms on the right side. Recall that in the derivation of the energy equation, the machine work was defined as work done by the fluid. Because 2,000 Nm/s of work is added here, this number must enter the equation with a minus sign, so that:

$$0 = \dot{Q}_{COND} - (-2,000) \Rightarrow \dot{Q}_{COND} = -2,000.$$

Because the $\dot{Q}_{COND}$ term defines the rate of heat transfer to the control volume as positive, the resulting negative sign in the solution means that 2,000 Joules/s of heat is removed from the pipe between the two stations.

Appraisal: Because work is being done on the flowing air between the two stations and there is no change in stagnation temperature, the result that heat is being removed agrees with intuition. Notice that the numerical value of the stagnation temperature was not used. The only important fact was that the stagnation temperature was constant. Thus, part of the given information was not required. For many problems encountered in the real world, not enough information is known (assumptions must be made) or too much information is known (irrelevant data must be discarded—but which?). This is where a physical understanding of the problem and the equations becomes vital.

EXAMPLE 3.7 *Given*: There is a steady flow of air at a rate of 15 lbm/s through the compressor of a stationary gas turbine. The air enters at average conditions of $T_1 = 80°F$ and $V_1 = 10$ ft/s and discharges at an average velocity of $V_2 = 90$ ft/s. The compressor receives 500 HP of work from the turbine, while the heat loss from the compressor to the surroundings is 300 BTU/s. Assume that the work on the control surface is due only to pressure and ignore changes in potential-energy and body-force effects.

Required: Predict the average enthalpy of the air at the exit of the compressor.

Approach: Choose a suitable control volume that encloses the compressor. Use the steady-flow integral-energy equation with certain terms omitted.

Solution: Select the control volume shown here:

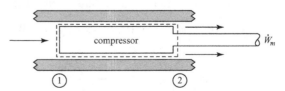

Apply Eq. 3.44; namely:

$$\iint_{\hat{S}} \rho \left(h + \frac{V^2}{2} \right) (\mathbf{V} \cdot \mathbf{n}) d\hat{S} = \dot{Q}_{COND} - \dot{W}_M$$

Now, $\rho(\mathbf{V} \cdot \mathbf{n})d\hat{S} = \dot{m}$; therefore, for one-dimensional flow,

$$\dot{m}\left(h + \frac{V^2}{2}\right)_2 - \dot{m}\left(h + \frac{V^2}{2}\right)_1 = \dot{Q}_{COND} - \dot{W}_M.$$

Recall that $h = c_p T$ and, from tables for air, $c_p = 0.24$ BTU/lbm°F. Thus, on the left side of the equation, $\dot{m}h$ has the units of [lbm/s][BTU/lbm] = BTU/s. Because all terms in the equation must have the same units, arbitrarily adopt BTU/s as the standard to be used in this problem. Examining the units of the other terms in the equation and supplying appropriate conversion factors, the units of the following terms are:

$$\begin{cases} \dot{m}V^2 = \left[\frac{\text{lb}_m}{s}\right] \cdot \left[\frac{\text{ft}^2}{s}\right] \cdot \left[\frac{\text{slug}}{32.2\,\text{lb}_m}\right] \cdot \left[\frac{\text{BTU}}{778\,\text{ft}-\text{lb}_f}\right] = \left[\frac{\text{BTU}}{s}\right] \\[2mm] \dot{W}_M = [\text{HP}] \cdot \left[\frac{550\,\text{ft}-\text{lb}_f}{\text{HP}}\right] \cdot \left[\frac{\text{BTU}}{778\,\text{ft}-\text{lb}_f}\right] = \left[\frac{\text{BTU}}{s}\right] \\[2mm] \dot{Q}_{COND} = \left[\frac{\text{BTU}}{s}\right] \end{cases}$$

Then, assemble these terms into Eq. 3.44, recognizing that heat is being transferred from the control volume and work is being done on the gas within the control volume. Thus,

$$\dot{m}\left\{\left[h_2 + \frac{V_2^2}{2}\right] - \left[h_1 + \frac{V_1^2}{2}\right]\right\} = \dot{Q}_{COND} - \dot{W}_M$$

$$15\left\{\left[h_2 + \frac{(90)^2}{2(32.2)(778)}\right] - \left[(.24)(80+460) + \frac{(10)^2}{2(32.2)(778)}\right]\right\} =$$

$$= -(300) - \left[-\frac{(500)(550)}{778}\right]$$

Solving, $h_2 = $ enthalpy of exit air $= 133$ BTU/lbm.

Appraisal: The resulting value for exit enthalpy implies an exit temperature of 94°F. The answer comes out positive (as it must be), and the magnitude of the exit temperature appears reasonable because it is greater than the inlet temperature. Note that we adopted the mass unit *lbm* for this problem, as often is done in thermodynamics textbooks. The student is reminded that the *slug* is the compatible mass unit so that the conversion factor, $g_c = 32.2$ lbm/slug, appears in the numerical evaluation.

Other Physical Laws in Control-Volume Form

We now have derived the control-volume equations that are needed throughout the book in numerous aerodynamics applications. Other physical laws are required only occasionally in these applications. For example, the Second Law of Thermo-dynamics, Eq. 3.13, can be expressed readily in control-volume form. The procedure follows closely that used for the First Law. Similarly, the part of Newton's Second Law pertaining to angular motion can be expressed readily in control-volume form following the same procedures described previously. Problems in the set of exercises at the end of the chapter provide the opportunity to test the student's understanding by carrying out the details for some of these additional physical laws.

3.5 Physical Laws in Differential-Equation Form

In examining the example problems describing the application of the physical laws in control-volume form, notice that evaluation of the surface and volume integrals often involve assuming average values of the variables. For example, averaged densities, velocities, and pressures were used to make the integrals tractable. That is, there was no detailed information on the actual point-to-point variations in any of the key variables. Clearly, in some situations, this would be a crude description of the fluid motion. It might be required, for example, to know how the pressure is distributed over the surface or how the velocity is distributed through a viscous boundary layer. In determining the forces on a wing, it is required to know not merely the magnitude of the lift and drag forces but also how they are distributed over the surface. Problems of this type require a sharper mathematical representation of the physical laws. This precision is available only if we convert the equations of fluid motion into differential form. Then, by solving the resulting set of *differential field equations*, we can determine in detail the behavior of the flow variables throughout the domain of the problem. In real problem solving, both the control volume and the differential-equation formulations often are used together to generate the required information.

The differential equations describing fluid flow can be derived directly by analysis of differential-control volumes. In this method, we define small control volumes that conform to the type of coordinate system to be used. For example, in Cartesian coordinates, a differential cube is appropriate. The procedure is repeated for each coordinate system of interest. However, a simpler way is to work from the control-volume formulation already worked out herein. This method involves converting all surface integrals and other terms into equivalent volume integrals by means of appropriate forms of *Green's Theorem*. Then, all terms are collected into a single-volume integral, which is equal to zero. Because the control volume is of arbitrary size and shape, a necessary condition for satisfaction of the resulting equation is that the integrand must be zero. The result is the required differential form of the original integral equation, which can be expressed in any desired coordinate system by knowing the vector operators such as the *gradient, divergence*, and *curl* in that coordinate system. This procedure is followed for each of the basic physical laws.

Differential-Continuity Equation

A derivation of the differential form of the continuity equation is carried out next. As explained, differential equations provide a tool for a more complete description of the flow field. In particular, a detailed description of the complete field behavior at any point in the flow then can be accomplished.

Recall from advanced calculus that a form of Green's Theorem, the *divergence theorem* (also called *Gauss's Theorem*), relates surface integrals to volume integrals —namely, if **A** is a vector, then:

$$\iint_{\hat{S}} (\mathbf{A} \cdot \mathbf{n}) d\hat{S} = \iiint_{\hat{V}} (\nabla \cdot \mathbf{A}) d\hat{V} = \iiint_{\hat{V}} div \mathbf{A} d\hat{V}. \tag{3.47}$$

In what follows, we use the gradient-operator notation ($\nabla \rightarrow$ gradient, $\nabla \cdot \rightarrow$ divergence, $\nabla \times \rightarrow$ curl) instead of *grad, div,* and *curl*. Now, by identifying the vector **A** in Eq. 3.47 with the vector ($\rho\mathbf{V}$) in Eq. 3.24, the surface integral in Eq. 3.24 may be changed to a volume integral by using the divergence theorem. Also, because the control volume that we are using is fixed in space, the limits on the volume integral in Eq. 3.24 do not vary with time. That being the case, the time derivative may be brought through the volume integral such that:

$$\frac{\partial}{\partial t} \iiint_{\hat{V}} \rho d\hat{V} = \iiint_{\hat{V}} \frac{\partial}{\partial t} (\rho d\hat{V}). \tag{3.48}$$

Then, using Eqs. 3.47 and 3.48 and putting both terms in Eq. 3.24 under the same volume integral:

$$\iiint_{\hat{V}} \left(\frac{\partial \rho}{\partial t} + \nabla \cdot (\rho\mathbf{V}) \right) d\hat{V} = 0. \tag{3.49}$$

Now, Eq. 3.49 must hold for any arbitrary control volume; therefore, the integrand must be zero. The integrand then represents the differential form of the continuity requirement.

The only way that Eq. 3.49 can be satisfied, then, is for the integrand to be zero. Thus,

$$\frac{\partial \rho}{\partial t} + \nabla \cdot (\rho\mathbf{V}) = 0. \tag{3.50}$$

This is the general continuity equation in differential-equation form, valid at any field point in an unsteady compressible flow. Again, as in the control-volume continuity formulation, there are special forms of Eq. 3.50 that fit cases of interest and simplify the equation. For steady compressible flow, for example, Eq. 3.50 becomes:

$$\nabla \cdot (\rho\mathbf{V}) = 0.$$

For incompressible flow, Eq. 3.50 becomes:

$$\nabla \cdot \mathbf{V} = 0,$$

which is valid for both steady and unsteady flow because in an incompressible flow, the density is constant and all of its derivatives therefore are zero. Table 3.5 summarizes the special cases of the continuity equation.

These differential equations can be written for any coordinate system by suitably expanding the vector operations. For example, in a three-dimensional Cartesian coordinate system with unit vectors **i**, **j**, and **k** in the coordinate directions $x, y,$ and z and with the velocity vector having components $\mathbf{V} = u\mathbf{i} + v\mathbf{j} + w\mathbf{k}$, Eq. 3.50 becomes:

Table 3.5. *Differential form of the continuity equation*

General form	$\dfrac{\partial \rho}{\partial t} + \nabla \cdot (\rho \mathbf{V}) = 0$	(3.50)
Steady compressible flow	$\nabla \cdot (\rho \mathbf{V}) = 0$	(3.51)
Incompressible flow	$\nabla \cdot \mathbf{V} = 0$	(3.52)

$$\frac{\partial \rho}{\partial t} + \frac{\partial}{\partial x}(\rho u) + \frac{\partial}{\partial y}(\rho v) + \frac{\partial}{\partial z}(\rho w) = 0. \tag{3.53}$$

For incompressible flow, Eq. 3.53 reduces to:

$$\frac{\partial u}{\partial x} + \frac{\partial v}{\partial y} + \frac{\partial w}{\partial z} = 0. \tag{3.54}$$

The first of the conservation laws now is expressed in terms of both an integral and a differential equation. The equivalent equations in other coordinates (e.g., cylindrical and spherical) are found by using the vector operators written in that system. For example, in three-dimensional cylindrical coordinates with unit vectors, $\mathbf{e}_r$, $\mathbf{e}_\theta$, and $\mathbf{e}_z$, the velocity vector can be written as $\mathbf{V} = u_r\mathbf{e}_r + u_\theta\mathbf{e}_\theta + u_z\mathbf{e}_z$, and the divergence of the velocity vector is:

$$\nabla \cdot \mathbf{V} = \frac{1}{r}\frac{\partial}{\partial r}(ru_r) + \frac{1}{r}\frac{\partial u_\theta}{\partial \theta} + \frac{\partial u_z}{\partial z}.$$

This expression is set to zero for continuity for an incompressible flow.
In the case of a compressible flow, the density multiplies each of the scalar velocity components as before. It is important to understand the process described here, to apply the equations in the coordinate system that best fits the geometry of the problem to be solved.

EXAMPLE 3.8 *Given*: A velocity field for an incompressible flow is described by:

$$\mathbf{V} = (2xy^2)\,\mathbf{i} + (2x^2y)\,\mathbf{j}.$$

Required: Determine if this flow field is physically possible.

Approach: To be physically possible, the flow field must satisfy the conservation of mass.

Solution: Test the flow field by checking whether it satisfies the continuity equation for an incompressible flow, Eq. 3.13. Then,

$$\frac{\partial u}{\partial x} + \frac{\partial v}{\partial y} = 2y^2 + 2x^2 \neq 0,$$

except at the origin. This flow field is not physically possible.

Appraisal: A flow field that is described as a given mathematical function or that is to be solved for any problem should always be checked to see whether it is physically possible—that is, whether it satisfies the continuity equation.

The Differential Form of the Momentum Equation

As in the case of the continuity equation, the starting point for the derivation of the differential form of the momentum equation is the integral form, Eq. 3.32:

$$\frac{\partial}{\partial t}\iiint_{\hat{V}}\rho \mathbf{V}d\hat{V}+\iint_{\hat{S}}\rho \mathbf{V}(\mathbf{V}\cdot\mathbf{n})d\hat{S}=\mathbf{F}=\iiint_{\hat{V}}\rho \mathbf{b}d\hat{V}+\iint_{\hat{S}}\tau d\hat{S}. \tag{3.55}$$

The surface force is decomposed into two parts: one due to normal pressure force only, τ_p; and the other containing both normal and shear components due to viscosity, τ_v, because there can be a viscous correction to the normal force due to the pressure. Thus, Eq. 3.55 becomes:

$$\frac{\partial}{\partial t}\iiint_{\hat{V}}\rho \mathbf{V}d\hat{V}+\iint_{\hat{S}}\rho \mathbf{V}(\mathbf{V}\cdot\mathbf{n})d\hat{S}=\iiint_{\hat{V}}\rho \mathbf{b}d\hat{V}+\iint_{\hat{S}}\tau_p d\hat{S}+\iint_{\hat{S}}\tau_v d\hat{S}. \tag{3.56}$$

In following the same procedure used for the continuity equation, we must assemble all of the terms under a single-volume integral. Work with the force components first; the body-force term is already in the required form. Because $\tau_p = -p\mathbf{n}d\hat{S}$, the surface integral can be converted from a surface to a volume integral by using the following form of the divergence theorem[*]:

$$\iint_{\hat{S}}G\mathbf{n}d\hat{S}=\iiint_{\hat{V}}\nabla G d\hat{V} \tag{3.57}$$

where G is any scalar (e.g., p in the present case) and the gradient of a scalar is a vector. The second term on the right side of Eq. 3.56 then becomes:

$$\iint_{\hat{S}}\tau_p d\hat{S}=\iint_{\hat{S}}-p\mathbf{n}d\hat{S}=-\iiint_{\hat{V}}\nabla p d\hat{V}. \tag{3.58}$$

---

[*] The derivation is shown here for convenience. Starting with the divergence theorem,

$$\iint_{\hat{S}}(\mathbf{A}\cdot\mathbf{n})d\hat{S}=\iiint_{\hat{V}}(\nabla\cdot\mathbf{A})d\hat{V}.$$

Put $\mathbf{A}=G\mathbf{B}$, where $\mathbf{B}$ = constant and G = scalar variable.

$$\iint_{\hat{S}}G\mathbf{B}\cdot\mathbf{n}d\hat{S}=\iiint_{\hat{V}}\nabla\cdot(G\mathbf{B})d\hat{V},$$

but $\nabla\cdot(G\mathbf{B})=G\nabla\cdot\mathbf{B}+\mathbf{B}\cdot\nabla G=\mathbf{B}\cdot\nabla G$

$$\iint_{\hat{S}}G\mathbf{B}\cdot\mathbf{n}d\hat{S}=\iiint_{\hat{V}}\mathbf{B}\cdot\nabla G d\hat{V}$$

$$\mathbf{B}\cdot\iint_{\hat{S}}G\mathbf{n}d\hat{S}=\mathbf{B}\cdot\iiint_{\hat{V}}\nabla G d\hat{V}, \text{ because } \mathbf{B}=\text{constant}.$$

Thus, $\iint_{\hat{S}}G\mathbf{n}d\hat{S}=\iiint_{\hat{V}}\nabla G d\hat{V}.$

The last term involving the viscous stresses may be rewritten in terms of a volume integral after first expressing τ_v in terms of its normal and tangential viscous stress components. Because details of the viscous-force modeling are the subject of Chapter 8, the volume integral corresponding to the viscous term is represented symbolically for the present as:

$$\iint_{\hat{S}} \tau_v d\hat{S} = \iiint_{\hat{V}} \mathbf{f}_v d\hat{V} \tag{3.59}$$

and is carried along as a placeholder. Regardless of the details, this term is zero if the flow is assumed to be inviscid—a simplification of which we often take advantage.

Working with the first acceleration term on the left side of Eq. 3.56, the time derivative may be passed through the integral sign because the limits of integration are independent of time; the control volume is fixed. Thus,

$$\frac{\partial}{\partial t} \iiint_{\hat{V}} \rho \mathbf{V} d\hat{V} = \iiint_{\hat{V}} \frac{\partial}{\partial t} (\rho \mathbf{V}) d\hat{V}. \tag{3.60}$$

The second term on the left side:

$$\text{convective acceleration} = \iint_{\hat{S}} \rho \mathbf{V} (\mathbf{V} \cdot \mathbf{n}) d\hat{S}, \tag{3.61}$$

requires special consideration. We again attempt to convert the surface integral of Eq. 3.61 to a volume integral by means of the divergence theorem,

$$\iint_{\hat{S}} (\mathbf{A} \cdot \mathbf{n}) d\hat{S} = \iiint_{\hat{V}} (\nabla \cdot \mathbf{A}) d\hat{V},$$

where $\mathbf{A}$ is any vector. Notice that there is an apparent difficulty here in converting the surface integral to a volume integral by using the divergence theorem in this form because $\rho \mathbf{V} (\mathbf{V} \cdot \mathbf{n})$ in Eq. 3.61 is *not* of the form $(\mathbf{A} \cdot \mathbf{n})$. In fact, $\rho \mathbf{V} (\mathbf{V} \cdot \mathbf{n})$ is a vector and $(\mathbf{A} \cdot \mathbf{n})$ is a scalar. The divergence theorem is not directly applicable to the conversion of Eq. 3.61 into a volume integral. Notice, however, that the problem is circumvented if we focus on any one scalar component of the velocity vector $\mathbf{V}$. Then, the procedure we follow is to work first with one component and then generalize the results to full vector form.

For simplicity, use Cartesian coordinates and focus on the x-component, u, of the velocity vector:

$$\mathbf{V} = u\mathbf{i} + v\mathbf{j} + w\mathbf{k}. \tag{3.62}$$

Then, the x-component of the vector surface integral in Eq. 3.61 is converted readily to a volume integral by means of the divergence theorem. We find:

$$\iint_{\hat{S}} \rho u (\mathbf{V} \cdot \mathbf{n}) d\hat{S} = \iiint_{\hat{V}} \nabla \cdot (\rho u \mathbf{V}) d\hat{V}.$$

Inserting this and the x-components of the other terms into Eq. 3.56:

$$\iiint_{\hat{V}} \frac{\partial}{\partial t} (\rho u) d\hat{V} + \iiint_{\hat{V}} \nabla \cdot (\rho u \mathbf{V}) d\hat{V} = \mathbf{i} \cdot \left[\iiint_{\hat{V}} \rho \mathbf{b} d\hat{V} - \iiint_{\hat{V}} \nabla p \, d\hat{V} + \iiint_{\hat{V}} \mathbf{f}_v d\hat{V} \right],$$

where the x-components of the vectors on the right side are found by dotting with the unit vector, $\mathbf{i}$. Collecting all of the terms under the same volume integral results in:

$$\iiint\limits_{V} \left\{ \frac{\partial}{\partial t}(\rho u) + \nabla \cdot (\rho u \mathbf{V}) - \mathbf{i} \cdot (\rho \mathbf{b} - \nabla p + \mathbf{f}_v) \right\} d\hat{V} = 0.$$

Thus, for an arbitrary control volume, we require that the integrand vanishes with the result that for the x-component,

$$\frac{\partial}{\partial t}(\rho u) + \nabla \cdot (\rho u \mathbf{V}) - \mathbf{i} \cdot (\rho \mathbf{b} - \nabla p + \mathbf{f}_v) = 0. \qquad (3.63)$$

This is the so-called conservation form of the differential-momentum equation (x-component). It is clear that the other two components can be found by a similar analysis. The three momentum-equation components are used commonly in conservation form to set up numerical solutions in *computational fluid dynamics* (CFD) procedures, discussed later in the book.

It is useful to expand the first term,

$$\frac{\partial}{\partial t}(\rho u) = \rho \frac{\partial u}{\partial t} + u \frac{\partial \rho}{\partial t},$$

and the second term by means of the vector identity $\nabla(\phi \mathbf{A}) = \mathbf{A} \cdot \nabla \phi + \phi \nabla \cdot \mathbf{A}$. Then,

$$\nabla \cdot (\rho u \mathbf{V}) = \rho \mathbf{V} \cdot \nabla u + u \nabla \cdot (\rho \mathbf{V}).$$

The combination of the first and second terms yields:

$$\frac{\partial}{\partial t}(\rho u) + \nabla \cdot (\rho u \mathbf{V}) = \rho \frac{\partial u}{\partial t} + u \frac{\partial \rho}{\partial t} + \rho \mathbf{V} \cdot \nabla u + u \nabla \cdot (\rho \mathbf{V}) =$$

$$= \rho \left(\frac{\partial u}{\partial t} + \mathbf{V} \cdot \nabla u \right) + u \left(\frac{\partial \rho}{\partial t} + \nabla \cdot (\rho \mathbf{V}) \right) = \rho \left(\frac{\partial u}{\partial t} + \mathbf{V} \cdot \nabla u \right),$$

where the part multiplied by u is the continuity equation and must be set to zero so that mass is conserved. Noting that the pressure gradient in Cartesian form is:

$$\nabla p = \frac{\partial p}{\partial x} \mathbf{i} + \frac{\partial p}{\partial y} \mathbf{j} + \frac{\partial p}{\partial z} \mathbf{k}, \qquad (3.64)$$

we can write the momentum equation in Cartesian component form as:

$$\begin{cases} \rho \left(\dfrac{\partial u}{\partial t} + \mathbf{V} \cdot \nabla u \right) = -\dfrac{\partial p}{\partial x} + \rho b_x + \mathbf{i} \cdot \mathbf{f}_v \\[2mm] \rho \left(\dfrac{\partial v}{\partial t} + \mathbf{V} \cdot \nabla v \right) = -\dfrac{\partial p}{\partial y} + \rho b_y + \mathbf{j} \cdot \mathbf{f}_v \\[2mm] \rho \left(\dfrac{\partial w}{\partial t} + \mathbf{V} \cdot \nabla w \right) = -\dfrac{\partial p}{\partial z} + \rho b_z + \mathbf{k} \cdot \mathbf{f}_v. \end{cases} \qquad (3.65)$$

This suggests that in compact vector form, the differential-momentum equation is:

$$\rho \left(\frac{\partial \mathbf{V}}{\partial t} + \mathbf{V} \cdot \nabla \mathbf{V} \right) = -\nabla p + \rho \mathbf{b} + \mathbf{f}_v, \qquad (3.66)$$

which is sometimes called the *primitive-variable* form of the momentum equation. In effect, we split off the continuity equation from the conservation form (i.e., Eq. 3.63

and the other two components) to arrive at the primitive form. Expressed in words, this equation states that "the mass per unit volume times the total acceleration equals the force per unit volume" at any point in the field of motion. In the usual particle-mechanics problem, the time derivative of the velocity vector is the acceleration. Because of our use of Eulerian-field point of view, the derivative of $\mathbf{V}$ is the *local acceleration* term; the second term, $\mathbf{V} \cdot \nabla \mathbf{V}$, is the *convective acceleration* term that corrects for the rate of change of fluid properties due to motion in the field.

Although the derivation was carried out for Cartesian coordinates, Eq. 3.66 is the correct result for any coordinate system provided that we properly define the convective accleration $\mathbf{V} \cdot \nabla \mathbf{V}$ and correctly interpret the implied mathematics. Note that unlike the Cartesian scalar-component form in Eq. 3.65, this term cannot be interpreted as simply $\mathbf{V}$ dotted with the gradient of $\mathbf{V}$ (the gradient of a vector is not defined!). A careful discussion and interpretation of this seemingly pathological term is required. Many errors in the application of the differential-momentum equation can be traced to an insufficient understanding of this term and its extension to other than simple rectangular Cartesian coordinate systems. Observe also that this term is nonlinear. That is, it involves products of the velocity components whereas, for comparison, the local acceleration term is linear in the sense that it contains no products (or transcendental functions) of the variables.

The operator $\mathbf{V} \cdot \nabla$ is sometimes called the *convective operator*. In Cartesian form, it is correct to interpret this as an operator:

$$\mathbf{V} \cdot \nabla = u \frac{\partial}{\partial x} + v \frac{\partial}{\partial y} + w \frac{\partial}{\partial z}$$

and then apply it to each scalar component of the vector on which it acts. For example, for the x-component of Eq. 3.65, we find that taking the dot product of the velocity vector with the gradient of u:

$$\mathbf{V} \cdot \nabla u = \mathbf{V} \cdot \left(\frac{\partial u}{\partial x} \mathbf{i} + \frac{\partial u}{\partial y} \mathbf{j} + \frac{\partial u}{\partial z} \mathbf{k} \right) = u \frac{\partial u}{\partial x} + v \frac{\partial u}{\partial y} + w \frac{\partial u}{\partial z}$$

is the same result found by treating $\mathbf{V} \cdot \nabla$ as an operator acting on u; thus,

$$\mathbf{V} \cdot \nabla(u) = \left(u \frac{\partial}{\partial x} + v \frac{\partial}{\partial y} + w \frac{\partial}{\partial z} \right)(u) = u \frac{\partial u}{\partial x} + v \frac{\partial u}{\partial y} + w \frac{\partial u}{\partial z}.$$

The convective element of the momentum balance must be treated with great care. It is best to treat the form, $\mathbf{V} \cdot \nabla$, as a shorthand notation that can be evaluated correctly and explicitly as demonstrated for Cartesian coordinates. To find the correct form for any coordinate system, it is necessary to use vector analysis.* The form that is correct in any coordinate system is:

---

* Using the vector identity:

$\nabla(\mathbf{u} \cdot \mathbf{v}) = \mathbf{u} \cdot \nabla \mathbf{v} + \mathbf{v} \cdot \nabla \mathbf{u} + \mathbf{u} \times (\nabla \times \mathbf{v}) + \mathbf{v} \times (\nabla \times \mathbf{u})$

and setting $\mathbf{u} = \mathbf{v}$,

$\nabla(\mathbf{V} \cdot \mathbf{V}) = 2\mathbf{V} \cdot \nabla \mathbf{V} + 2\mathbf{V} \times (\nabla \times \mathbf{V}).$

Thus,

$\mathbf{V} \cdot \nabla \mathbf{V} = \nabla\left(\frac{1}{2} \mathbf{V} \cdot \mathbf{V}\right) - \mathbf{V} \times (\nabla \times \mathbf{V}) = \nabla\left(\frac{1}{2} V^2\right) - \mathbf{V} \times (\nabla \times \mathbf{V}).$

$$(\mathbf{V} \cdot \nabla)\mathbf{V} = \nabla \left(\frac{\mathbf{V} \cdot \mathbf{V}}{2} \right) - \mathbf{V} \times \nabla \times \mathbf{V}. \tag{3.67}$$

Then, whenever it is necessary to write the momentum equation for cylindrical or spherical coordinates, we use Eq. 3.67 to arrive at the correct equations. This expression is demonstrated herein when we evaluate the momentum equation in component form for cylindrical coordinates (see Eq. 3.71). As is typical, mathematical results such as Eq. 3.67 often have important physical implications. For example, the first term contains the square of the local flow speed, which suggests a connection to the kinetic energy of the flow. The second term involves the curl of the velocity vector, $\nabla \times \mathbf{V}$, which is associated in subsequent chapters with important properties of the flow field, such as *rotation* and *vorticity*.

Now that the general form for the momentum balance, Eq. 3.66, is deduced, we can proceed as usual to write reduced forms that suit situations of special interest. In the complete form, the viscous-force term, $\mathbf{f}_v$, is replaced by a suitable model that relates the components of the viscous force to the geometry of the flow. The Newtonian model for viscous force introduced in Chapter 2 (see Eq. 2.4) is generalized in Chapter 8 for this purpose. The resulting equation (or equations, if written in component form) is called the *Navier–Stokes Equation* after Navier (French) and Stokes (English), who developed the equations independently in the mid-1800s. Famous names are associated with other special forms of the momentum balance. For example, an important case is Euler's equations, in which the viscous forces are dropped. The inviscid approximation has many applications in aerodynamics problems. Table 3.6 summarizes important special forms of the differential-momentum equation.

The second set of equations (Eq. 3.68) in Table 3.6 represents the Cartesian form of the Navier–Stokes equations for an incompressible fluid with a Newtonian model for the viscous forces. The viscosity coefficient, μ, is assumed to be constant (see Chapter 2). The details of the viscous model used here are elaborated on in Chapter 8. Notice that in the other cases shown in the table, there is no assumption of incompressibility. Equations 3.69–3.71 are valid for compressible flow because after splitting the continuity equation off of the original momentum balance, there are no derivatives of the density remaining in the general equation. However, if viscous forces are retained, density gradients may have important consequences.

Equation 3.70, the two-dimensional Euler's equations for Cartesian coordinates, is used frequently and should be compared to the set shown in Eq. 3.71 in Table 3.6, written in polar coordinates. Students should test their understanding of these results by reducing the complete vector-momentum equation (Eq. 3.66) to the special cases shown in the table. It is important to see that treatment of the convective acceleration by using the $\mathbf{V} \cdot \nabla$ operator approach (correct only for the Cartesian case) does not work for cylindrical coordinates. It is necessary to use the full form of the convective accleration in vector form (Eq. 3.67) to derive the result shown in Eq. 3.71.

Because the momentum equation is a vector equation, the result is that three more of the required six equations needed for the general fluid-mechanics problem now are established. If the flow is assumed to be incompressible, then the continuity and momentum equations are sufficient to define the problem. This simplification is used to advantage in subsequent chapters.

Table 3.6. *Differential form of the momentum equation*

General Navier–Stokes in vector form (Eq. 3.66)	$\rho\left(\dfrac{\partial \mathbf{V}}{\partial t}+\mathbf{V}\cdot\nabla\mathbf{V}\right)=-\nabla p+\rho\mathbf{b}+\mathbf{f}_v$
Navier–Stokes equation for incompressible fluid with constant viscosity, no body forces, three-dimensional Cartesian coordinates, unsteady flow	$\begin{cases}\rho\left(\dfrac{\partial u}{\partial t}+u\dfrac{\partial u}{\partial x}+v\dfrac{\partial u}{\partial y}+w\dfrac{\partial u}{\partial z}\right)=-\dfrac{\partial p}{\partial x}+\mu\nabla^2 u \\[2mm] \rho\left(\dfrac{\partial v}{\partial t}+u\dfrac{\partial v}{\partial x}+v\dfrac{\partial v}{\partial y}+w\dfrac{\partial v}{\partial z}\right)=-\dfrac{\partial p}{\partial y}+\mu\nabla^2 v \qquad (3.68) \\[2mm] \rho\left(\dfrac{\partial w}{\partial t}+u\dfrac{\partial w}{\partial x}+v\dfrac{\partial w}{\partial y}+w\dfrac{\partial w}{\partial z}\right)=-\dfrac{\partial p}{\partial z}+\mu\nabla^2 w\end{cases}$
Euler's equations in three-dimensional Cartesian coordinates; no body forces; steady, inviscid flow	$\begin{cases}\rho\left(u\dfrac{\partial u}{\partial x}+v\dfrac{\partial u}{\partial y}+w\dfrac{\partial u}{\partial z}\right)=-\dfrac{\partial p}{\partial x} \\[2mm] \rho\left(u\dfrac{\partial v}{\partial x}+v\dfrac{\partial v}{\partial y}+w\dfrac{\partial v}{\partial z}\right)=-\dfrac{\partial p}{\partial y} \qquad (3.69) \\[2mm] \rho\left(u\dfrac{\partial w}{\partial x}+v\dfrac{\partial w}{\partial y}+w\dfrac{\partial w}{\partial z}\right)=-\dfrac{\partial p}{\partial z}\end{cases}$
Euler's equations in two-dimensional Cartesian coordinates; no body forces; steady, inviscid flow	$\begin{cases}\rho\left(u\dfrac{\partial u}{\partial x}+v\dfrac{\partial u}{\partial y}\right)=-\dfrac{\partial p}{\partial x} \\[2mm] \rho\left(u\dfrac{\partial v}{\partial x}+v\dfrac{\partial v}{\partial y}\right)=-\dfrac{\partial p}{\partial y}\end{cases} \qquad (3.70)$
Euler's equations in two-dimensional cylindrical coordinates; no body forces; steady, inviscid flow	$\begin{cases}\rho\left(u_r\dfrac{\partial u_r}{\partial r}+\dfrac{u_\theta}{r}\dfrac{\partial u_r}{\partial \theta}-\dfrac{u_\theta^2}{r}\right)=-\dfrac{\partial p}{\partial r} \\[2mm] \rho\left(u_r\dfrac{\partial u_\theta}{\partial r}+\dfrac{u_\theta}{r}\dfrac{\partial u_\theta}{\partial \theta}+\dfrac{u_r u_\theta}{r}\right)=-\dfrac{1}{r}\dfrac{\partial p}{\partial \theta}\end{cases} \qquad (3.71)$

EXAMPLE 3.9 *Given*: Consider the following flow field. It is steady and inviscid with no body forces. The flow field is described by $\mathbf{V} = (2x)\mathbf{i} + (3x\text{-}2y)\mathbf{j}$. Assume the density (constant) to be $\rho = 0.2$ slug/ft^3.

Required: Find the pressure gradient in this flow field in the x-direction at a certain point specified by $(x = 3, y = 2)$.

Approach: Because the x-component of the pressure gradient (i.e., $\partial p/\partial x$) is required, use the x-component of the differential-momentum equation (Eq. 3.70) with constant density.

Solution: First, is this flow physically possible? Check using Eq. 3.52: Yes (verify this). Next, substitute Eq. 3.70 and evaluate:

$$u\frac{\partial u}{\partial x}+v\frac{\partial u}{\partial y}=4x+0=-\frac{1}{\rho}\frac{\partial p}{\partial x}, \quad \text{and solving,} \quad \frac{\partial p}{\partial x}=-2.4 \text{ lbf/ft}^2/\text{ft.}$$

Appraisal: For this particular flow field, the pressure gradient in the x-direction is independent of y. This, of course, is not a general result. Note that this example does not represent a solution of the differential-momentum equation but rather is an exercise in applying the equation.

EXAMPLE 3.10 *Required*: Derive Euler's equation in two-dimensional polar coordinates by direct evaluation of the general momentum equation (Eq. 3.66) and Eq. 3.67.

Approach: Assume steady, inviscid flow without body forces and insert Eq. 3.67 for the convective acceleration so that the momentum equation is valid for any coordinate system. Then, use the vector expressions for the gradient and curl in polar coordinates.

Solution: Euler's equation in vector form valid for any coordinates becomes:

$$\rho\left(\nabla\left(\frac{\mathbf{V}\cdot\mathbf{V}}{2}\right)-\mathbf{V}\times\nabla\times\mathbf{V}\right)=\nabla p.$$

Carrying out the required vector operations,

$$
\begin{cases}
\mathbf{V}=u_r\mathbf{e}_r+u_\theta\mathbf{e}_\theta \qquad \mathbf{V}\cdot\mathbf{V}=u_r^2+u_\theta^2 \\[2mm]
\nabla p=\dfrac{\partial p}{\partial r}\mathbf{e}_r+\dfrac{1}{r}\dfrac{\partial p}{\partial \theta}\mathbf{e}_\theta \\[3mm]
\nabla\left(\dfrac{\mathbf{V}\cdot\mathbf{V}}{2}\right)=\left(u_r\dfrac{\partial u_r}{\partial r}+u_\theta\dfrac{\partial u_\theta}{\partial r}\right)\mathbf{e}_r+\left(\dfrac{u_r}{r}\dfrac{\partial u_r}{\partial \theta}+\dfrac{u_\theta}{r}\dfrac{\partial u_\theta}{\partial \theta}\right)\mathbf{e}_\theta \\[3mm]
\nabla\times\mathbf{V}=\dfrac{1}{r}\begin{vmatrix}\mathbf{e}_r & r\mathbf{e}_\theta & \mathbf{e}_z \\[1mm] \dfrac{\partial}{\partial r} & \dfrac{\partial}{\partial \theta} & \dfrac{\partial}{\partial z} \\[1mm] u_r & ru_\theta & 0\end{vmatrix}=\left(\dfrac{1}{r}\dfrac{\partial}{\partial r}(ru_\theta)-\dfrac{1}{r}\dfrac{\partial u_r}{\partial \theta}\right)\mathbf{e}_z \\[3mm]
\mathbf{V}\times\nabla\times\mathbf{V}=u_\theta\left(\dfrac{1}{r}\dfrac{\partial}{\partial r}(ru_\theta)-\dfrac{1}{r}\dfrac{\partial u_r}{\partial \theta}\right)\mathbf{e}_r-u_r\left(\dfrac{1}{r}\dfrac{\partial}{\partial r}(ru_\theta)-\dfrac{1}{r}\dfrac{\partial u_r}{\partial \theta}\right)\mathbf{e}_\theta \\[3mm]
\nabla\left(\dfrac{\mathbf{V}\cdot\mathbf{V}}{2}\right)-\mathbf{V}\times\nabla\times\mathbf{V}=\left(u_r\dfrac{\partial u_r}{\partial_r}-\dfrac{u_\theta^2}{r}+\dfrac{u_\theta}{r}\dfrac{\partial u_r}{\partial_\theta}\right)\mathbf{e}_r+ \\[3mm]
\qquad\qquad +\left(\dfrac{u_\theta}{r}\dfrac{\partial u_\theta}{\partial \theta}+u_r\dfrac{\partial u_\theta}{\partial r}+\dfrac{u_r u_\theta}{r}\right)\mathbf{e}_\theta.
\end{cases}
$$

Inserting these expressions into the vector Euler's equations and writing the radial and transverse components yields Eq. 3.71:

$$
\begin{cases}
\rho\left(u_r\dfrac{\partial u_r}{\partial r}-\dfrac{u_\theta^2}{r}+\dfrac{u_\theta}{r}\dfrac{\partial u_r}{\partial \theta}\right)=-\dfrac{\partial p}{\partial r} \\[3mm]
\rho\left(\dfrac{u_\theta}{r}\dfrac{\partial u_\theta}{\partial \theta}+u_r\dfrac{\partial u_\theta}{\partial r}+\dfrac{u_r u_\theta}{r}\right)=-\dfrac{1}{r}\dfrac{\partial p}{\partial \theta},
\end{cases}
$$

the Euler's equations in plane-polar coordinates.

Acceleration of a Fluid Particle: The Total Derivative

It is instructive to analyze the dynamics of fluid motion by means of differential elements rather than the somewhat mechanical procedure of converting the finite

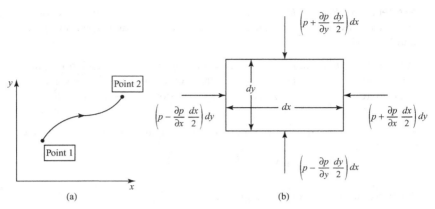

Figure 3.3. Motion of infinitesimal fluid particle.

control-volume formulation into differential form. By this means, we can gain a deeper understanding of important features of the differential equations displayed in Table 3.6. We do this analysis in this subsection for the case of a two-dimensional flow in a rectangular coordinate grid to illustrate the procedure.

The Conservation of Momentum Principle is now applied to an *infinitesimal* system moving with the flow (i.e., to a fluid particle). For simplicity, the following derivation assumes an inviscid flow with negligible body-force effects. Details of the derivation are carried out in a two-dimensional Cartesian coordinate system for simplicity. The first step is to define a special way to look at the time derivative in a moving fluid, the so-called *Eulerian, total, substantial,* or *substantive* derivative. All of these terms are used interchangeably throughout the literature of fluid mechanics.

We illustrate the ideas by considering the time rate of change of the velocity (i.e., the acceleration) of a moving fluid particle. This special time rate of change is composed of the following two parts:

- *unsteady acceleration:* the rate of change due to changing conditions at a reference point in the field.
- *convective acceleration:* the rate of change due to the motion of the particle as it passes through the reference point

To illustrate these two different types of accelerations, imagine we are observers-moving with a fluid particle, as illustrated in Fig. 3.3.

Suppose that we reach a certain location (i.e., Point 1 in Fig. 3.3) in the flow at a certain time and at that same instant, the particle velocity changes because the flow is unsteady (i.e., the flow properties depend on time). This change in velocity represents acceleration of the particle. Now, what happens if the flow velocity is independent of time at a particular point? That is, any particle passing through that point always has the same velocity. Because the flow field is not necessarily uniform, then as the particle moves from Point 1 to Point 2, the moving observer notes that the velocity at Point 2 may be different from that experienced at Point 1 and that this change in velocity has occurred over the transit time from Point 1 to Point 2. This corresponds to a change in velocity with respect to time (i.e., an apparent acceleration) due to the motion (i.e., convection) of the particle through the flow—hence, the descriptive term *convective acceleration* introduced in the previous subsection. The *total* time rate of change of velocity of a fluid particle traveling through an unsteady flow field then consists of two parts: the local acceleration and the convective acceleration. This combination often is referred to as the *total derivative*; other terminology also is used. With these physical ideas in mind, define:

$$\text{Eulerian derivative = total derivative = substantial derivative} = \frac{D\mathbf{V}}{Dt}$$

This notation represents the sum of both the unsteady (local) and the convective acceleration terms.

Recall that the velocity vector is a field property; it can be represented by a mathematical function of the spatial coordinates and time. That is, in general, the velocity vector can be written as $\mathbf{V} = \mathbf{V}(x,y,z,t)$. For the two-dimensional case considered here, we write $\mathbf{V} = \mathbf{V}(x,y,t)$. Then, if two points, Point 1 and Point 2, are differential distance $\Delta\mathbf{r} = \Delta x\mathbf{i} + \Delta y\mathbf{j}$ apart in the x-y plane and Δt apart in time, we can write:

$$\frac{D\mathbf{V}}{Dt} = \lim_{\Delta t \to 0}\left\{\frac{\mathbf{V}(x+\Delta x, y+\Delta y, t+\Delta t) - \mathbf{V}(x,y,t)}{\Delta t}\right\}.$$

Use a Taylor series expansion to represent the first term in brackets. Notice that the partial derivatives of $\mathbf{V}$ in all three coordinate directions and time are required:

$$\frac{D\mathbf{V}}{Dt} = \lim_{\Delta t \to 0}\left\{\frac{1}{\Delta t}\left[\left(\begin{array}{c}\mathbf{V}(x,y,t)+\dfrac{\partial\mathbf{V}}{\partial x}\Delta x+\dfrac{\partial\mathbf{V}}{\partial y}\Delta y+\dfrac{\partial\mathbf{V}}{\partial t}\Delta t+\\[2mm]+\dfrac{\partial^2\mathbf{V}}{\partial x^2}\dfrac{(\Delta x)^2}{2}+\dots\end{array}\right)-\mathbf{V}(x,y,t)\right]\right\}.$$

In the limit of a small time increment, the terms involving the squares (and higher) of the differential displacements become negligible. Then, passing to the limit as $\Delta t \to 0$ gives:

$$\frac{D\mathbf{V}}{Dt} = \frac{\partial\mathbf{V}}{\partial x}\frac{dx}{dt}+\frac{\partial\mathbf{V}}{\partial y}\frac{dy}{dt}+\frac{\partial\mathbf{V}}{\partial t}\frac{dt}{dt}.$$

Now, $dx/dt = u$ because dx is the distance that the particle travels in the x-direction over the time interval dt. That is, dx is the x-projection of the infinitesimal distance between Point 1 and Point 2 that the particle traveled. Similarly, $dy/dt = v$. Thus, in this two-dimensional example:

$$\frac{D\mathbf{V}}{Dt} = \frac{\partial\mathbf{V}}{\partial t}+u\frac{\partial\mathbf{V}}{\partial x}+v\frac{\partial\mathbf{V}}{\partial y}.$$

In vector form, this equation may be written in shorthand notation as follows:

$$\frac{D\mathbf{V}}{Dt} = \frac{\partial\mathbf{V}}{\partial t}+(\mathbf{V}\cdot\nabla)\mathbf{V}. \tag{3.72}$$

Notice that this is precisely the representation for the rate of change of the velocity vector deduced in Eq. 3.66. Again, for emphasis, the first term represents the local acceleration and the second term is the convective acceleration. It must be emphasized that the convective acceleration can be expressed in component form by direct substitution of the velocity vector in Eq. 3.72 only if rectangular coordinates are used. Thus, it is better to write:

$$\frac{D\mathbf{V}}{Dt} = \frac{\partial \mathbf{V}}{\partial t} + \nabla\left(\frac{\mathbf{V}\cdot\mathbf{V}}{2}\right) - \mathbf{V}\times\nabla\times\mathbf{V}, \tag{3.73}$$

which is valid for any curvilinear coordinate system.

It is now useful to use these ideas to check our derivation of the momentum equation. As before, the conservation of Momentum Principle can be expressed as Newton's Second Law (i.e., sum of forces, $\sum F$ = time rate of change of momentum, $\rho\mathbf{V}$) applied to a fluid particle (i.e., a system). Thus, the net force acting on the particle (due to pressure only, by assumption) is equal to the product of the mass of the particle (a constant) and the acceleration of the particle—that is, the product of the mass of the particle and the time rate of change of the particle velocity.

The net pressure force on the particle may be derived by taking a snapshot of the moving particle at an instant of time. At the time the picture is taken, the particle is assumed to be a cube (i.e., mathematically convenient; no loss of generality). Considering a two-dimensional Cartesian coordinate system, the cube appears as a square and the pressure force is pA, where the area is the length of a side times unity out of the page and the pressure acts perpendicular to all four sides. This is shown in Fig. 3.3(b).

The pressure is arbitrarily assumed to have the reference value p at the center of the particle—that is, at the point to which the element shrinks in the limit of a small mass element. With the positive coordinate directions assumed to be the positive force directions, the net pressure force, $\mathbf{F}$, on the fluid particle is given by:

$$\left(-\frac{\partial p}{\partial x}dxdy\right)\mathbf{i} + \left(-\frac{\partial p}{\partial y}dydx\right)\mathbf{j} = \mathbf{F}.$$

If there are no body forces or viscous forces, then, according to the Newton's Second Law:

$$\mathbf{F} = \frac{D}{Dt}(m\mathbf{V}) = m\frac{D\mathbf{V}}{Dt} = \rho(dx)(dy)(1)\frac{D\mathbf{V}}{Dt},$$

where the total acceleration can be represented by the total derivative, Eq. 3.72. Assembling all of the terms that enter into the momentum change/force balance,

$$\mathbf{F} = \rho(dxdy)\frac{D\mathbf{V}}{Dt},$$

it follows (for rectangular coordinates in two dimensions) that:

$$-\frac{\partial p}{\partial x}\mathbf{i} - \frac{\partial p}{\partial y}\mathbf{j} = \rho\frac{D\mathbf{V}}{Dt} = \rho\frac{\partial \mathbf{V}}{\partial t} + \rho u\frac{\partial \mathbf{V}}{\partial x} + \rho v\frac{\partial \mathbf{V}}{\partial y}.$$

Now, recall that $\mathbf{V} = u\mathbf{i} + v\mathbf{j}$ and write the two-component equations that follow from this vector equation; namely:

$$\begin{cases} \rho\dfrac{Du}{Dt} = \rho\dfrac{\partial u}{\partial t} + \rho u\dfrac{\partial u}{\partial x} + \rho v\dfrac{\partial u}{\partial y} = -\dfrac{\partial p}{\partial x} \\[2mm] \rho\dfrac{Dv}{Dt} = \rho\dfrac{\partial v}{\partial t} + \rho u\dfrac{\partial v}{\partial x} + \rho v\dfrac{\partial v}{\partial y} = -\dfrac{\partial p}{\partial y} \end{cases}. \qquad (3.74)$$

This is the momentum balance for a two-dimensional, inviscid, unsteady flow with no body forces written in Cartesian coordinate form. Compare Eq. 3.74 with Eq. 3.70, which originated in the statement of momentum conservation for a fixed control volume. Because steady flow was assumed in Eq. 3.70, the time derivative terms are not present. It is clear, however, that the two methods of analysis yield identical results, as they must. Thus, starting with either a fixed control volume or an infinitesimal fluid particle, the derivation of the differential equation expressing conservation of momentum leads to the same result. Note that although we illustrated the procedure in two dimensions, it is extended readily to three. The student is invited to repeat these steps for a three-dimensional element to verify that the method leads to the correct three-dimensional result (e.g., Eq. 3.69).

The other governing equations can be derived similarly. The differential-element technique is sometimes used to deduce the equations of motion in other than Cartesian coordinates by defining the appropriate infinitesimal element in terms of that coordinate system. For example, Fig. 3.4 shows an element that is appropriate for a three-dimensional cylindrical coordinate system. If the same procedure used to determine the Cartesian momentum balance of Eq. 3.73 were used with this element, the equivalent of Eq. 3.71 would be found. The details are left for the student in an exercise at the end of the chapter. This method is more lengthy than the one used previously; that is, by direct use of the vector operations appropriate to the coordinate system. However, it has the advantage of being more geometrically motivated and intuitive than the direct evaluation of the general vector expressions (e.g., Eq. 3.73) in dealing with other coordinate systems.

Comments Regarding the Total Derivative

The total derivative appears naturally in many places in the differential analysis of fluid motion. Examine the general continuity equation, Eq. 3.50. By expanding the divergence term using the appropriate vector identity, we find that:

$$\frac{\partial \rho}{\partial t} + \nabla \cdot (\rho\mathbf{V}) = \frac{\partial \rho}{\partial t} + \mathbf{V} \cdot \nabla\rho + \rho\nabla \cdot \mathbf{V} = 0; \qquad (3.75)$$

therefore, the differential-continuity equation can be expressed in terms of the total rate of change of the density:

$$\frac{D\rho}{Dt} + \rho\nabla \cdot \mathbf{V} = 0. \qquad (3.76)$$

Thus, the Eulerian derivative D/Dt applies to changes in other properties of the moving fluid and has the same physical interpretation we attached to its use with

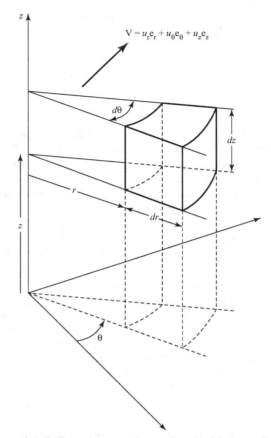

Figure 3.4. Differential-mass element for cylindrical coordinates.

the acceleration of the fluid. The rate of change of *any* scalar property of the moving fluid particle is composed of an unsteady part and a convective part, so that:

$$\frac{DH}{Dt} = \frac{\partial H}{\partial t} + \mathbf{V} \cdot \nabla H, \tag{3.77}$$

where H represents any scalar property of the field such as pressure, p; density, ρ; and temperature, T. Only when the property being differentiated is a vector, such as the velocity, $\mathbf{V}$, is it necessary to use the special form for the total derivative given in Eq. 3.73. For emphasis, note that it is not necessary to use the special form of the total derivative expressed in Eq. 3.73 if the derivative of a scalar variable is to be assumed. That is, ∇H is a proper vector operation. Then, Eq. 3.77 can be evaluated for *any* curvilinear coordinate system in explicit form.

The notation $d\mathbf{V}/dt$ sometimes is used to indicate the Eulerian derivative in a flow field that is *independent of time* (i.e., steady flow). In this situation, $d\mathbf{V}/dt$ only represents the convective acceleration.

EXAMPLE 3.11 *Given*: An unsteady, incompressible, two-dimensional inviscid flow with no body forces has a flow field described by $\mathbf{V} = (2x + 4t^2)\mathbf{i} + (3x - 2y)\mathbf{j}$. Assume that the units of velocity are m/s.

Required: Determine the total acceleration of a fluid particle in this flow field in the *y*-direction at point $(x,y,t) = (3,2,1)$.

Approach: Use the total acceleration expression Eq. 3.72.

Solution: From Eq. 3.72:

$$\frac{D\mathbf{V}}{Dt} = \frac{D}{Dt}(u\mathbf{i} + v\mathbf{j}) = \frac{\partial \mathbf{V}}{\partial t} + (\mathbf{V} \cdot \nabla)\mathbf{V} = \frac{\partial}{\partial t}(u\mathbf{i} + v\mathbf{j}) + \left(u\frac{\partial}{\partial x} + v\frac{\partial}{\partial y}\right)(u\mathbf{i} + v\mathbf{j}).$$

Equating the terms associated with the unit vector $\mathbf{j}$ (i.e., the terms in the *y*-direction):

$$\frac{Dv}{Dt} = \frac{\partial v}{\partial t} + u\frac{\partial v}{\partial x} + v\frac{\partial v}{\partial y} = 0 + (2x + 4t^2)(3) + (3x - 2y)(-2).$$

Evaluating, $Dv/Dt = 20$ m/s^2.

Appraisal: The Eulerian derivative depends on both space and time in this example.

The Differential Form of the Energy Equation

Because the integral energy equation is a scalar equation, the derivation of the differential form parallels the development of the differential-continuity equation. Begin with the integral form of the energy equation, Eq. 3.42:

$$\frac{\partial}{\partial t}\iiint_{\hat{V}} \rho\left(e + \frac{V^2}{2}\right)d\hat{V} + \iint_{\hat{S}} \rho\left(e + \frac{V^2}{2}\right)(\mathbf{V} \cdot \mathbf{n})\,d\hat{S} =$$

$$= \begin{pmatrix} \dot{Q}_{\text{INT}} + \dot{Q}_{\text{COND}} - \dot{W}_M + \dot{W}_V - \\ -\iint_{\hat{S}} p(\mathbf{V} \cdot \mathbf{n})\,d\hat{S} + \iiint_{\hat{V}} \rho(\mathbf{V} \cdot \mathbf{b})\,d\hat{V} \end{pmatrix}. \tag{3.42}$$

Next:

1. Recall that the time derivative may be put inside the volume integral as previously for the continuity equation.
2. Use the divergence theorem to change the two surface integrals to volume integrals.
3. Assume no machine work or chemical combustion and assume that potential energy changes are negligible.
4. Recognize that the heat-conduction term and the viscous-work-rate term each represent integrated effects over the entire surface of the fixed-control volume.

Thus, for example:

$$\dot{Q}_{COND} = \iint_{S} k \frac{\partial T}{\partial n} d\hat{S} = \iint_{S} k\mathbf{n} \cdot \nabla T \, d\hat{S} = \iiint_{V} k(\nabla \cdot \nabla T) d\hat{V} \tag{3.78}$$

from the Fourier Law of heat conduction, where k is the thermal conductivity (taken here to be a constant property of the fluid) and $\partial T/\partial n$ is the temperature gradient everywhere normal to the control surface. This is converted readily to control-volume form by means of the divergence theorem, as shown in the last term in Eq. 3.72. The work rate due to viscosity in Eq. 3.42 usually is represented by a mechanical or viscous dissipation function, Φ'. These viscous effects, as well as the heat conduction, are assumed negligible in most of the aerodynamics applications that follow (i.e., the fluid often can be assumed to be inviscid and non-heat-conducting). This is clearly inappropriate in a hypersonic flow—for example, wherein there are high-temperature gradients and important viscous interactions. These two terms are evaluated in appropriate detail when needed.

Incorporating all of the assumptions and manipulations denoted previously into Eq. 3.42, collecting all of the terms under a single-volume integral, and arguing that the integrand must be zero for an arbitrary control volume, it follows that:

$$\frac{\partial}{\partial t} \rho \left(e + \frac{\mathbf{V}^2}{2} \right) + \nabla \cdot \left[\rho \left(e + \frac{\mathbf{V}^2}{2} \right) \mathbf{V} \right] + \nabla \cdot (p\mathbf{V}) = k\nabla^2 T + \Phi' + \rho(\mathbf{b} \cdot \mathbf{V}). \tag{3.79}$$

Following the terminology used in development of the momentum equation, we refer to this equation as the *conservation form* of the energy equation. This means that it is the differential equivalent to the conservation of energy written in either system or integral control-volume form. In this form, the variable in the differential equation is the energy (i.e., internal energy plus kinetic energy) rather than a single variable. As with the momentum equation, further manipulations can be made to write the equation so that only a single variable such as the internal energy or enthalpy appears in the derivatives; this is referred to as the *primitive form*. Although Eq. 3.79 is not the most general case (we ignored internal heat generation and radiation heat transfer), it exhibits most of the energy-balance interactions needed in aerodynamic applications. Equation 3.79 often is used in this (conservation) form to represent the energy balance in CFD machine solutions for complicated flow problems because experience shows that the numerical errors are smaller.

Equation 3.79 simplifies under the assumptions of inviscid, adiabatic (i.e., no heat transfer) flow, and negligible body-force effects to:

$$\frac{\partial}{\partial t} \rho \left(e + \frac{1}{2} V^2 \right) + \nabla \cdot \left[\rho \left(e + \frac{1}{2} V^2 \right) \mathbf{V} \right] + \nabla \cdot (p\mathbf{V}) = 0. \tag{3.80}$$

This equation is important in the treatment of compressible flow.

Further manipulation of Eq. 3.79 leads to considerable simplification. Working with the second term on the left side and using the vector identity:

$$\nabla \cdot (\phi \mathbf{A}) = \mathbf{A} \cdot \nabla \phi + \phi \nabla \cdot \mathbf{A} \tag{3.81}$$

(i.e., the divergence of the product of a scalar times a vector), we find:

$$\nabla \cdot \left[\rho\left(e+\frac{1}{2}V^2\right)\mathbf{V}\right] = \rho\mathbf{V}\cdot\nabla\left(e+\frac{1}{2}V^2\right)+\left(e+\frac{1}{2}V^2\right)\nabla\cdot(\rho\mathbf{V}), \qquad (3.82)$$

where the vector in the identity equation (Eq. 3.81) was chosen to be $\rho\mathbf{V}$ because this combination (i.e., momentum per unit volume) appeared several times before. In fact, the last term in the resulting expression clearly is related to the continuity equation. To take advantage of this, expand the derivative in the first term in Eq. 3.79 to give:

$$\frac{\partial}{\partial t}\rho\left(e+\frac{1}{2}V^2\right)=\left(e+\frac{1}{2}V^2\right)\frac{\partial\rho}{\partial t}+\rho\frac{\partial}{\partial t}\left(e+\frac{1}{2}V^2\right). \qquad (3.83)$$

Then, adding Eqs. 3.82 and 3.83, we find:

$$\frac{\partial}{\partial t}\rho\left(e+\frac{1}{2}V^2\right)+\nabla\cdot\left[\rho\left(e+\frac{1}{2}V^2\right)\mathbf{V}\right]=$$

$$=\left(e+\frac{1}{2}V^2\right)\left(\frac{\partial\rho}{\partial t}+\nabla\cdot(\rho\mathbf{V})\right)+\rho\frac{\partial}{\partial t}\left(e+\frac{1}{2}V^2\right)+\rho\mathbf{V}\cdot\nabla\left(e+\frac{1}{2}V^2\right).$$

The first term on the right side of this expression must vanish so that the flow satisfies the continuity equation (Eq. 3.50). Then, remembering the definition of the total derivative, we write:

$$\frac{\partial}{\partial t}\rho\left(e+\frac{1}{2}V^2\right)+\nabla\cdot\left[\rho\left(e+\frac{1}{2}V^2\right)\mathbf{V}\right]=\rho\frac{D}{Dt}\left(e+\frac{1}{2}V^2\right). \qquad (3.84)$$

Further simplifications result if we separate the kinetic and internal energy so that:

$$\frac{D}{Dt}\left(e+\frac{1}{2}V^2\right)=\frac{De}{Dt}+\frac{D}{Dt}\left(\frac{1}{2}\mathbf{V}\cdot\mathbf{V}\right)=\frac{De}{Dt}+\mathbf{V}\cdot\frac{D\mathbf{V}}{Dt},$$

where by means of the momentum equation, Eq. 3.66,

$$\rho\left(\frac{\partial\mathbf{V}}{\partial t}+\mathbf{V}\cdot\nabla\mathbf{V}\right)=\rho\frac{D\mathbf{V}}{Dt}=-\nabla p+\rho\mathbf{b}+\mathbf{f}_v.$$

We see that

$$\rho\frac{D}{Dt}\left(e+\frac{1}{2}V^2\right)=\rho\frac{De}{Dt}+\mathbf{V}\cdot(-\nabla p+\rho\mathbf{b}+\mathbf{f}_v).$$

It also is useful to expand the term representing the rate of work done by the pressure force. Again, using the identity for expanding the divergence of the product of a vector times a scalar (Eq. 3.81), we find:

$$\nabla\cdot(p\mathbf{V})=\nabla\cdot(p\mathbf{V})=\mathbf{V}\cdot\nabla p+p\nabla\cdot\mathbf{V}.$$

Inserting the several expanded terms into the energy equation, Eq. 3.79, we find that the body-force term from the momentum equation cancels what appears on the left side of Eq. 3.79. Part of the term involving the pressure also is canceled, leaving:

$$\rho\frac{De}{Dt}+p\nabla\cdot\mathbf{V}=k\nabla^2T+\Phi, \qquad (3.85)$$

where the work-rate term involving the viscous force, $\mathbf{V} \cdot \mathbf{f}_v$, was incorporated in the viscous-dissipation function, $\Phi = \Phi' - \mathbf{V} \cdot \mathbf{f}_v$. Equation 3.85 is the primitive variable form of the complete energy-balance equation. The only terms not displayed here are those representing the effect of internal heat generation (e.g., by combustion) and radiation heat transfer. It also must be remarked that the conduction-heat-transfer term was written here for the case of constant heat conduction, k. There are cases in which this assumption may be inappropriate. Because k may depend on the temperature, its variation with position may affect the outcome of calculations involving flows with strong temperature gradients. These matters as well as the dissipation function Φ are described in more detail in texts on high-speed flows, in which high temperatures and viscous effects may be quite important.

Table 3.7. *Special forms of the differential energy equation*

General energy equation (Eq. 3.85) (no radiation heat transfer or internal heat generation)	$\rho \dfrac{De}{Dt} + p\nabla \cdot \mathbf{V} = k\nabla^2 T + \Phi$
Steady, adiabatic, inviscid flow (Eq. 3.88) along streamlines (no body forces or internal heat generation; no radiation heat transfer)	$h + \dfrac{V^2}{2} = h_0 = \left\{ \begin{array}{l}\text{constant along} \\ \text{streamlines}\end{array}\right\}$
Steady, adiabatic, inviscid flow (Eq. 3.88) with uniform temperature upstream (no body forces or internal heat generation; no radiation heat transfer)	$h + \dfrac{V^2}{2} = h_0 = \left\{ \begin{array}{l}\text{constant at any} \\ \text{point in the field}\end{array}\right\}$

There are many useful special forms of the energy equation; those most useful are summarized in Table 3.7. The so-called steady-flow form of the energy equation shown in the table can be deduced by rewriting Eq. 3.79 in primitive form by means of Eq. 3.84. Then,

$$\rho \frac{D}{Dt}\left(e + \frac{1}{2}V^2\right) + \nabla \cdot (p\mathbf{V}) = 0, \qquad (3.86)$$

which is valid for inviscid, adiabatic flow with negligible body forces. The divergence term can be expanded by means of the vector identity, Eq. 3.78. We find:

$$\nabla \cdot (p\mathbf{V}) = \nabla \cdot \left(\frac{p}{\rho}\rho\mathbf{V}\right) = \rho\mathbf{V} \cdot \nabla\left(\frac{p}{\rho}\right) + \frac{p}{\rho}\nabla \cdot (\rho\mathbf{V}),$$

and if the flow is steady, the last term vanishes. The first term on the right side is the density times the total derivative (assuming steady flow) of p/ρ. Then, the two total derivative terms can be combined to give:

$$\rho \frac{D}{Dt}\left(e + \frac{1}{2}V^2 + \frac{p}{\rho}\right) = 0. \qquad (3.87)$$

Because the total (Eulerian) derivative applies to a fluid particle moving with the fluid and the flow is steady, Eq. 3.87 indicates that the combination of terms:

$$e + \frac{1}{2}V^2 + \frac{p}{\rho}$$

is constant along a streamline. Notice the appearance of the familiar thermodynamic property, h, the specific enthalpy:

$$h = e + \frac{p}{\rho} = e + pv,$$

where v is the specific volume. The energy equation, Eq. 3.87, often is written in terms of the enthalpy in place of the internal energy in thermodynamics problems involving fluid motion. Discussions related to the flow work and its connection to the pv combination are found in any good basic thermodynamics textbook, and are not included here.

It follows from Eq. 3.87 that the quantity:

$$h + \frac{V^2}{2}$$

is constant along a streamline—that is, along the path of the particle through the steady flow field. Evaluating the constant at a stagnation point where the velocity is zero and the flow was brought to rest adiabatically:

$$h + \frac{V^2}{2} = h_0, \tag{3.88}$$

where h is the static enthalpy, V is the magnitude of the local velocity vector, and h_0 is the stagnation enthalpy. Equation 3.85 states that the stagnation enthalpy is constant along a streamline. If the flow from upstream infinity is uniform (i.e., all of the upstream flow has a constant temperature, which is the usual case), then the constant in Eq. 3.85 (i.e. the stagnation enthalpy) is the same *everywhere* in the flow field.

The energy equation of Eq. 3.85 is the simplest form of the conservation of energy principle because it is an algebraic equation. It is convenient to use this form instead of a differential equation when the assumptions for a particular problem allow it. Thus, if the flow is steady, adiabatic, inviscid, non-heat-conducting, and experiences no body forces, then Eq. 3.85 is all that is needed to ensure that energy conservation is accounted for properly.

If the random kinetic energy, e, is constant along a streamline, then:

$$\frac{1}{2}V^2 + \frac{p}{\rho} = \text{constant.}$$

If the flow is also incompressible, then the density is constant and we find:

$$\frac{1}{2}\rho V^2 + p = \text{constant.}$$

This is the famous Bernoulli Equation, which has considerable practical value because it directly relates changes in velocity to the local pressure. The derivation just presented is not displayed often because it does not adequately illuminate all of the underlying assumptions. It is considered better to derive it as a special form of the momentum balance, and we discuss it from that point of view in the next chapter.

3.6 Properties of the Defining Equations

The following observations are made concerning the equations that are derived in this chapter. We assume for now that only reversible thermodynamic processes are involved in the main part of the flow field of interest. If this is not so, additional equations may be required. For example, if it is required to study flow regions in which there are rapid changes in temperature, density, and pressure (e.g., within shock waves in compressible flow), then it may be necessary to supplement the equations already derived with an additional one containing consequences of the Second Law of Thermodynamics (Eq. 3.13). It may be necessary to include an additional variable like the entropy, s, to properly represent such situations. Similarly, there are cases in which noninertial control volumes and angular-momentum effects may be important. Because this does not happen often in this textbook, propose to introduce such complexities only where they are necessary. If students have studied the material carefully, they will see that all of the mathematical techniques were presented to enable them to easily write any additional equations that may be necessary.

With this in mind, the problems we treat have the following characteristics from a mathematical point of view:

1. Six conservation equations—continuity (one), momentum (three), energy (one), and the equation of state (one)—provide the necessary set of equations to solve for the six unknowns of a general flow field—namely, u, v, w, p, ρ, and T, where u, v, and w can be replaced by any three orthogonal velocity components appropriate to the coordinate system most convenient to the geometry of the problem.
2. The integral conservation equations are useful mainly when the flow is steady. For unsteady flow, evaluation of the volume integral in the equations requires detailed knowledge of the physical properties of the fluid inside the control volume.
3. Even without the considerable complication of including viscous forces and thermal-conductivity effects, the general solution of the complete differential-conservation equations is a formidable task. In fact, it is usually pointed out that general solutions do not exist because solution of a set of nonlinear partial-differential equations is required. Notice that several nonlinear terms are present; in particular, the convective acceleration terms are nonlinear. For example, quantities such as $u\dfrac{\partial u}{\partial x}$ appear that involve products of the variables and their derivatives. The mathematical consequence is that there exist no general solutions of the complete set of equations. Analytical solutions require sets of assumptions that usually lead to some form of linearization. Much of this book is devoted to devising useful solutions of this type.

3.7 Boundary Conditions

Before further discussion of aerodynamic problem solving, it is important to realize that problem formulation requires a careful mathematical statement of the *boundary conditions* in addition to the set of governing differential equations that occupies much of our attention in this chapter. This section is a brief discussion of

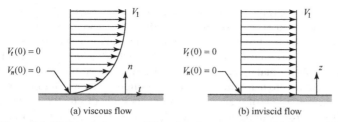

Figure 3.5. Wall boundary conditions for viscous and inviscid-flow models.

key features of aerodynamic boundary conditions, which must always be considered when applying the equations in particular applications.

The set of defining equations that must be solved is the same for many diverse problems in aerodynamics. The mathematical solution that correctly describes a specific problem is obtained by applying the appropriate boundary conditions. Boundary conditions represent physical constraints that the solution must obey; they usually involve the physical behavior of the flow and the particular geometry of the problem.

As observed in subsequent chapters, a primary boundary condition is one that specifies what the flow must do at the surface of an object in the flow or on surfaces bounding the flow field. In a viscous-flow model, the boundary condition is that there is no slip at the surface—that is, the viscous flow adheres to the surface (see Chapter 2), so that both the normal and the tangential velocity components are zero at the surface. Of course, the normal velocity may not be zero if the surface is porous with fluid entering the region of interest through the boundary.

In an inviscid-flow model, only the *tangency boundary condition* can be invoked, which states that the flow must be everywhere parallel to the body surface or, equivalently, that there can be no velocity component perpendicular to a solid surface. Thus, in the inviscid-flow model, the normal component of velocity is zero at the surface but the tangential component is not! These two situations are contrasted in Fig. 3.5. Other types of boundary conditions may be encountered—for example, if mass transfer through a porous body surface forms part of the flow-field boundary.

3.8 Solution Procedures

Now that we are confronted by an intimidating set of partial-differential equations and complex boundary conditions, it is appropriate to discuss strategies for finding solutions. One method that is doomed to fail is to mount a frontal attack in an attempt to find a general solution to the problem. The student has no doubt already noticed that considerable effort has gone into examining special cases for the governing equations as set forth in the tables in this chapter. For example, equations pertaining to steady, incompressible, or inviscid flows are carefully worked out. It should be obvious that our intention is to approach each problem from the standpoint of which simplifications can be made to reduce it to the most understandable and mathematically tractable form. In many cases, this results in a problem that can be solved fairly easily even without recourse to numerical means. Some of the most elegant and useful solutions were created using this approach.

We call this process the *art of approximation*. It was practiced in the past by the most innovative workers in aeronautics, and the various solutions are named after their creators. Confronted by the difficult nonlinear equations we now have derived, these investigators carefully determined where simplifying assumptions could be used to make the problem mathematically tractable. Students of these remarkable individuals were required to learn this approach and often went on to improve the results or to create new methods of their own. As students study this textbook, they are presented with numerous examples that show how this approach can be used to bring about extraordinary physical insight, as well as practical solutions, to difficult problems.

Many types of approximations can be introduced. These may take the form of appropriate physical assumptions such as the flow being incompressible or inviscid. They also may be based on linearization techniques such as those used in supersonic small-disturbance theory. These methods of approximation all receive careful attention throughout the book.

Unfortunately, the analytical approach is not used as often at present as it was in the past. Some aerodynamicists believe that all of the "simple results have already been deduced." Too often, we observe a tendency—when confronted by a difficult aerodynamics problem—for an investigator to go directly to an experimental approach, or to what is equivalent to an experiment—the numerical solution by means of computational methods such as CFD. The latter approach is based on the ready availability of powerful and fast digital computers that can solve complex sets of differential equations (usually written by representing derivatives in an algebraic difference form) by iterative methods. In effect, this replaces the problem formulation by a general-purpose "black box." In some complex situations, this is the only approach that can lead to the necessary information. However, it is important to understand that nothing can take the place of a thorough physical understanding of a problem. Therefore, it is of the utmost importance to practice solving problems in an *approximate* way by applying a set of appropriate simplifying assumptions. Then, even if the final solution requires relaxing some of those assumptions, and a brute-force numerical attack cannot be avoided, the means to check the numerical computations are available. Many costly errors have been made because this principle was not clearly understood. Therefore, it is a major goal of this textbook to help students develop a thorough understanding of the available methods of solution, their relative value in typical real-life situations, and their relationships to one another.

Practical Methods of Problem Solution

Three approaches can be used to seek solutions to problems in aerodynamics of the type defined in this chapter as follows:

- analytical methods based on assumptions regarding the flow regime or the presence of small parameters in the problem
- experimental methods that seek to illuminate the problem by measurements and flow visualization in a laboratory environment or by instrumenting actual flight vehicles
- numerical methods that solve the appropriate sets of equations and boundary conditions by high-speed computation.

It is important to understand that these methods always should be used in a mutually supportive manner. Too often, we see experimentalists who steadfastly insist that their (sometimes incorrect) experimental data are a better representation of reality than what is predicted by a theoretical model. Similarly, theorists who have grown too close to their favorite analysis may criticize the work of an experimentalist as incorrect because it does not conform to their predictions. In contrast, those who practice the "black art" of CFD sometimes insist that their numerical computations represent the "correct" solution without checking whether all constraints and physical models used in the "black box" are truly appropriate.

Therefore, we start our study of aerodynamics with a strong bias in favor of methods that use application of simplifying assumptions. To make this work, we must be sure of the understanding of basic principles. This method has produced many of the most useful physical insights into the underlying behavior of aerodynamic flow fields. Nevertheless, we frequently see situations in which such approximations fail. It is important to learn the appropriate strategies for addressing situations that do not readily yield to the familiar and comfortable assumptions that may have worked in a similar problem solved previously. It also is imperative to learn not to misuse simple theoretical results in cases in which the underlying simplifying assumptions may be violated.

It is shown by example how to determine when an experiment is necessary or when resort should be made to strictly numerical methods. Again, our approach is to show how all of the available techniques can be used in concert to produce a powerful, adaptable problem-solving strategy. Our first encounter with the need for experimental verification is in Chapter 5 as we learn about important applications of airfoil theory in wing design. The value of the theory in formulating experiments and the need for them in validating the theoretical predictions is an oft-repeated scenario.

A valuable application of computational methods is the ability to represent complex results in easily interpreted graphical form. Therefore, even in problems that can be well represented by an approximate analytical solution, numerical evaluation of the results and plotting them in graphical form is needed frequently. As part of the text, students are provided with an integrated set of numerical tools to use in the manner described. Numerical methods are used frequently to extend analytical models that can be used only in their original form in simplified geometries, and so on.

Examples of the Art of Approximation

A significant breakthrough in applying the equations of fluid dynamics in an approximate way to represent the solution to an important aerodynamics problem is due to Ludwig Prandtl. His genius in realizing that many complex flow problems of practical interest may be decomposed into (1) an inviscid irrotational part, *potential flow*, and (2) a viscosity-dominated part, *boundary layer flow*, revolutionized aeronautical thinking in the early years of the twentieth century. The equations that describe the component parts are vastly simplified and can be solved separately. The two separate solutions are "matched" so that all of the boundary conditions are accommodated. This yields a remarkably satisfactory solution for many practical problems in aerodynamics. This decomposition into simpler parts, along with the subsequent synthesis into a complete solution, is the basis of a powerful approach to many engineering problems. It also is the foundation of an analytical technique called the *perturbation method* that has been used to produce many of the standard solutions used by all practicing engineers. Various versions of this approach are used to linearize the sometimes nonlinear governing equations in a given aerodynamics problem.

We examine many such problems and their solutions in this book. The perturbation approach can be used when a key parameter occurring in the formulation is either very small or very large. A famous example is the solution for compressible flow over a slender wing or body: the *small-disturbance theory*. In that situation, the size of the object measured perpendicular to the flow (e.g., h) can be considered small compared to the length, L, along the direction of the flow. Then, the ratio of these two physical lengths results in a small perturbation parameter—call it ε:

$$\varepsilon = \frac{h}{L},$$

which can be used to greatly simplify the governing equations. For example, if a particular term in one equation is proportional to the square (ε^2) or cube (ε^3) of this small parameter, such a term often can be dropped. This is because it is implied that it is very small compared to terms that either do not involve the small parameter directly or contain it only to the first power. The equations remaining when the presumed small parts are discarded lead to what is called the *linearized formulation* of the problem.

If analytical solutions of the defining partial-differential equations are sought, general simplifications must be made, even after the problem has been decomposed. They may involve dropping irrelevant terms and/or neglecting certain terms as negligible compared to other terms in an equation in a particular physical situation—thus, the importance of understanding the physical basis for the equation and the physical reason for the appearance of each term. Only then can a reasonable approximation be carried out that results in an approximate theory yielding satisfactorily accurate results.

3.9 Summary

Because there are many special forms of the three equations (i.e., continuity, momentum, and energy) that govern an aerodynamic flow, they are tabulated in concise form throughout the chapter. The tables are arranged to enable the student to readily find the form needed in a particular situation and to check the limitations that apply and the degree of approximation being made.

PROBLEMS

3.1 Two streams of water (ρ = constant = 2 slug/ft^3) enter a duct as shown here. The flow is parallel to the duct centerline. Stream A enters with a uniform velocity of 5 ft/s and Stream B enters with a velocity of 25 ft/s. Both streams have a constant area of 1 ft^2 and the duct has a constant area of 2 ft^2. The two flows mix thoroughly, resulting in a uniform velocity at Station 2. Assume a steady flow. What is the velocity at Station 2?

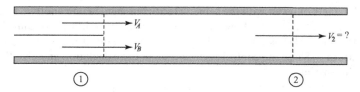

Problem 3.1

3.2 An incompressible jet flows into and out of a bent pipe at atmospheric pressure. The pipe is mounted on a frictionless roller and is restrained by a spring. Two pipe bend angles are available, as shown. Determine which configuration causes the smallest deflection of the spring. Assume a steady, inviscid flow.

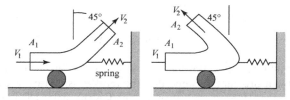

Problem 3.2

3.3 Rework Example Problem 3.5 using the specified control volume and absolute pressure. Recognize that when the control surface cuts through the support strut, there is a force imposed on the control surface at that point.

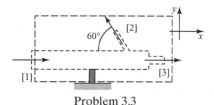

Problem 3.3

3.4 A long cylindrical body of revolution with a cross-sectional area of 2 ft^2 is installed in a constant-area duct of 5 ft^2, as shown. The flow is steady and incompressible. Assume that the viscous effects are negligible and the flow is one-dimensional. Find the pressure force acting on the nose of the body in the streamwise direction. Assume $\rho = 0.2$ slugs/ft^3. Show the control volume used in the calculation and give the units of the answer.

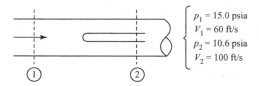

$$p_1 = 15.0 \text{ psia}$$
$$V_1 = 60 \text{ ft/s}$$
$$p_2 = 10.6 \text{ psia}$$
$$V_2 = 100 \text{ ft/s}$$

Problem 3.4

3.5 Two jets of water with the same velocity impinge on one another, forming two other water jets, as shown. Assume a steady, inviscid flow of unit depth into the page. Also assume that all of the jets have a uniform flow. Calculate the jet width t_2. Show the control surface used in the calculation and give the units of the answer.

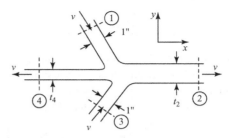

Problem 3.5

3.6 A free-jet flow passes through a fixed cascade of turbine blades in a test stand at ambient pressure. Assume that the jet flow is inviscid (i.e., no losses) and steady and that the cascade turns the jet flow through a 45° angle. Calculate the lift and drag on the cascade. The flow may be taken as two-dimensional. Show the control volume used in the calculation and give the units for the answer.

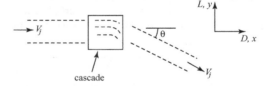

Problem 3.6

3.7 A liquid ($\rho = 64.4$ lbm/ft^3) flows through a vent pipe and discharges to the atmosphere through two jets, as shown ($p_2 = p_3 =$ ambient pressure). At Station 1, the vent pipe is bolted to the supply pipe by means of a flange. The static pressure at Station 1 is 12 psi above ambient pressure. Assume that the flow is steady and incompressible; ignore viscous and gravity effects. Show the control volume used in the calculation and give the units for the answer. Calculate the x-component of force on the coupling. Are the bolts in tension or compression?

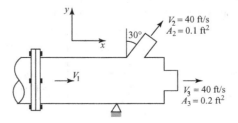

Problem 3.7

3.8 A pipe receives a steady flow of liquid ($\rho = 2$ slug/ft^3) through Station 1 and discharges into two other pipes at Stations 2 and 3. The volume flow rate through Station 2 is measured as 40 ft^3/s. The pipe is restrained by a support, as shown. The couplings cannot support any load. Assume an inviscid one-dimensional incompressible flow. Ignore gravity effects and the weight of the pipe. Show the control volume used in the calculation and give the units for the answers.

Find: (a) exit area A_3

(b) y-component of force on the support: magnitude direction (up, down).

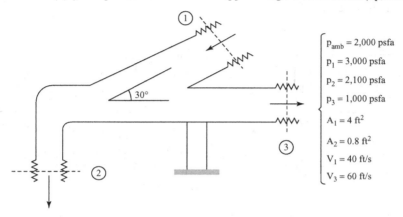

Problem 3.8

3.9 An air compressor takes in air at standard conditions (use 14.7 psia and 70°F) at a negligible velocity. The air is discharged at 115 psia and 125°F with a density of 0.5 lb$_m$/ft^3. The mass flow rate out of the 0.5 ft^2 discharge pipe is 12 lb$_m$/s. If the compressor input horsepower is 800, find the magnitude and direction (in words) of the heat transfer to/from the compressor in units of BTU/s. Assume a steady flow. Draw a control surface for the calculation.

3.10 A steady flow of a gas passes into and out of a certain device. At the inlet, Station 1, $T_1 = 140$°F and $V_1 = 20$ ft/s. At the outlet, Station 2, $T_2 = 600$°R and $V_2 = 30$ ft/s. In passing through the device, 2 BTU/s of heat is extracted from the flowing medium and 4 HP of machine work is done on the flowing medium by an outside source. Calculate the mass flow rate through the device (slug/s). Ignore gravity effects and assume an inviscid flow. Draw a control surface for the calculation.

3.11 A steady flow of air at 5 slug/s passes through a certain device, entering at 100 ft/s and exiting at 200 ft/s. The entrance and exit static enthalpy of the air is the same, and 10 HP is supplied to the device. What amount of heat (BTU/s) is being added to (or subtracted from) the device? State the heat flux and direction. Assume an inviscid flow and ignore gravity effects. Draw a control surface for the calculation.

3.12 Consider the turboprop engine shown here. The intake and exhaust airflow rates are 100 lb$_m$/s and the stagnation enthalpy of the entering air is 120 BTU/lb$_m$. The turbine, compressor, and propeller HP are 20,000, 10,000, and 10,000, respectively. The turbine drives the compressor and the propeller. Assume that the propeller does no work on the intake air; however, it does work on the air outside of the engine. Compute the jet stagnation enthalpy, h_{oj}, assuming a uniform and steady flow. Use the suggested control volume and give the units for the answer. Then, work the problem again by doing it in two steps: the first step based on a control surface extending from upstream of the propeller to just downstream of the compressor, and the second step using a control surface extending from just downstream of the compressor to well downstream in the jet exhaust.

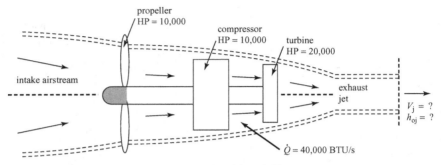

Problem 3.12

3.13 The velocity components u and v in a two-dimensional flow field are given by:
$$u = 4yt \text{ ft/s}, \qquad v = 4xt \text{ ft/s},$$

where t is time. What is the time rate of change of the velocity vector $\mathbf{V}$ (i.e., the acceleration vector) for a fluid particle at $x = 1$ ft and $y = 1$ ft at time $t = 1$ second?

3.14 Consider a two-dimensional, incompressible, inviscid flow field given by:
$$\mathbf{V} = (3x + 4t^2)\mathbf{i} + (2x - 3y)\mathbf{j}$$
a. What is the pressure gradient in the x-direction at a point $(x = 2, y = 3)$ at $t = 1/2$ second? Assume the density of the flow to be 0.2 slug/ft^3.
b. What is the convective acceleration of a particle in this flow field in the y-direction at $(x,y,t) = (3,2,1)$?

3.15 The two-dimensional channel shown here has a linear variation of area between Stations 1 and 2 and has a steady volume flow rate of 20 ft^3/s per unit width of channel. What is the convective acceleration of a fluid particle at $x = 2$ ft? Give the units for the answer.

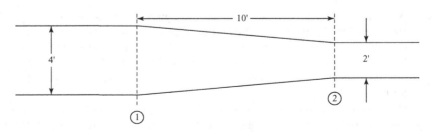

Problem 3.15

3.16 Follow the same method used to derive Eq. 3.74 to determine the three-dimensional form of the differential-momentum equation.

3.17 Using the vector form of the differential-momentum equation, deduce the three scalar components of the momentum equation in cylindrical coordinates (see Fig. 3.4).

3.18 Use the differential-element method to find the momentum equation in three-dimensional polar coordinates (see Fig. 3.4). Reduce to the two-dimensional form (i.e., no dependence on the z-coordinate) and compare to the results for Problem 3.17.

3.19 Use the differential-element method to find the momentum equation in three-dimensional spherical coordinates. Compare the results to the appropriate tables in the appendix.

REFERENCES AND SUGGESTED READING

Johnson, Richard W., *The Handbook of Fluid Dynamics*, CRC Press, LLC, 1998.

Kuethe, A. M., and Schetzer, J. D., *Foundations of Aerodynamics*, 2nd ed., John Wiley & Sons, New York, 1961.

Schlichting, Hermann, *Boundary Layer Theory*, 7th ed., McGraw-Hill Book Company, New York, 1979.

Serrin J., "Mathematical Principles of Classic Fluid Dynamics," Flugge S. (ed.), *Handbuch der Physik* VIII/1, Springer-Verlag, Berlin, Heidelberg, New York, pp. 125–263, 1959.

Wilcox, David C., *Basic Fluid Mechanics*, Third Edition, DCW Industries, Inc., 2007.

Fundamentals of Steady,
Incompressible, Inviscid Flows

4.1 Introduction

In this chapter, solutions of the conservation equations in partial-differential equation form are sought for a simple case—namely, steady, incompressible, inviscid two-dimensional flow. Each of these crucial assumptions is discussed in detail and their applicability as models of real flow-field situations are justified. Body forces such as gravity effects are neglected because they are negligible in most aerodynamics problems. Simple geometries are considered first. The analysis is then extended so that finally it is possible to represent the complex flow field around realistic airfoil shapes, such as those needed to efficiently produce lift forces for flight vehicles. Chapter 5 is a detailed treatment of two-dimensional airfoil flows.

The intention here is to obtain solutions valid throughout the entire flow field; hence, the differential-conservation equations are integrated so as to work from the small (i.e., the differential element) to the large (i.e., the flow field). In this regard, the integral form of the conservation equations is not a useful starting point because in steady flow, the integral equations describe events over the surface of only some fixed control volume. We are seeking detailed information regarding the pressure and velocity fields at any point in the flow. What are the implications of each assumption listed previously?

1. *Steady flow.* The assumption of steady flow enables the definition of a streamline as the path traced by a fluid particle moving in the flow field, from which it follows that a streamline is a line in the flow that is everywhere tangent to the local velocity vector. Also, all time-derivative terms in the governing equations can be dropped; this results in a much simpler formulation.
2. *Incompressible flow.* The assumption of incompressible flow means that the density is assumed to be constant. As shown herein, and as the conservation equations in Chapter 3 indicate, the assumption of incompressibility in a problem leads to enormous simplifications. The obvious one is that terms in the equations containing derivatives of density are zero. The other major simplification is that the number of equations to be solved is reduced. If the density is constant, then there cannot be large variations in temperature, and the temperature may be assumed to be constant as well. With density and temperature no

longer variables, the equation of state and the energy equation may be set to one side and the continuity and momentum equations solved for the remaining variables—namely, velocity and pressure.

In other words, for incompressible flows, the equation of state and the energy equation may be uncoupled from the continuity and momentum equations. It is true that no fluid (liquid or gas) is absolutely incompressible; however, at low speeds, the variation in density of an airflow is small and can be considered essentially incompressible. For example, considerations of compressible flow show that at a Mach number of 0.3 (a velocity of 335 ft/s, or 228 mph, at sea level), the maximum possible change in density in a flow field is about 6 percent and the maximum change in temperature of the flow is less than 2 percent. For flows of this velocity or less, the incompressible assumption is good. However, at Mach number 0.5 (558 ft/s, or 380 mph, at sea level), the maximum change in density in a flow field is almost 19 percent. An incompressible-flow assumption for such a case leads to prohibitive errors.

Results from an assumed incompressible flow around thin airfoils or wings and around slender bodies provide a foundation for the prediction of the flow around these bodies at higher, compressible-flow Mach numbers (i.e., less than unity). It turns out that the effects of compressibility on pressure distribution, lift, and moment at flow Mach numbers less than 1 can be expressed as a correction factor times a related incompressible flow value. Thus, results using the incompressible model are useful not only for low-speed flight, they also provide a database for the accurate prediction of vehicle operation at much higher (but subsonic) speeds.

3. *Inviscid flow.* The inviscid-flow assumption means physically that viscous-shear and normal stresses are negligible. Thus, all of the viscous shear-stress terms on the force side of the momentum equations drop out, as well as the normal stresses due to viscosity. As a result, the only stresses acting on the body surface are the normal stresses due to pressure. Recall from Chapter 3 that when considering incompressible viscous-flow theory (see Chapter 8), the viscous-shear stresses are assumed to be proportional to the rate of strain of a fluid particle, with the constant of proportionality as the coefficient of viscosity. Thus, an assumption equivalent to that of negligible viscous stresses is the assumption that the coefficient of viscosity is essentially zero. Such a flow is termed *inviscid* (i.e., of zero viscosity). In effect, the boundary layer on the surface of the body is deleted by this assumption. This implies that the boundary layer must be very thin compared to a dimension of the body and that the presence or absence of the boundary layer has a negligible effect relative to modifications to the body geometry as "seen" by the flow.

The inviscid, incompressible-fluid model is often termed a *perfect fluid* (not to be confused with a perfect or ideal gas as defined in Chapter 1). The boundary layer in many practical situations is extremely thin compared to a typical dimension of the body under study such that the body shape that a viscous flow "sees" is essentially the geometric shape. The exception is where the flow separates and the boundary layer leaves the body, resulting in a major change in the effective geometry of the body. Such separated regions occur on wings, for example, at large angles of attack. However, the wing angle of attack of a vehicle at a cruise condition is only a few degrees so that the effects of separation are minimal. Thus, the inviscid-flow assumption provides useful results that match closely the experiment for conditions corresponding to cruise, and the inviscid-flow model breaks down when large regions of separated flow occur.

Because the presence of the boundary layer is neglected in perfect-fluid theory, the theory does not predict the frictional drag of a body; that must be left to viscous-flow theory. However, within the framework of incompressible inviscid flow, predictions for low-speed pressure distribution, lift, and pitching moment are valid and useful.

4. *Two-dimensional flow.* The assumption of two-dimensional flow is a simplifying assumption in that it reduces the vector-component momentum equations from three to two. *Two-dimensional* simply means that the flow (and the body shape) is identical in all planes parallel to, say, a page of this book; there are no variations in any quantity in a direction normal to the plane. Consider a cylinder or wing extending into and out of this page, with each cross section of the body exactly the same as any other. The flow around the body in all planes parallel to the page are then identical. It follows that the cylinder in two-dimensional flow has an infinite axis length and the wing has an infinite span. Any cross section of this wing of infinite span is termed an *airfoil section*. Theoretical predictions for such an airfoil may be validated by experiments in a wind tunnel in which the wing model extends from one wall to the opposite wall. If the wing/wall interfaces are properly sealed, the model then behaves as if it were a wing of infinite span—that is, as if it has no wing tips around which there would be a flow due to the difference in pressure between the top and bottom surfaces of the wing. Theoretical results for an airfoil (i.e., a two-dimensional problem) form the basis for predicting the behavior of wings of finite span (i.e., a three-dimensional problem) because each cross section (i.e., airfoil section) of the finite wing is assumed to behave as if the flow around it were locally two-dimensional (see Chapters 5 and 6). Thus, two-dimensional results have considerable value. Most (but not all) of the concepts discussed in this chapter may be extended to three dimensions and/or to compressible flow. Such extensions are introduced at appropriate points.

4.2 Basic Building Blocks

Incompressible, inviscid, two-dimensional flow is now defined, and it was emphasized that many results of wide practical value and crucial importance are forthcoming from a theory that, at first glance, may seem to be so restricted by assumptions as to be of extremely limited value.

Before examining the defining differential equations and their solution, concepts regarding the kinematics of the flow are explained and defined. These concepts will prove useful as the theory is developed. In this section, several basic building blocks are established (some of which are independent of one another) and then set aside for later use. Patience is required in mastering these conceptual building blocks: The importance of the material can be appreciated only in retrospect after each building block plays its role in the analysis.

Streamline and Stream Tube

These terms are defined in Section 2.3. For steady flows—the subject of this chapter—a streamline is the path that a fluid particle traces as it traverses the flow field. Another way to express this is that streamlines are lines in the flow that are everywhere tangent to the local velocity vector. Notice that because there can be no flow across a streamline, a solid wall can be interpreted as a limiting streamline. From this, it can be seen that any flow streamline can be "cross-hatched" and thought of as a solid wall if such an interpretation proves useful. In a two-dimensional flow, the tangency property of a streamline requires that

$$\frac{v}{u} = \left(\frac{dy}{dx}\right)_{streamline}, \tag{4.1}$$

where (x,y) are the coordinate directions and (u,v) are the respective velocity components. Notice that Eq. 4.1 is equally valid for a compressible flow. A single streamline indicates velocity direction, not magnitude. However, because there is no flow across streamlines, the relative spacing between two nearby streamlines at several points in a flow field in an incompressible flow is a measure of the relative flow-velocity magnitudes at these points because the mass flux (ρV) is constant between streamlines and the density is constant as well. A stream tube in a two-dimensional flow is defined by the distance between two adjacent streamlines and a dimension that is an arbitrary length perpendicular to the plane of the flow and usually set to unity.

EXAMPLE 4.1 *Given:* A steady, incompressible, two-dimensional flow has a velocity field with velocity components given by $u = y, v = x^2$.

Required: Find the equation of the streamline passing through the point $(3,2)$.

Approach: Appeal to the definition of a streamline.

Solution: Using Eq. 4.1, the slope of the streamline is given by:

$$\frac{dy}{dx} = \frac{x^2}{y} \rightarrow ydy = x^2dx. \text{ Integrating, } \frac{y^2}{2} - \frac{x^3}{3} = \text{constant}.$$

Therefore, the equation of the stramline passing through point $(3,2)$ is given by:

$$\frac{y^2}{2} - \frac{x^3}{3} = 7$$

Appraisal: As a preliminary, the differential form of the continuity equation (Eq. 3.13) could have been used to test whether the "given" is a physically possible flow field. (It is, and the student should confirm this.)

EXAMPLE 4.2 *Given:* A two-dimensional, incompressible flow is described by the magnitude of the local velocity vector and the equation of the streamlines—namely:

$$|V| = (y^2 + x^4)^{1/2}, \quad \frac{y^2}{2} - \frac{x^3}{3} = \text{constant}.$$

Required: Find the velocity components, u and v.

Approach: Appeal to the definition of a vector in terms of its components and then use the streamline definition, Eq. 4.1.

Solution: $|\mathbf{V}| = \sqrt{u^2 + v^2} = u\sqrt{1 + \dfrac{v^2}{u^2}} = \sqrt{y^2 + x^4}$. Next, taking the derivative of the given equation of the streamlines, $ydy - x^2dx = 0$. From this, it follows that: $\dfrac{dy}{dx} = \dfrac{x^2}{y} = \dfrac{v}{u}$ (see the following appraisal).

Substituting the quantity x^2/y for v/u in the square root expression and squaring both sides: $u^2\left(1 + \dfrac{x^4}{y^2}\right) = y^2 + x^4 \rightarrow u^2 = \dfrac{y^2(y^2 + x^4)}{(y^2 + x^4)}$. From this, it follows that $u = \pm y$. Finally,

$$\frac{v}{u} = \frac{v}{\pm y} = \frac{x^2}{y} \Rightarrow v = \pm x^2.$$

Hence, the required velocity components are found.

Appraisal: There is a plus/minus in the answer because the flow direction is not specified. Thus, the flow could be either up and to the right or down and to the left. Note also that although $v/u = x^2/y$, we cannot state that $v = x^2$ and $u = y$ (because the v/u expression corresponds to one equation in two unknowns). To see that taking v and u as simply the numerator and denominator of the quotient is not correct, rework this problem with the same equation for the streamlines but with the magnitude of the velocity vector given by $2\sqrt{y^2 + x^4}$.

Angular Velocity and Strain

In general, as a fluid particle moves from point to point in a flow field, it rotates and its shape is distorted (i.e., *strain*). In this section, the *rotation* or angular velocity of a moving fluid particle is defined in terms of derivatives of the velocity components of the particle. Any deformation of the particle shape is due to viscous-shear effects, which are absent in the perfect-fluid model considered in this chapter. An expression for strain is needed later in the development of the viscous-flow terms in the conservation equations (see Chapter 8) and is easily obtained simultaneously with the rotation effects. Hence, the strain terms are derived here and then set aside for later use.

Rotation is defined as a property of the fluid particle and not of the global flow field. Thus, for example, a flow field with circular streamlines under certain circumstances might have zero rotation if individual particles move along the streamlines without spin. In considering the rotation of a solid body (e.g., a spinning flywheel), it is sufficient to determine angular velocity simply by making a straight line on the flywheel and a reference line on a fixed surface. The time rate of change of the angle made by the line marked on the wheel with respect to the fixed reference line is the angular velocity of the flywheel.

A more general definition is necessary for a moving fluid particle because it can deform. Of course, the general definition of rotation for a fluid particle when

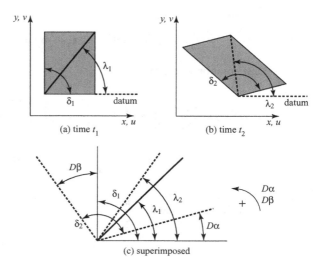

Figure 4.1. Deformation and
rotation of a moving fluid particle.

(a) time t_1

(b) time t_2

(c) superimposed

applied to a solid-body rotation problem must produce the familiar answer. Consider a moving fluid particle of infinitesimal size in a two-dimensional flow in the (x,y) plane at an arbitrary initial time t_1. For example, assume that at time t_1, the particle is a rectangle. The assumption of a rectangular shape is, of course, not a requirement; use of a more general shape leads to the same results. Now, imagine that the molecules making up the straight edges of the particle are painted and that a straight diagonal is drawn between two corners of the rectangle (Fig. 4.1(a)).

Take a snapshot of this particle and, on the photographic negative, define the angle δ_1 as the included angle in the corner (here, 90°) and the angle λ_1 to be the angle that the diagonal makes with a horizontal datum parallel to the x-axis. Now, imagine that the fluid particle moves and distorts and that a second snapshot is taken at a later time, t_2 (Fig. 4.1(b)). Although the particle distorts, the lines defining the angles still may be considered straight because the element is infinitesimally small. Next, mark each negative with a dot at the lower left corner of the fluid particle, superpose the two negatives, and put a pin through the dots (Fig. 4.1(c)). Such a superposition eliminates the translation of the particle between snapshots.

Now define:

1. *rotation or angular* velocity $\equiv \dfrac{D\lambda}{Dt} \equiv \omega_z$

2. *strain* $\equiv \dfrac{D\delta}{Dt} \equiv -\varepsilon_{xy}$

Notice that the time rates of change of the angles are written with an uppercase D because these are rates of change with respect to a moving fluid particle. The subscript xy indicates that this is the strain in the x-y plane, and the subscript z indicates that this is rotation in a plane perpendicular to the z-axis. These subscripts are necessary because two other planes in three-space could be drawn and other appropriate expressions defined.

Figure 4.1 shows that:

$$\lambda_1 = \frac{1}{2}\delta_1 = \frac{\pi}{4} \text{ and } \lambda_2 = D\alpha + \frac{1}{2}\delta_2 = D\alpha + \frac{1}{2}\left(\frac{\pi}{2} + D\beta - D\alpha\right).$$

Thus, rotation ω_z is given by:

$$\omega_z = \frac{D\lambda}{\Delta t} = \lim_{Dt \to 0}\left\{\frac{\lambda_2 - \lambda_1}{\Delta t}\right\} = \lim_{Dt \to 0}\left\{\frac{\left(D\alpha + \dfrac{\pi}{4} + \dfrac{D\beta}{2} - \dfrac{D\alpha}{2}\right) - \dfrac{\pi}{4}}{Dt}\right\}$$

$$\frac{D\lambda}{Dt} = \omega_z = \text{rotation} = \frac{1}{2}\left(\frac{D\beta}{Dt} + \frac{D\alpha}{Dt}\right).$$

Similarly, $\delta_1 = \dfrac{\pi}{2}$ and $\delta_2 = \dfrac{\pi}{2} + D\beta - D\alpha$. Thus:

$$\frac{D\delta}{Dt} = -\varepsilon_{xy} = \lim_{\Delta t \to 0}\left\{\frac{\delta_2 - \delta_1}{\Delta t}\right\} = \lim_{Dt \to 0}\left\{\frac{\left(\dfrac{\pi}{2} + D\beta - D\alpha\right) - \dfrac{\pi}{2}}{Dt}\right\}$$

$$\frac{D\delta}{Dt} = \frac{D\beta}{Dt} - \frac{D\alpha}{Dt} \Rightarrow \varepsilon_{xy} = \text{strain} = \frac{D\alpha}{Dt} - \frac{D\beta}{Dt}.$$

At first glance, it appears that not much was accomplished because rotation and strain were expressed in terms of other variables and in a more complicated way. However, these formulations allow rotation and strain to be expressed in terms of velocity-component derivatives.

Next, recast Fig. 4.1(c) as Fig. 4.2 and recognize why the angles $D\alpha$ and $D\beta$ are generated. It is because the "painted" molecules that comprise the edges of the fluid particle are moving with increasing speed in both the x- and y-directions the farther the molecule is located from the corner. (Recall that the gross translation components u and v of the particle were eliminated in the superposition in Fig. 4.1(c).)

Because Fig. 4.2 describes the limit as $Dt \to 0$, the paths of the painted molecules may be represented accurately as straight lines even though they are actually circular arcs. In this short time interval, assume that the molecule farthest from the center along the x-axis has moved a distance h. Then, with the tangent of the angle being equal to the angle itself for a small angle:

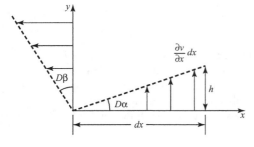

Figure 4.2. Differential motion of sides of a fluid particle.

$$D\alpha = \frac{h}{dx} = \frac{(\text{relative velocity})(\text{time interval})}{dx} = \frac{\left(\frac{\partial v}{\partial x}dx\right)(Dt)}{dx},$$

from which it follows that:

$$\frac{D\alpha}{Dt} = \frac{\partial v}{\partial x}. \tag{4.1a}$$

The student should prove by a similar argument that:

$$\frac{D\beta}{Dt} = -\frac{\partial u}{\partial y}. \tag{4.1b}$$

Note from Fig. 4.1(c) that an arbitrary counterclockwise positive sign was given to $D\alpha$ and $D\beta$. Because Fig. 4.2 indicates that $D\beta/Dt$ is positive but that the rate of change of u with respect to y is negative, a compensating minus sign must be inserted into Eq. 4.1b, as shown.

Finally, substituting Eq. 4.1 into the definitions of *rotation* and *strain*, the desired result is obtained—namely, that these definitions are expressed now in terms of definable and measurable local properties of the flow field. Thus, the rotation in the x-y plane is given by:

$$\omega_z = \frac{1}{2}\left(\frac{\partial v}{\partial x} - \frac{\partial u}{\partial y}\right). \tag{4.2}$$

Substituting Eq. 4.1 into the definition for *strain* yields:

$$\varepsilon_{xy} = \frac{\partial v}{\partial x} + \frac{\partial u}{\partial y}. \tag{4.3}$$

Expressions similar to Eq. 4.3 likewise can be derived for ε_{xz} and ε_{yz}. The concept of strain is discussed in more detail and applied in Chapter 8. Eq. 4.3 is not needed at this time.

Analogous expressions for rotation ω_x and ω_y can be derived similarly in the other two orthogonal planes. These three expressions for rotation represent the components of the angular velocity of a fluid particle in three-space and may be thought of as the three components of an angular velocity vector. Thus,

$$\omega = \omega_x \mathbf{i} + \omega_y \mathbf{j} + \omega_z \mathbf{k} = \frac{1}{2}\left[\left(\frac{\partial w}{\partial y} - \frac{\partial v}{\partial z}\right)\mathbf{i} + \left(\frac{\partial u}{\partial z} - \frac{\partial w}{\partial x}\right)\mathbf{j} + \left(\frac{\partial v}{\partial x} - \frac{\partial u}{\partial y}\right)\mathbf{k}\right]. \tag{4.4}$$

A flow is said to be *irrotational* if the magnitude of this angular velocity vector is identically zero; that is, all three components of the rotation vector are zero. Flows that are irrotational are demonstrated as far simpler to analyze than those that are not. It might seem that irrotational flows would be such special cases as to be of little practical importance. If we consider the velocity profile within a boundary layer in two dimensions (cf. Fig 2.3, where $\partial u/\partial y >> \partial v/\partial x$) or realize that the large shearing action within the boundary layer causes the fluid particles present there to "spin," then it is clear that a viscous boundary layer is a region of rotational flow. However, it happens that outside the thin boundary layer on a body, the flow can be

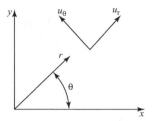

Figure 4.3. Polar-coordinate notation.

irrotational; hence, inviscid flow is irrotational. An exception occurs when entropy gradients may be present, such as behind a curved shock wave in a supersonic flow. Of course, the entire flow field near the body can be rotational if there is a source of vorticity that causes the flow field to be rotational far upstream. The word *vorticity* is used often instead of *rotation*; it differs by only a factor of two, as discussed in the next subsection.

The expression for angular velocity can be written readily in polar coordinates. Such coordinates are frequently of great value in analyzing flow fields, especially those involving curved streamlines. A detailed list of the fundamental fluid equations in various coordinate systems is included in the appendixes. The two velocity-vector components at a point in a polar-coordinate frame are shown in Fig. 4.3.

In polar coordinates (Fig. 4.3), the angular velocity expression is:

$$\omega_z = \frac{1}{2}\left[\frac{\partial u_\theta}{\partial r} + \frac{1}{r}\left(u_\theta - \frac{\partial u_r}{\partial \theta}\right)\right]. \tag{4.5}$$

In this derivation, both sides of the benchmark fluid particle considered in Fig. 4.1 are assumed to rotate with time in a counterclockwise direction. The student should verify that the same results for ω_z and ε_{xy} are obtained if the rotations of both sides are assumed to be clockwise or if one is taken to be clockwise and the other counterclockwise.

EXAMPLE 4.3 *Given*: A physically possible flow field is steady, incompressible, and two-dimensional and it has circular streamlines; it is also irrotational.

Required: Determine the behavior of the tangential velocity component of this flow field.

Approach: Because the flow field is physically possible, continuity must be satisfied. Thus, Eq. 3.52 is needed, as well as the irrotationality condition that $\omega_z = 0$ in Eq. 4.5.

Solution: If the flow has circular streamlines, it follows that $u_r = 0$ and that $u_\theta = Kr^n$, where K is a constant and exponent n is to be determined. Next, write the vector continuity equation (Eq. 3.52) in polar-coordinate form:

$$\frac{\partial u_r}{\partial r} + \frac{1}{r}\left[u_r - \frac{\partial u_\theta}{\partial \theta}\right] = 0.$$

Then, substitute these expressions for the velocity components u_r and u_θ into the continuity equation. This leads to the requirement that $\frac{\partial}{\partial \theta}(Kr^n) = 0$, which is true for any value of the exponent n. Thus, the use of the continuity equation did not determine n; any value leads to a physically real flow.

In addition, we required that the flow must be irrotational, meaning that Eq. 4.4 must apply with $\omega_z = 0$. Substituting the expressions for u_r and u_θ into Eq. 4.4 and performing the differentiation, irrotationality demands that

$$nKr^{n-1} + Kr^{n-1} = 0 = r^{n-1}K(n+1).$$

This states that if $n = -1$, the flow field is irrotational but that the expression for the rotation, ω_z, is indeterminate at the origin ($r = 0$). Thus, the flow field described by $u_r = 0$ and $u_\theta = K/r$ satisfies continuity and is irrotational everywhere except at the origin. To settle the question of what happens at the origin requires the development of additional building blocks (see Section 4.6). An irrotational flow with circular streamlines is an important special case called a *vortex*.

Appraisal: This flow has circular streamlines, yet the fluid particles in the flow exhibit zero rotation. The fluid particles are behaving like the seats on a Ferris wheel in that the angle between a horizontal datum and the diagonal in Fig. 4.1 does not vary with time. If the flow field had circular streamlines and was exhibiting solid-body rotation—as if it were instantaneously frozen and rotating like a wheel—then $u_\theta = Kr$ and Eq. 4.5 yields $\omega_z = K$. This simply states that the general angular velocity expression, Eq. 4.4, gives the expected answer for a simpler and more familiar special case.

Vorticity and Curl

The rotation or angular velocity, ω, plays a crucial role in the analysis of flow fields in fluid mechanics. In the analysis equations, the rotation term appears so often together with a factor of 2 that this quantity has been given a special name and notation; thus, vorticity $\equiv \xi = 2\omega$. The factor of 2 enters the definition only for convenience in relating the vorticity to the curl-vector operation.

In Eq. 4.4, multiply both sides by the factor of 2. The resultant expression is $2\omega = \xi$. The right side of the equation is identified from vector analysis as a fundamental vector operation—namely, the curl of a vector. Thus, vorticity = curl $\mathbf{V}$ = $\nabla \times \mathbf{V}$. From this, it follows that if a flow is irrotational, then the vorticity is zero and the curl of the velocity vector likewise must be zero.

The expression for vorticity in Cartesian coordinates is:

$$\zeta = \zeta_x \mathbf{i} + \zeta_y \mathbf{j} + \zeta_z \mathbf{k} = \text{Curl } \mathbf{V} = \left(\frac{\partial w}{\partial y} - \frac{\partial v}{\partial z}\right)\mathbf{i} + \left(\frac{\partial u}{\partial z} - \frac{\partial w}{\partial x}\right)\mathbf{j} + \left(\frac{\partial v}{\partial x} - \frac{\partial u}{\partial y}\right)\mathbf{k}, \quad (4.6a)$$

whereas in plane-polar coordinates, the vorticity vector is given by:

$$\zeta = \nabla \times \mathbf{V} = \zeta_z \mathbf{k} = \left[\frac{\partial u_\theta}{\partial r} + \frac{1}{r}\left(u_\theta - \frac{\partial u_r}{\partial \theta}\right)\right]\mathbf{k}. \quad (4.6b)$$

Note that the vorticity vector is normal to the plane of the flow field in a two-dimensional flow.

For a two-dimensional planar field, a flow is irrotational if:

$$\frac{\partial v}{\partial x} - \frac{\partial u}{\partial y} = 0, \tag{4.7a}$$

whereas in planar-polar coordinates, a flow is irrotational if:

$$\frac{\partial u_\theta}{\partial r} + \frac{1}{r}\left(u_\theta - \frac{\partial u_r}{\partial \theta}\right) = 0. \tag{4.7b}$$

If the differential form of the momentum equation, Eq. 3.66, is written for a two-dimensional incompressible flow with all of the viscous terms detailed on the right side, it may be combined with the continuity equation and the definition of vorticity, Eq. 4.6, and written as a single equation for the time rate of change of vorticity of a moving fluid particle, $D\zeta/Dt$. The resulting equation for $D\zeta/Dt$ is called the *vorticity-transport equation*, which shows that $D\zeta/Dt$ is proportional to the coefficient of viscosity of the fluid. This means that the vorticity or rotation of a particle moving in a fluid changes because of the presence of viscosity. If attention is focused outside of the boundary layer, where the flow may be represented as inviscid (i.e., a flow with zero viscosity coefficient), the time rate of change of vorticity in this flow is zero. Now, the usual viewpoint for aerodynamics problems is to assume that the vehicle is at rest and that the flow is issuing from upstream infinity as a uniform flow. If the flow from upstream infinity is uniform, then none of the fluid particles coming from infinity has any vorticity because all of the velocity-component derivatives are zero. In an inviscid flow, the vorticity continues to be zero, as discussed previously. We conclude that outside of a thin boundary layer, a subsonic flow may be considered irrotational or to have zero vorticity and rotation.

Circulation

Circulation is defined as the negative of the line integral (defined in mathematics in the counterclockwise sense) of the tangential component of the velocity vector around an arbitrary but fixed closed path in the velocity field, as illustrated in Fig. 4.4. By definition, then:

$$\text{circulation} \equiv \Gamma \equiv -\oint_C \mathbf{V} \cdot \mathbf{ds}. \tag{4.8}$$

Thus, a negative sign is introduced into the definition of circulation in Eq. 4.8 because, although mathematicians associate a positive result with a counterclockwise integration, it is more convenient in aerodynamics applications to define positive circulation as clockwise, as becomes apparent when applications are discussed later. Some authors omit the minus sign in Eq. 4.8 and assume the line integration to be in the clockwise sense. Thus, it is important to specify the direction of integration before evaluating the circulation for a given problem.

The evaluation of the line integral starts at an arbitrary point on the closed path and "marches" around that path in a counterclockwise direction. With each

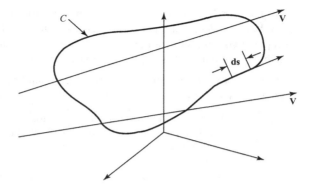

Figure 4.4. Contour C and
directed-line segment **ds**.

incremental step of length **ds** along the path, the velocity component in the direction
of the line segment **ds** is multiplied by the magnitude of **ds** and the product noted.
The line integral is the sum of all of these products. The integrand in the line integral,
Eq. 4.8, may be expanded as:

$$\mathbf{V} \cdot \mathbf{ds} = |\mathbf{V}||\mathbf{ds}| \cos (\text{angle between}).$$

As reviewed in Chapter 3, the vector-dot product is a component-taking operation.
Thus, in the integrand of Eq. 4.8, the component of **V** in the direction of **ds** is multiplied by the magnitude of **ds** as required by the line-integral definition. Note that **ds**
here is an incremental line segment and *not* a unit vector. That is, $\mathbf{ds} = dx\,\mathbf{i} + dy\,\mathbf{j} + dz\,\mathbf{k}$.
  In Cartesian coordinates:

$$\Gamma = -\oint_C (u\mathbf{i} + v\mathbf{j} + w\mathbf{k}) \cdot (dx\mathbf{i} + dy\mathbf{j} + dz\mathbf{k})$$
$$\text{or} \quad \Gamma = -\oint_C [u\,dx + v\,dy + w\,dz]. \tag{4.9}$$

  The circulation may be linked to the curl of the velocity vector and then to the
vorticity in a flow by appealing to Stokes' Theorem, which relates line integrals and
surface integrals. By Stokes' Theorem, if **A** is any vector, then:

$$\oint_C \mathbf{A} \cdot \mathbf{ds} = \iint_{\hat{S}} (\nabla \times \mathbf{A}) \cdot \mathbf{n}\, d\hat{S},$$

where **n** is a unit vector everywhere perpendicular to $\hat{S}$ and the surface $\hat{S}$ is the area
bounded by the curve C. Think of a closed wire frame, C, oriented in three-space.
The surface $\hat{S}$ would be a diaphragm with edges along C; it need not be in the plane
of C and may be of any arbitrary shape.
  Letting **A** be the velocity vector **V**, Stokes' Theorem states that:

$$\text{circulation} = \Gamma = -\oint_C \mathbf{V} \cdot \mathbf{ds} = -\iint_{\hat{S}} (\mathbf{V} \cdot \mathbf{n})\, d\hat{S}. \tag{4.10}$$

Equation 4.10 links vorticity (i.e., particle rotation) to the magnitude of circulation. Later, a close link between circulation and airfoil lift is demonstrated. The concepts of vorticity, circulation, and curl apply to two- or three-dimensional, compressible or incompressible, viscous or inviscid flows.

EXAMPLE 4.4 *Given*: Consider a two-dimensional flow field given by
$\mathbf{V} = (2y)\,\mathbf{i} + (4x)\,\mathbf{j}$.

Required: Find the circulation around the closed path *A-B-C* shown here by evaluating the line integral and by using Stokes' Theorem.

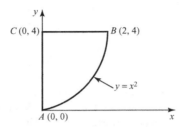

Approach: Because two different methods are required, both the line-integral definition and Stokes' Theorem are used.

Solution: Evaluating the line integral around the given closed path starting at point *A*:

$$A - B: \ \int_A^B (udx + vdy) = \int_0^2 (2y)dx + \int_0^4 (4x)dy = +\frac{80}{3}$$

because along *A-B*, $y = x^2$ and so $(2y)dx = (2x^2)dx$ and $(4xdy) = (4y^{1/2})dy$:

$$B - C: \ \int_B^C (udx + vdy) = \int_2^0 (2y)dx + \int_4^4 (4x)(0)dy = -16, \text{ because along } B\text{-}C, y = 4$$

$$\text{and } C = A: \ \int_C^A (udx + vdy) = \int_0^0 (2y)(0) + \int_4^0 (4x)dy = 0, \text{ because along } C\text{-}A, x = 0.$$

Now, utilizing Stokes' Theorem:

$$\nabla \times \mathbf{V} = \left(\frac{\partial v}{\partial x} - \frac{\partial u}{\partial y}\right)\mathbf{k} = 2\mathbf{k}, \text{ and because } \mathbf{n} = \mathbf{K}, \text{ then:}$$

$$\iint_{\hat{S}} (\nabla \times \mathbf{V}) \cdot \mathbf{k}d\hat{S} = \iint_{\hat{S}} 2d\hat{S} = \frac{32}{3} \text{ so } \Gamma = -\frac{32}{3}.$$

Appraisal: The two results agree, as they should. Note that this flow is rotational. Also note that here, the curl of the velocity vector for this velocity field is a constant independent of *x* and *y*, so that the evaluation of the double integral simply amounted to calculating the area of the figure *A-B-C*. It is important to note that this is a special result; the value of the curl need not always be a constant independent of *x* and *y*.

Stream Function and Velocity Potential

Both the stream function and the velocity potential are scalar functions of the velocity-field coordinates. The derivatives of both of these functions are related to the velocity components of the flow, so that if the function can be found, the velocity field is determined. The existence of the two functions does not depend on the same condition. Thus, one function may exist and be useful in a problem whereas the other function may not exist at all. These functions make the analysis of certain flow problems much easier, as shown herein, so that it is important to know when they may be used and when they do not apply.

Stream Function. A unique stream function exists for two-dimensional planar flows and for axisymmetric flows. For a three-dimensional flow, a stream function may be defined for each of two intersecting surfaces; the intersection of these two surfaces determines a streamline. The two- or three-dimensional flows may be compressible or incompressible, viscous or inviscid. If a stream function is defined for a flow, the flow must satisfy the continuity equation (i.e., the flow must be physically possible). Only two-dimensional planar-incompressible flows are considered here.

Consider the continuity equation, Eq. 3.52, for incompressible, two-dimensional planar flow; namely:

$$\frac{\partial u}{\partial x} + \frac{\partial v}{\partial y} = 0.$$

Then ask: Is it possible to define a scalar function of the coordinates (x,y) that identically satisfy this equation? If so, call this function $\psi(x,y)$. After some thought, we conclude that there is indeed such a scalar function and that in its simplest form, it must be a function such that:

$$u = \frac{\partial \psi}{\partial y} \text{ and } v = -\frac{\partial \psi}{\partial x}. \tag{4.11}$$

If Eq. 4.11 is true, then the continuity equation is identically satisfied because:

$$\frac{\partial^2 \psi}{\partial x \partial y} - \frac{\partial^2 \psi}{\partial y \partial x} = 0.$$

(The order of differentiation is unimportant because there are no discontinuities in the flow.) The minus sign is introduced with the v term rather than the u term in Eq. 4.11 because it is observed that the derivative of the stream function yields a velocity component that is orthogonal to the derivative direction. Thus, a negative Δx with a minus sign yields a positive v component. The student should not be uneasy about Eq. 4.11 being derived here by inspection; the same equation may be derived in a fully rigorous mathematical manner if desired.

It may be shown that the derivative of the stream function in any direction n yields the magnitude of the velocity in a direction perpendicular to n. What is important is that if $\psi(x,y)$ can be determined or specified, then the velocity components of the flow field follow directly by differentiation (this is important later).

It has been shown that if a two-dimensional flow satisfies the continuity equation (i.e., is physically possible), then a stream function exists. What can be said about the properties of such a stream function?

Recall from vector analysis that the gradient of a scalar is a vector and, further, that the gradient vector is at right angles to *isolines*, which are lines along which the scalar is a constant. With this in mind, form the following:

$$\mathbf{V} \cdot \nabla \psi = (u\mathbf{i} + v\mathbf{j}) \cdot \left(\frac{\partial \psi}{\partial x}\mathbf{i} + \frac{\partial \psi}{\partial y}\mathbf{j} \right) = (u\mathbf{i} + v\mathbf{j}) \cdot (-v\mathbf{i} + u\mathbf{j}) = 0.$$

Now, if the dot product of two vectors is zero, then the two vectors must be everywhere at right angles to one another. It follows that the velocity vector is everywhere perpendicular to the vector gradient of the stream function. However, the gradient vector is everywhere perpendicular to isolines ψ = constant. However, if the velocity vector is perpendicular to a vector that is perpendicular to isolines, it follows that the velocity vector must be everywhere *tangent* to the isolines ψ = constant. This means that lines ψ = constant are streamlines; if a stream function can be found, then the streamlines of the flow field are determined as well.

Because no flow can cross a streamline, the volume flow (or mass flow for a compressible flow field) in a stream tube defined by any two streamlines and a unit height out of the paper must be constant. By choosing appropriate values of the constant for ψ = constant, streamlines may be labeled by the volumetric flow rate between them and an arbitrarily designated "base" streamline, as shown in Fig. 4.5.

A stream function also may be defined for polar coordinates in the same manner as for Cartesian coordinates. Write the incompressible flow continuity equation, Eq. 3.11, in polar form by suitably expressing the divergence of the velocity vector in polar coordinates—namely:

$$\nabla \cdot \mathbf{V} = \frac{\partial u_r}{\partial r} + \frac{1}{r}\left[u_r + \frac{\partial u_\theta}{\partial \theta} \right] = 0, \tag{4.12}$$

where the polar-coordinate notation is defined in Fig. 4.3. It may be determined from inspection that a stream function in this coordinate system must be defined as:

$$u_r = \frac{1}{r}\frac{\partial \psi}{\partial \theta} \quad \text{and} \quad u_\theta = -\frac{\partial \psi}{\partial r} \tag{4.13}$$

if the requirement of continuity as expressed in Eq. 4.12 is to be satisfied.

To summarize:

1. A stream function exists for any steady flow that satisfies the continuity equation (i.e., physically possible).

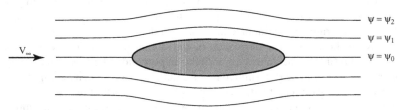

Figure 4.5. Streamlines: lines of ψ = constant.

2. The derivative of the stream function in any direction yields the velocity components of the flow field at right angles to that direction.
3. Explicit expressions for velocity components in terms of derivatives of the stream function were developed for two-dimensional planar flows in Eqs. 4.11 and 4.13.
4. Lines of ψ = constant are streamlines.

EXAMPLE 4.5 *Given*: The equation of a family of streamlines in the physically possible flow of a perfect fluid is given by:

$$xy^2 - \frac{x^3}{3} = \text{constant.}$$

Find the slope of the streamline passing through the point (1,2) and verify that the expression for the stream function as given by Eq. 4.11 is consistent with this answer.

Approach: Appeal to Eq. 4.1 for the definition of a streamline; then, find the slope from this result and also by evaluating Eq. 4.11 for this flow field.

Solution: From Eq. 4.1,

$$\frac{v}{u} = \left[\frac{dy}{dx}\right]_{\text{streamline.}}$$

Differentiating the given equation for the streamline, $y^2 dx + 2xy\,dy - x^2 dx = 0$. Solving,

$$\frac{dy}{dx} = \frac{x^2 - y^2}{2xy} \text{ and at}(1.2),$$

$$\frac{dy}{dx} = \frac{v}{u} = -\frac{3}{4}.$$

This is the required streamline slope at (1,2). Now, because the stream function is constant along a streamline, the equation of the streamline also must be the equation of the stream function—namely,

$$\psi = xy^2 - \frac{x^3}{3}.$$

Applying Eq. 4.11 at the point (1,2), $u = 4$ and $v = -3$ so that the ratio of the velocity components is $v/u = -3/4$, as before. Hence, the slope of the streamline agrees with the prior result.

Appraisal: This example reinforces the validity of the derivation leading to Eq. 4.11. Because the problem was framed as a "physically possible flow field," no check using the continuity equation was performed (neither was it necessary) before the problem was begun.

EXAMPLE 4.6 *Given*: A velocity field is given in polar coordinates for a perfect fluid flow as:

$$u_r = \left(\frac{\theta^2}{r} - 1\right) \text{ and } u_\theta = (\theta - 2r).$$

Required: Find the stream function for this flow.

Approach: Use Eq. 4.13 to link velocity components with the stream function in polar coordinates.

Solution: Using Eq. 4.13,

$$u_r = \frac{1}{r}\frac{\partial\psi}{\partial\theta} = \frac{\theta^2}{r} - 1 \Rightarrow \frac{\partial\psi}{\partial\theta} = \theta^2 - r.$$

Integrating this *partial* derivative,

$$\psi = \frac{\theta^3}{3} - r\theta + f(r),$$

where $f(r)$ is an arbitrary function of r (recall that there is not simply a constant of integration here because a partial derivative is being integrated). Now, using this result for ψ, form:

$$u_\theta = -\frac{\partial\psi}{\partial r} = 0 + \theta - f'(r) = \theta - 2r$$

from the given information. Solving for $f'(r)$ and integrating,
$f'(r) = 2r \Rightarrow f(r) = r^2 + \text{constant}.$

From this, it follows that:

$$\psi = \frac{\theta^3}{3} - r\theta + r^2 + \text{constant},$$

which is the required stream function.

Appraisal: The resulting stream function also is the equation for the streamlines for this flow, with different values of the integration constant corresponding to different streamlines. If it is only required to find velocity components from a given stream function, the value of this constant is not important because it drops out on differentiation for the velocity components. Note that it is well to check the given velocity field to see that it satisfied the continuity equation (*i.e.*, Eq. 3.52) before the stream function was sought.

Velocity Potential: This function of the field coordinates exists *only* for irrotational flows; the flows may be compressible or incompressible, two- or three-dimensional. Only inviscid, incompressible flows are treated here. As discussed, an inviscid subsonic flow is irrotational unless vorticity is created in the flow upstream of the region being investigated.

From Section 4.2, if a flow is irrotational, then this implies that the vorticity is zero and that the curl of the velocity vector is also zero—or, in vector notation, $\nabla \times \mathbf{V} = 0$. Now, there is a vector identity that says $\nabla \times (\nabla \phi) = 0$, where ϕ is any scalar function. From this identity, it follows that the velocity vector may be defined in terms of the gradient of a scalar function of the field coordinates if the flow is irrotational. This function is called the *velocity potential* and is usually given the symbol $\phi(x,y)$. Thus,

$$\mathbf{V} = \nabla\{\phi(x,y)\} \Rightarrow u\mathbf{i} + v\mathbf{j} = \frac{\partial \phi}{\partial x}\mathbf{i} + \frac{\partial \phi}{\partial y}\mathbf{j},$$

and, equating vector components:

$$u = \frac{\partial \phi}{\partial x}, \quad v = \frac{\partial \phi}{\partial y}. \tag{4.14}$$

Because the density is not involved in the derivation, the same expression holds true for an inviscid compressible flow. If the compressible flow is supersonic, there may be regions of the flow that are not irrotational by reason of entropy gradients (as behind curved shock waves), even though the flow is assumed to be inviscid. A third Cartesian dimension simply adds a third term in the development and the expression for the third velocity component follows directly.

A parallel development using the vector expressions proper for polar coordinates leads to the result:

$$u_r = \frac{\partial \phi}{\partial r}, \quad u_\theta = \frac{1}{r}\frac{\partial \phi}{\partial \theta}. \tag{4.15}$$

Notice that the derivative of the velocity potential with respect to a coordinate direction yields the velocity component in the *same* direction.

The velocity potential has the same role as the stream function in that if the velocity potential can be determined or specified, then the velocity components of the flow field follow by differentiation. The velocity potential often is useful as well in the treatment of simultaneous partial-differential equations for a flow field. The introduction of the velocity potential allows the equations to be combined in terms of a single dependent variable (i.e., the velocity potential) although the single equation is of higher order than the simultaneous equations.

Lines of constant velocity potential are *not* streamlines. However, they have a certain property with respect to streamlines that may be observed by forming the vector-dot product $[\nabla\{\phi(x,y)\}]\cdot[\nabla\psi\{(x,y)\}]$. Evaluation of this product yields zero for any flow field for which the two scalar functions exist. This states that the two gradient vectors are perpendicular to one another; however, these two vectors, in turn, are each perpendicular to isolines of their particular function. The conclusion, then, is that isolines of constant-velocity potential (i.e., equipotential lines) are everywhere perpendicular to isolines of constant-stream function (i.e., streamlines). Thus, equipotential lines form an orthogonal mesh with streamlines everywhere in a flow field—provided, of course, that the potential exists (i.e., the flow is irrotational).

Henceforth, it is assumed that any flow field under consideration satisfies the continuity equation so that a stream function exists. However, a particular physically possible flow may or may not be irrotational, depending on the circumstances.

To summarize:

1. A velocity potential exists for any flow that is irrotational.
2. The derivative of the velocity potential in any direction yields the velocity component of the flow field in that same direction.
3. Explicit expressions for velocity components in terms of derivatives of the velocity potential were developed for two-dimensional planar flows in Eqs. 4.14 and 4.15.
4. Equipotential lines are everywhere orthogonal to streamlines.

EXAMPLE 4.7 *Given*: A two-dimensional flow field is described by:

$$\mathbf{V} = (2x+y+1)\,\mathbf{i} + (-2y)\,\mathbf{j}.$$

Required: Determine the velocity potential for this flow.

Approach: Before trying to find the velocity potential, check first whether one exists.

Solution: If the velocity potential exists, then the flow must be irrotational and the curl of the velocity vector must be zero. For this two-dimensional flow, this means that the coefficient of the unit vector $\mathbf{k}$ in Eq. 4.6 must be zero. Evaluating,

$$\frac{\partial v}{\partial x} - \frac{\partial u}{\partial y} = (0) - (1) \neq 0.$$

Thus, the flow is not irrotational and no velocity potential exists.

Appraisal: Care must be taken to establish the existence of the velocity potential.

EXAMPLE 4.8 *Given*: A physically possible irrotational flow is
$\mathbf{V} = (2x + 1)\,\mathbf{i} - (2y)\,\mathbf{j}$.

Required: Find the velocity potential for this flow.

Approach: Use Eq. 4.14 and integrate to find the velocity potential.

Solution: Using Eq. 4.14,

$$u = \frac{\partial \phi}{\partial x} = 2x+1 \Rightarrow \phi = x^2 + x + f(y) \qquad \frac{\partial \phi}{\partial y} = f'(y).$$

Also, however, $v = \dfrac{\partial \phi}{\partial y} = -2y \Rightarrow f'(y) = -2y \Rightarrow f(y) = -y^2 + \text{constant}.$

Substituting for $f(y)$ in the previous expression for the velocity potential, $\phi = x^2 + x - y^2 + \text{constant}$, where again the evaluation of the constant of integration is of no practical importance because the primary concern is with the derivatives of the velocity potential—that is, with the velocity components.

Appraisal: The determination of the velocity potential knowing the velocity components of a flow field is straightforward and parallels the finding of the stream function from given velocity components.

4.3 Special Solutions of the Conservation Equations

Certain special solutions of the basic equations describing an incompressible flow are of great value in applications. We start by looking for an important integral of the equations that provides information regarding the connection between the local pressure and flow speed.

From Chapter 3, the defining equations in Cartesian coordinates for a two-dimensional planar, incompressible, inviscid steady flow are:

$$\text{continuity:} \qquad \frac{\partial u}{\partial x} + \frac{\partial v}{\partial y} = 0 \qquad (3.52)$$

$$x\text{-momentum:} \qquad \rho u \frac{\partial u}{\partial x} + \rho v \frac{\partial u}{\partial y} = -\frac{\partial p}{\partial x} + \rho b_x$$

$$y\text{-momentum:} \qquad \rho u \frac{\partial v}{\partial x} + \rho v \frac{\partial v}{\partial y} = -\frac{\partial p}{\partial y} + \rho b_y, \qquad (3.70)$$

where the density is constant and a body-force term was added to Eq. 3.2 as two components b_x and b_y to make a point (the body-force term is shown to be negligible in most aerodynamics applications and later dropped). The task, then, is to solve this set of three simultaneous partial-differential equations for the three unknowns u, v, and pressure p. In fact, this task is not as difficult as it may first appear.

Consider first the momentum-conservation equations. Rather than write them as two-component equations, as previously, write the vector equation, Eq. 3.66, for a steady, two-dimensional, inviscid flow. Then, Eq. 3.66 becomes:

$$\rho u \frac{\partial \mathbf{V}}{\partial x} + \rho v \frac{\partial \mathbf{V}}{\partial y} = -\nabla p - \rho \mathbf{b}. \qquad (4.16)$$

The body force is assumed to be the gravity force acting in the downward direction (i.e., the y direction in two dimensions); this is the usual case. Thus, $\mathbf{b} = \mathbf{g} = |\mathbf{g}| \mathbf{J} = -g\mathbf{j}$, where g is the acceleration due to gravity and $\mathbf{j}$ is the unit vector in the y-coordinate (upward) direction. Because rectangular coordinates are used, Eq. 4.16 can be written in a more compact vector form as:

$$\rho(\mathbf{V} \cdot \nabla)\mathbf{V} = -\nabla p - \rho \mathbf{g}. \qquad (4.17)$$

The objective is to directly integrate this form of the momentum equation. Integration can be carried out in a straightforward way under two different assumptions, each of which leads to the same final form of integrated equation. However, the equations actually apply to different sets of physical conditions. Both

assumptions are demonstrated here to emphasize the difference in the applicability of the resulting integrated equations. Because the full vector form of the
momentum equation is used, the two approaches that follow also are valid in three
dimensions.

The first approach is to integrate Eq. 4.17 along a streamline; in this case, the
flow may be rotational. The second approach considers integration along any arbitrary path through the flow field; however, the flow then must be irrotational as
demonstrated here.

1. *Integration along a streamline:*
   First expand Eq. 4.17 in standard vector notation, as follows:

$$\rho(\mathbf{V}\cdot\nabla)(u\mathbf{i}+v\mathbf{j}) = -\frac{\partial p}{\partial x}\mathbf{i} - \frac{\partial p}{\partial y}\mathbf{j} + \rho g\mathbf{j}$$

$$\rho\left[u\frac{\partial u}{\partial x}+v\frac{\partial u}{\partial y}\right]\mathbf{i}+\rho\left[u\frac{\partial v}{\partial x}+v\frac{\partial v}{\partial y}\right]\mathbf{j} = -\frac{\partial p}{\partial x}\mathbf{i} - \frac{\partial p}{\partial y}\mathbf{j} + \rho g\mathbf{j}.$$

Consider now a general incremental displacement vector $\mathbf{ds} = dx\,\mathbf{i} + dy\,\mathbf{j}$. Taking
the vector-dot product of $\mathbf{ds}$ with the vector-momentum equation results in:

$$\rho\left[u\frac{\partial u}{\partial x}dx+v\frac{\partial u}{\partial y}dx+u\frac{\partial v}{\partial x}dy+v\frac{\partial v}{\partial y}dy\right] = -\frac{\partial p}{\partial x}dx-\frac{\partial p}{\partial y}dy+\rho g dy, \qquad (4.18)$$

which is valid along any line in the flow field. Focus on the square bracket on
the left side of this equation. Now, require that the length $\mathbf{ds}$ is not any arbitrary
length but rather is a length $\mathbf{ds}$ along a streamline. Recall from the definition of a
streamline that a certain relationship between u and v must hold along a streamline—namely, that:

$$\frac{dy}{dx} = \frac{v}{u} \Rightarrow u\,dy = v\,dx.$$

Use this relation in the second and third terms within the square bracket. Substitute $(u\,dy)$ for $(v\,dx)$ in the second term and $(v\,dx)$ for $(u\,dy)$ in the third term. The
result is:

$$\rho\left[\left(u\frac{\partial u}{\partial x}dx+u\frac{\partial u}{\partial y}dy\right)+\left(v\frac{\partial v}{\partial x}dx+v\frac{\partial v}{\partial y}dy\right)\right].$$

Now, from calculus, if $u = u\,(x,y)$, then:

$$du = \frac{\partial u}{\partial x}dx+\frac{\partial u}{\partial y}dy.$$

The same thing is true for $v(x,y)$ and $p(x,y)$. Thus, the full equation, Eq. 4.18, may
be rewritten along a streamline as follows:

$$\rho[u\,du+v\,dv]=-dp+\rho g dy$$

But
$$u\,du = d(u^2/2),$$

and
$$v\,dv = d(v^2/2),$$

so that on dividing through by ρ, Eq. 4.18 finally becomes:

$$d\left(\frac{u^2}{2}\right) + d\left(\frac{v^2}{2}\right) = \frac{1}{2}d(V^2) = \frac{dp}{\rho} + gdy. \tag{4.19a}$$

In Eq. 4.19a, the quantity V is the *magnitude* of the velocity vector (sometimes referred to as the speed) and the square of V is equal to the sum of the squares of the orthogonal velocity components.

Eq. 4.19a is set aside to derive this same equation in another way. The equation then is integrated. Note that Eq. 4.19a is easy to integrate if the density, ρ, and the gravitational acceleration are constant.

2. *Integration throughout the flow field (irrotational flow only)*:
The starting point is again Eq. 4.17. Begin by using a standard vector identity that states:

$$(\mathbf{V}\cdot\nabla)\mathbf{V} = \nabla\left(\frac{|\mathbf{V}|^2}{2}\right) - \mathbf{V}\times(\nabla\times\mathbf{V}).$$

Now, for an irrotational flow, the curl of the velocity vector, $\nabla\times\mathbf{V} \equiv \zeta$, is zero by definition. Hence, for an irrotational flow, the left side of Eq. 4.17 simplifies and the equation becomes:

$$\rho\nabla\left(\frac{V^2}{2}\right) = -\nabla p + \rho g\mathbf{j},$$

where again, V is the magnitude of the velocity vector. Taking the vector-dot product of this equation with an arbitrary line segment $d\mathbf{s} = dx\mathbf{i} + dy\mathbf{j}$ yields Eq. 4.17 in the following form:

$$\rho\left[\frac{\partial}{\partial x}\left(\frac{V^2}{2}\right)dx + \frac{\partial}{\partial y}\left(\frac{V^2}{2}\right)dy\right] = -\frac{\partial p}{\partial x}dx - \frac{\partial p}{\partial y}dy + \rho gdy.$$

Again using the basic ideas of calculus, this may be written as follows:

$$d\left(\frac{V^2}{2}\right) = -\frac{dp}{\rho} + gdy. \tag{4.19b}$$

Eqs. 4.19a and 4.19b are identical but arise from different assumptions. Notice that under either assumption, the two partial-differential momentum equations in component form have reduced to a single equation, Eq. 4.19a or 4.19b, which may be integrated easily *providing that the density and gravity are constant*.

The Bernoulli Equation

Recall that the flows considered here are incompressible. It follows that:

$$\int\frac{dp}{\rho} = \frac{1}{\rho}\int dp.$$

Also, assume that the magnitude of the acceleration due to gravity, g, is a constant (to be justified later). Then, integrate Eq. 4.19 as follows:

$$\int d\left(\frac{V^2}{2}\right) = -\frac{1}{\rho}\int dp + g\int dy =$$

$$= \frac{1}{2}V^2 + \frac{p}{\rho} - gy = \text{constant.}$$

(4.20)

The integration follows directly because each integral operation simply undoes the differential operation that follows it. Eq. 4.20 is a simple algebraic equation. Considering Eq. 4.19 to be derived along a streamline as in Eq. 4.19a, the most that can be said is that the sum of the three terms in Eq. 4.20 is equal to a constant at any point along a particular streamline. It is not a requirement that this constant is the same among streamlines. However, if irrotationality is assumed as in Eq. 4.19b, then the constant must be the same on every streamline and, hence, throughout the flow field.

Taking two points in the flow field, either along a streamline or in an irrotational flow field, Eq. 4.20 states:

$$\frac{1}{2}\rho V_1^2 + p_1 - gy_1 = \frac{1}{2}\rho V_2^2 + p_2 - gy_2.$$

In ordinary aerodynamic problems, the extent of the region of interest around an airfoil or fuselage is not large in the vertical (y) direction, so that ($y_1 - y_2$) is small. (This may not be true for a dirigible with large dimensions in the vertical direction!) Thus, in the usual problems of flows around flight vehicles, the gravity-force contribution is negligible and this term, henceforth, is neglected. Eq. 4.20 is written as follows:

$$p + \frac{1}{2}\rho V^2 = \text{constant} = p_o.$$

(4.21)

Eq. 4.21 is the celebrated Bernoulli Equation.* It is emphasized that *this equation may be used only for an incompressible flow*. Serious errors have been made by forgetting this important limitation. It was the incompressible-flow assumption that allowed the term $\int \frac{dp}{\rho}$ to be written as $\frac{1}{\rho}\int dp$ and the integration of the exact differentials followed directly. If the flow is compressible, other similar but somewhat more complicated equations relating the flow variables may be derived, as accomplished later. However, for emphasis, the Bernoulli Equation as derived here *may not* be used to make calculations for a compressible flow.

Now, examine the Bernoulli Equation. If two points are taken in an irrotational flow field, the equation states that the sum of the static-pressure term and the term $\rho V^2/2$ is equal to the same constant at those two points. If the second point is taken to be a location where the velocity is zero, the pressure term alone appears in the sum at that point. This term is the pressure when the velocity is zero (i.e., when the flow is brought to rest or stagnates), and this pressure is give a special name and

---

* Named after Daniel Bernoulli, a famous mathematician and physicist in the 1700s who made many important contributions to the study of hydrodynamics.

symbol: *stagnation pressure* = p_0 = pressure where the flow velocity is zero. This reference pressure often is referred to as the *total pressure* (because it is evaluated by summing two terms) or the "total head" (because pressure is sometimes measured as the height of a liquid column).

Recalling the two different approaches in the derivation of the Bernoulli Equation, it follows that if a flow is rotational (e.g., within a boundary layer), then the stagnation pressure is constant along any streamline within the boundary layer, but the constant is not the same value among streamlines. If a flow is irrotational, the stagnation pressure is the same everywhere in the flow field.

All of the terms in the Bernoulli Equation have the units of pressure because:

$$\rho V^2 = \left[\frac{slug}{ft^3}\right] \cdot \left[\frac{ft^2}{s^2}\right] = \left[\frac{lb_f s^2}{ft^4}\right] \cdot \left[\frac{ft^2}{s^2}\right] = \left[\frac{lb_f}{ft^2}\right].$$

Thus, each term in the Bernoulli Equation can be given a descriptive pressure name as follows:

$p \equiv$ *static pressure* = the pressure in a flow field that would be measured by an instrument moving with the same velocity that the flow is moving at the point in question
$\rho V^2/2 \equiv$ *dynamic pressure* = the "pressure" due to fluid motion
$p_0 \equiv$ *stagnation pressure* = the pressure at some point in the flow where the velocity is zero, as at the stagnation point at the nose of an airfoil or vehicle

These three definitions are fundamental and appear throughout this book. The definitions must be committed to memory and a clear understanding of their physical meaning is vital.

Static pressure at a point must be measured in a way that does not disturb the flow because if the flow is locally slowed or accelerated by the measuring device, then a true measurement cannot be made. One way to measure the static pressure of a flow in a wind-tunnel test section, for example, is to make a small hole (i.e., a static-pressure tap) in the wall of the test section (Fig. 4.6a). The hole should be smooth and small enough to ensure a local measurement, and it is connected by tubing to some suitable pressure-measuring device. Of course, there is a boundary layer growing on the test-section wall, but it is shown in Chapter 8 that the variation of static pressure through this boundary layer in a direction normal to the wall essentially is zero. Hence, the static pressure as measured at the wall is that of the external,

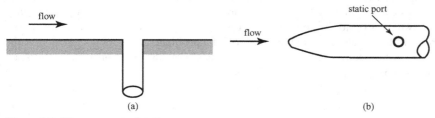

(a)   (b)

Figure 4.6. Measurement of static pressure.

Figure 4.7. Stagnation pressure probe.

undisturbed flow at a particular streamwise station. If the static pressure is not constant across the test section, a local measurement in the flow may be made by using a static-pressure probe (Fig. 4.6b), which is a small tube aligned with the flow and streamlined at the upstream end so as to produce as small a flow disturbance as possible. A static-pressure tap is made in the tube wall several diameters downstream of the front of the probe to allow any local-flow accelerations around the nose to die out. This pressure tap is connected to a suitable pressure-measuring device.

The dynamic pressure cannot be measured directly because it is not a physical quantity but rather is only a variable having the units of pressure. The dynamic pressure can be measured indirectly as the difference between the stagnation pressure and the static pressure.

The stagnation pressure in a flow can be measured by using a stagnation-pressure probe (also called a pitot probe or a total head probe), shown in Fig. 4.7. This is simply an open-ended tube aligned with the flow and connected to a pressure-measuring device, which effectively "dead-ends" the tube. The oncoming flow slows and then comes to a rest at the entrance to the tube. The resulting stagnation pressure can be measured by means of an aneroid device or a liquid column.

The simultaneous measurement of stagnation and static pressure allows for the determination of flow velocity through the use of the Bernoulli Equation (providing, or course, that the density is known or can be measured).

This would be a truly local measurement only if the static and stagnation-pressure probes were located close to one another. Hence, they are often combined into one instrument, called a pitot-static probe (Fig. 4.8).

The differential (i.e., stagnation minus static) pressure is most often measured by using such a probe because this is more accurate than measuring each pressure independently and then finding the difference between the two values. Again, the static-pressure tap must be located far enough from the nose of the probe that the flow has returned to essentially undisturbed conditions.

The stagnation pressure exists as a physical reality anywhere that the flow is brought to rest (e.g., at the end of a pitot probe or at the stagnation point on an airfoil or wing). However, the stagnation pressure can be defined at any point in the flow field, whether or not the flow there is at rest. Thus, for purposes of calculation, the flow can be thought of as being brought to rest at any convenient point in the flow (whether or not it is physically at rest there) by means of an "imaginary" pitot probe, if this is helpful during the solution of a problem.

It is often convenient to express a physical pressure in terms of a nondimensional ratio called the *pressure coefficient*:

$$C_p \equiv \frac{p - p_\infty}{\frac{1}{2}\rho_\infty V_\infty^2}, \qquad (4.22)$$

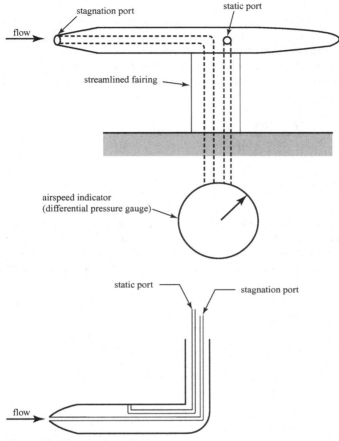

Figure 4.8. Pitot-static probes.

where p is the local static pressure in question and the subscript ∞ refers to values of static pressure, density, and velocity in the undisturbed freestream. The density usually is written without a subscript in material pertaining to incompressible flow because for incompressible flow, the density is constant everywhere in the flow field. The dynamic pressure in the denominator of Eq. 4.22 always is formed with the freestream values of velocity and density (which are constants) rather than with local values, which vary throughout a flow. Expressing a pressure in coefficient form conveys considerably more information than just a quoted pressure magnitude because knowledge of the nondimensional coefficient value allows the prediction of the pressure magnitude in a corresponding flow that could have a different freestream velocity and/or freestream static-pressure level.

This definition of the pressure coefficient also is valid for a compressible flow (with the density now carrying an "infinity" subscript), including supersonic flow. Care must be taken in examining plots of chordwise pressure distribution on airfoils because the convention is to graph the pressure-coefficient ordinate as negative upward. Notice that the pressure coefficient can be either positive or negative, depending on the relative values of the local and freestream static pressures. This is in contrast to the absolute pressure, which never can be negative.

In the case of incompressible, inviscid flow in which the Bernoulli Equation holds and the stagnation pressure is constant everywhere, the pressure coefficient may be written as:

$$C_p = \frac{\left(p_0 - \frac{1}{2}\rho V^2\right) - \left(p_{0_\infty} - \frac{1}{2}\rho V_\infty^2\right)}{\frac{1}{2}\rho V_\infty^2} = 1 - \frac{V^2}{V_\infty^2}. \tag{4.23}$$

Pressure distributions often are tabulated as the ratio $\dfrac{V^2}{V_\infty^2}$ for incompressible flows.

EXAMPLE 4.9 *Given*: Air is drawn from the atmosphere at standard sea-level conditions ($\rho = 1.225\ Kg/m^3, p = 1.013 \times 10^5\ N/m^2$) into a wind tunnel by means of a fan at the discharge end (see illustration). It is known that at a certain fan speed, the test-section velocity is 60 m/s. Assume the flow to be steady, incompressible, and inviscid.

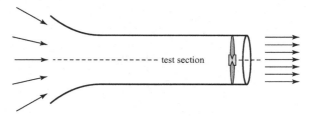

Required: Predict (a) the pressure at the stagnation point of a model in the test section, and (b) the static pressure in the test section with the model removed.

Approach: Because the flow is at low speed (and assumed to be incompressible), use the Bernoulli Equation.

Solution: (a) Consider any fluid particle suitably far from the wind-tunnel entrance so that it is initially in the atmosphere at rest and then is sucked into the test section by the inflow. The path of this particle is a streamline along which the stagnation pressure is constant. Far from the entrance, the pressure of the air at rest is atmospheric pressure, so that the stagnation pressure along that streamline is atmospheric pressure. The same is true for any streamline because the flow is assumed to be inviscid and, hence, irrotational. Thus, when the fluid particle is brought to rest at the stagnation point of the model, the pressure there is the stagnation pressure along the stagnation streamline, which in this wind tunnel is equal to the atmospheric pressure. No calculation is necessary.

(b) Consider any streamline and imagine bringing the flow to rest somewhere in the empty test section. The stagnation pressure there is atmospheric pressure, as noted in (a). Apply the Bernoulli Equation at this point:

$$p + \frac{1}{2}\rho V^2 = p_0 = 1.013 \times 10^5 = p + \frac{1}{2}(1.225)(60)^2.$$

Solving, the static pressure in the empty test section is 9.915×10^4 N/m^2.

Appraisal: The static pressure in the test section must be less than the stagnation (atmospheric) pressure, as indicated by the answer. Careful thought is necessary in this problem to correctly identify static and stagnation pressure (contrast the role of atmospheric pressure in this problem with Example 4.10). In (b), the ability to "imagine" the flow being brought to rest in the test section—even though there is no stagnation point physically present there—provides the key to the application of the Bernoulli Equation. Now, suppose that air from the atmosphere is pulled into the wind tunnel, but the fan is relocated so as to be just downstream of the entrance. Would the answers to the questions be changed? Yes. (The student should give this conclusion careful consideration.) In this latter case, work is done on the fluid particles as they pass through the fan, so that the pressure at a stagnation point on the model would not be the same as the atmospheric pressure!

EXAMPLE 4.10 *Given*: An airplane is in steady flight at 150 mph at an altitude of 15,000 feet ($\rho = 0.0015$ slugs/ft^3, $p = 1,194$ lbf/ft^2 absolute). The pressure at a certain Point A on the wing is measured at 8.0 psia. Assume a steady, incompressible, inviscid flow.

Required: (a) What is the pressure at a stagnation point on the airplane, and (b) what is the velocity at Point A?

Approach: Because the flow is incompressible, use the Bernoulli Equation. Because the flow is assumed to be inviscid, the stagnation pressure is the same everywhere in the flow field.

Solution: The problem is posed from the viewpoint of an observer on the ground watching an airplane fly past. Apply a coordinate transformation such that the airplane is fixed and the air is flowing by. Because the airplane is in steady flight, this transformation is made by imagining that the observer now is riding on the airplane. As the observer looks around at the airplane, it appears to be at rest and there is a wind of 150 mph blowing against the observer's face (which is the uniform flow from upstream infinity). The atmospheric pressure is unchanged by this shift in observer location. Thus, in this problem with the airplane apparently at rest (i.e., body-fixed coordinates), the static pressure of the oncoming flow is the atmospheric pressure and there is a dynamic pressure due to the velocity of the oncoming flow.

(a) From the Bernoulli Equation:

$$p_0 = p_{0_\infty} = p_{atm} + \frac{1}{2}\rho V_\infty^2 = 1,194 + \frac{1}{2}(0.0015)[(150)(1.467)]^2 = 1,230.3 \text{ psfa},$$

where consistent units require the velocity to be expressed in ft/s.

(b) Again applying the Bernoulli Equation at Point A with p_o a constant everywhere:

$$p_o = p_{o_A} = p_A + \frac{1}{2}\rho V_A^2 \Rightarrow V_A = 323 \text{ ft/s}.$$

Appraisal: The ambient pressure has a completely different role in this example than in the prior one; the two examples should be carefully reviewed until this becomes clear. Steady-flow problems normally are easiest to solve in a "body-axis" system—that is, in which the observer rides with the moving vehicle so that it appears to be at rest.

Return now to the Bernoulli Equation as the integral solution of the momentum equation. The Bernoulli Equation is a simple scalar equation. In particular, if the magnitude of the velocity is known at any point of interest, then the pressure at that point follows directly.

This suggests a strategy for solving the incompressible-flow problem that was posed initially as described by three partial-differential equations. Suppose that the velocity components and, hence, the magnitude of the velocity, at any field point could be found by working with the continuity equation. Then, the substitution of this velocity into the scalar Bernoulli Equation (which represents the integrated effect of the vector-momentum balance) yields the pressure at that point. If needed, the pressure distribution over a body then could be found and integrated to determine important forces, such as the lift on an airfoil.

At first glance, this strategy does not appear to be feasible because two velocity components must be found using only one (continuity) equation. This is where the velocity potential and the stream function become useful. In the next section, the continuity equation is examined with the scalar functions, the velocity potential, and the stream function. It is shown that it is possible to establish a powerful method for evaluating the magnitude of the velocity at any point in the flow field by starting with the continuity equation.

4.4 Solving the Conservation Equations

In this section, we continue the quest started in Section 4.3 for special forms of the governing equations that can be solved using standard mathematical tools. We found that a special form of the momentum equation could be solved by simple integration leading to the powerful Bernoulli Equation. We continue our examination of the governing equations for the case of incompressible steady flow.

Laplace's Equation

Return now to the continuity equation, Eq. 3.52, which is one of the defining equations for a two-dimensional incompressible-flow problem. The continuity equation may be written in terms of a single dependent variable—either the velocity potential or the stream function—provided that the flow is irrotational. To show this in both

Cartesian and polar coordinates, write the continuity equation in vector form, as in Eq. 3.52—namely, $\nabla \cdot \mathbf{V} = 0$. Now, for an irrotational flow, $\mathbf{V} = \nabla \phi$, and substituting this into the continuity equation yields:

$$\nabla \cdot (\nabla \phi) = \nabla^2 \phi = 0. \tag{4.24}$$

Eq. 4.24 is Laplace's Equation, an equation that appears in many fields including heat transfer and electrodynamics. Eq. 4.24 in Cartesian coordinates is:

$$\frac{\partial^2 \phi}{\partial x^2} + \frac{\partial^2 \phi}{\partial y^2} = 0 \tag{4.25}$$

and in polar coordinates, Laplace's Equation is:

$$\nabla^2 \phi = \frac{\partial^2 \phi}{\partial r^2} + \frac{1}{r^2} \frac{\partial^2 \phi}{\partial \theta^2} + \frac{1}{r} \frac{\partial \phi}{\partial r} = 0. \tag{4.26}$$

These two equations each represent a single equation in terms of a single dependent variable, which suggests that a solution is possible.

Regarding the stream function, recall that for irrotational flow, it follows from Eq. 4.3 that $\dfrac{\partial u}{\partial y} = \dfrac{\partial v}{\partial x}$ and, from the definition of the stream function, Eq. 4.11, that:

$$u = \frac{\partial \psi}{\partial y} \text{ and } v = -\frac{\partial \psi}{\partial x}.$$

Substituting the second relation into the first yields:

$$\frac{\partial^2 \psi}{\partial x^2} + \frac{\partial^2 \psi}{\partial y^2} = \nabla^2 \psi = 0, \tag{4.27}$$

which is the Laplace's Equation in Cartesian coordinates.

A parallel development for the stream function shows that it satisfies the Laplace's Equation in cylindrical coordinates—namely:

$$\frac{\partial^2 \psi}{\partial r^2} + \frac{1}{r^2} \frac{\partial^2 \psi}{\partial \theta^2} + \frac{1}{r} \frac{\partial \psi}{\partial r} = 0. \tag{4.28}$$

Notice that the original set of three simultaneous partial-differential equations describing a two-dimensional, incompressible, inviscid flow was replaced by the Laplace's Equation and a scalar equation, the Bernoulli Equation. This represents a considerable mathematical advantage because it is not necessary to deal with any vector variables. A strategy for predicting the pressure distribution in a steady two-dimensional inviscid, irrotational flow then is as follows:

1. Solve the Laplace's Equation for the dependent variable $\phi(x,y)$ or $\psi(x,y)$.
2. Suitably differentiate this expression to determine the velocity components at any point.

3. Knowing the velocity components, determine the magnitude of the local flow velocity.
4. Use the Bernoulli Equation to solve for the static pressure at any point in the flow field. The stagnation pressure is known or can be measured for a given flow.

Solving Laplace's Equation

The Laplace's Equation $\nabla^2 \phi = \nabla^2 \psi = 0$, Eq. 4.24, must be solved for the single dependent variable, either ϕ or ψ, subject to the appropriate boundary conditions for the problem under consideration. This ensures that the solution describes the proper physical reality. The second-order Laplace's Equation requires two such boundary conditions. For the aerodynamic problems, these are (1) the disturbance due to the presence of the body in the flow must die out far away from the body in the freestream; and (2) the velocity component normal to the body surface must be zero because the body is a solid. These two boundary conditions are expressed in terms of the scalar functions, as follows:

$$\text{far from the body, } u = \frac{\partial \phi}{\partial x} = \frac{\partial \psi}{\partial y} = V_\infty; \quad v = 0$$

and

$$\text{at the body surface, } \frac{\partial \phi}{\partial n}, \frac{\partial \psi}{\partial s} \equiv 0,$$

where n and s are the directions normal and tangential to the body surface, respectively. In other words, the flow must be tangent to the body surface.

The Laplace's Equation has a property that may be used to simplify enormously the solution—namely, that the equation is linear. Linearity means that in the equation, there are no products or terms taken to a power that contains the dependent variable. Thus, $\frac{\partial^2 \phi}{\partial x^2}$ is a linear term whereas $\phi \frac{\partial^2 \phi}{\partial x^2}$ and $\left(\frac{\partial^2 \phi}{\partial x^2}\right)^2$ are not. It also means that no transcendental functions of the dependent variable appear. This linearity property means that *superposition* of solutions may be used. Thus, if ϕ_1 and ϕ_2 are each a solution to the Laplace's Equation, then the linear combination $\phi = a\phi_1 + b\phi_2$ (where a and b are arbitrary constants) also is a solution.* This may be verified by direct substitution into Laplace's Equation. Of course, the appropriate boundary conditions must be satisfied by solution ϕ.

The far-reaching significance of superposition is that simple (or elementary) flow solutions may be added together to form solutions corresponding to more complicated flows. This means that it may not be necessary to solve Laplace's Equation to carry out the solution strategy described previously. A much simpler strategy can be followed. Consider the following steps:

1. Specify the velocity components for simple flows by inspection (e.g., for a uniform flow, the velocity components are $u = V_\infty$ and $v = 0$).

---

* The constant coefficients that appear in these elementary solutions (often referred to as *strengths*) are assigned convenient arbitrary values when illustrating superposition in this chapter. However, in Chapter 5, in which it is required that superposition correspond to a specified body shape, the coefficients must be adjusted so that the composite solution matches the required boundary conditions.

2. Integrate to find ϕ or ψ for each simple flow as in Examples 4.6 and 4.8. The constant of integration may be set equal to zero or to any convenient value because the final variables of interest are the velocity components, the derivatives of ϕ or ψ, and the constant of integration drops out.
3. Superpose (add) the simple (elementary) solutions for ϕ or ψ corresponding to these simple flows.
4. Differentiate this superposed solution to obtain the velocity components for a new, more complicated flow and then use these components to determine the velocity magnitude.
5. Use the Bernoulli Equation to solve for the magnitude of the corresponding static pressure.

The next section explains how elementary solutions to the Laplace's Equation for simple flows are found. These elementary solutions then are superposed in Section 4.6 to construct more complex and realistic flow-field solutions.

4.5 Elementary Solutions

Recalling the findings of the last section, we proceed to construct a set of elementary solutions to Laplace's Equation. The corresponding stream-function and velocity potentials are determined, which then form the basis for a powerful problem-solving technique using superposition of the elementary solutions. We start with the simplest and proceed to the more complex flow patterns as we are guided by both physical and geometrical considerations.

Uniform Flow

For *uniform flow*, $u = V_\infty$, $v = 0$. Thus,

$$u = V_\infty = \frac{\partial \psi}{\partial y} \Rightarrow \psi = \int V_\infty \, dy = V_\infty y + f(x) + \text{constant}$$

$$v = 0 = -\frac{\partial \psi}{\partial x} \Rightarrow \psi = g(y) + \text{constant}. \quad \text{Comparing, } \psi_{UF} = V_\infty y.$$

Because $f(x)$ must be zero, the x-axis has been made the zero streamline (i.e., at $y = 0$, $\psi = 0$), so that the constant of integration is zero. The subscript UF designates uniform flow. Thus, by a similar procedure, $\phi_{UF} = V_\infty x$.

Collecting the results, in Cartesian coordinates,

$$\begin{cases} \psi_{UF} = V_\infty y \\ \phi_{UF} = V_\infty x. \end{cases} \tag{4.29}$$

For many problems, the polar-coordinate form is useful. In this case,

$$\begin{cases} \psi_{UF} = V_\infty r \sin \theta \\ \phi_{UF} = V_\infty r \cos \theta. \end{cases} \tag{4.30}$$

The student should verify that uniform flow satisfies continuity and is irrotational and that Eqs. 4.29 and 4.30 both satisfy the Laplace's Equation on substitution.

In what follows, the uniform flow usually is taken as coincident with the x-axis. The results may be extended to uniform flow at an arbitrary angle by letting $u = V_\infty \cos\alpha$ and $v = V_\infty \sin\alpha$.

Source Flow

Source flow is flow that is assumed to emanate from a point source. In two-dimensional flow, a line source of infinite length pierces the x-y plane at a point (Fig. 4.9a). This point source in the x-y plane corresponds to outflow from the axis of a cylinder (Fig. 4.9b). In three-dimensional flow, a point source corresponds to outflow from the center of a sphere (see Chapter 7). Source flow is emitted in a purely radial direction; there is no tangential component of velocity. Thus, the streamlines are all straight lines originating at a point.

Assume two-dimensional planar flow and consider a point source at the origin of coordinates. Let Λ be defined as the volume flow rate from the source per unit time per unit depth out of the page. The magnitude of this quantity is the "strength" of the source, which can be adjusted as necessary to match boundary conditions during analysis. If Conservation of Mass is to be satisfied, then for any circle of arbitrary radius r, about the origin (Fig. 4.9c),

$$\rho\Lambda = \text{mass flux out of source} = \rho(2\pi r)(1)u_r \Rightarrow u_r = \frac{\Lambda}{2\pi r},$$

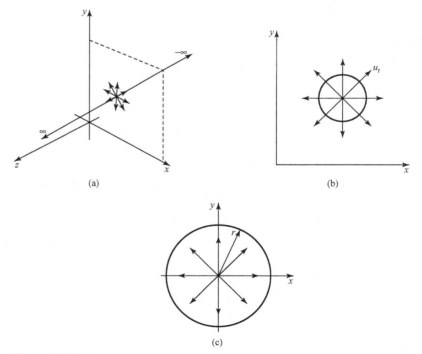

(a)   (b)

(c)

Figure 4.9. Point source.

where 2π is the circumference of the circle of radius r and the product of this with unity normal to the x-y plane yields the area through which the outflow occurs. By definition, the circumferential velocity $u_\theta = 0$. Notice that continuity requires that the radial velocity be proportional to $1/r$ so that at the source ($r = 0$), the velocity is infinite. This singularity is not troublesome in constructing flows by super position, as shown later.

Integrating the velocity components yields the following in polar coordinates (i.e., the most convenient coordinates for this flow):

$$u_r = \frac{\Lambda}{2\pi r} = \frac{1}{r}\frac{\partial \psi}{\partial \theta} \quad \Rightarrow \quad \psi = \int \frac{\Lambda}{2\pi} d\theta = \frac{\Lambda}{2\pi}\theta + f(r) + \text{constant}$$

$$u_\theta = 0 = -\frac{\partial \psi}{\partial r} \quad \Rightarrow \quad \psi = g(\theta) + \text{constant}.$$

Comparing the two expressions for ψ, it follows that $f(r) = 0$. The constant of integration may be set to zero for convenience. Similarly,

$$u_r = \frac{\partial \phi}{\partial r} = \frac{\Lambda}{2\pi r} \quad \Rightarrow \quad \phi = \int \frac{\Lambda}{2\pi r} dr = \frac{\Lambda}{2\pi}\ln r + f(\theta) + \text{constant}$$

$$u_\theta = \frac{1}{r}\frac{\partial \phi}{\partial \theta} = 0 \quad \Rightarrow \quad \phi = g(r) + \text{constant}.$$

Comparing, $f(\theta) = 0$ and the constant of integration may be set to zero. Summarizing for polar coordinates:

$$\psi_s = \frac{\Lambda}{2\pi}\theta$$

$$\phi_s = \frac{\Lambda}{2\pi}\ln r$$

(4.31)

and converting to rectangular coordinates:

$$\psi_s = \frac{\Lambda}{2\pi}\tan^{-1}\left(\frac{y}{x}\right)$$

$$\phi_s = \frac{\Lambda}{2\pi}\ln\left(\sqrt{x^2 + y^2}\right),$$

(4.32)

where the subscript s designates a source.

Again, substitution shows that Eqs. 4.31 and 4.32 both satisfy the Laplace's Equation. Considering a circular path enclosing the origin, the circulation around that closed path is zero because there is no component of velocity tangential to the path ($u_\theta = 0$). Thus, by Stokes' Theorem, the source flow is an irrotational flow.

A *sink* is simply the opposite of a source; that is, the strength Λ is negative instead of positive. The radial velocity is inward (negative u_r) and the fluid flows into the origin rather than being emitted from the origin. Eqs. 4.31 and 4.32 apply, with a change in sign.

Vortex

The elementary solution called a *vortex* was encountered in Example 4.3, where it was shown that a physically possible flow with circular streamlines must have a

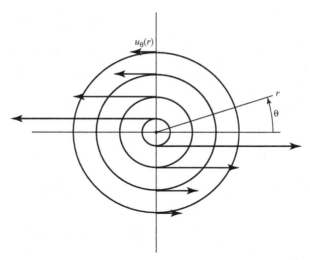

Figure 4.10. Point vortex at the origin of a polar-coordinate system.

tangential velocity that varies inversely as the distance from the center if the flow field is to be irrotational. Figure 4.10 illustrates such a flow field. Thus, $u_\theta = K/r$ and $u_r = 0$. It was not possible to state previously whether the flow is rotational at the exact center. The answer to this question had to await development of the concept of circulation. Because this now is accomplished, we apply the definition of circulation by taking the line integral around a closed path, which is a circle of arbitrary radius r centered at the origin. Then:

$$\Gamma = -\int V.\,ds = -\int_0^{2\pi} u_\theta\,(rd\theta) = -\int_0^{2\pi} \frac{K}{r}(rd\theta)$$

and, integrating:

$$\Gamma = -2\pi K \Rightarrow K = -\frac{\Gamma}{2\pi}.$$

(4.33)

Next, we appeal to Stokes' Theorem, Eq. 4.10—namely:

$$\Gamma = -\int_C \mathbf{V} \cdot \mathbf{ds} = -\iint_S (\nabla \times \mathbf{V}) \cdot \mathbf{n} d\hat{S}.$$

Take the surface integral to be over the surface enclosed by the circle and lying in the x-y plane, so that $d\hat{S}$ in Eq. 4.10 is an element of an area of a circle and the unit vector perpendicular to the surface in the x-y plane is $\mathbf{n} = \mathbf{k}$. Furthermore, in the x-y plane, the curl of the velocity vector, $\nabla \times \mathbf{V}$, is a vector with a single component perpendicular to the plane and with a magnitude $|\nabla \times \mathbf{V}|$. Then, it follows that:

$$-\Gamma = 2\pi K = \iint_{\hat{S}} (\nabla \times \mathbf{V}) \cdot \mathbf{n} d\hat{S} = \iint_{\hat{S}} (|\nabla \times \mathbf{V}|) \mathbf{k} \cdot \mathbf{k} d\hat{S} = \iint_{\hat{S}} |\nabla \times \mathbf{V}| d\hat{S}$$

Because within the double integral, the dot product $\mathbf{k} \cdot \mathbf{k}$ is unity.

Consider now a small circle enclosing the origin. The circulation on the path enclosing this circle is unchanged because for this flow, Γ does not depend on r. If the area is small, then the magnitude $|\nabla \times \mathbf{V}|$ may be assumed to be an average value over the area, and this constant quantity may be taken through the surface integral. Thus, the previous equation becomes:

$$-\Gamma = 2\pi K = \iint_{\hat{S}} |\nabla \times \mathbf{V}| d\hat{S} = |\nabla \times \mathbf{V}| \iint_{\hat{S}} d\hat{S} = |\nabla \times \mathbf{V}| (\pi r^2).$$

From this, it follows that:

$$|\nabla \times \mathbf{V}| = \frac{2K}{r^2},$$

which indicates that as $r \to 0$, the curl of the velocity vector (i.e., the vorticity) becomes infinite.

The conclusion drawn from this application of Stokes' Theorem is that a vortex is irrotational everywhere except at the origin, where it is rotational and, in fact, the vorticity is infinite. Note that for any closed curve that *does not enclose* the origin, the circulation is zero. For any closed curve that *does enclose* the origin (regardless of shape), the circulation is a constant.

It is demonstrated in Example 4.3 and Eq. 4.33 that for a point vortex:

$$u_r = 0 \quad \text{and} \quad u_\theta = \frac{K}{r} = -\frac{\Gamma}{2\pi r}.$$

Then:

$$\begin{cases} u_\theta = -\dfrac{\partial \psi}{\partial r} = -\dfrac{\Gamma}{2\pi r} \Rightarrow \psi = \dfrac{\Gamma}{2\pi} \ln r + f(\theta) + c_1 \\ u_r = \dfrac{1}{r}\dfrac{\partial \psi}{\partial \theta} = 0 \Rightarrow \psi = g(r) + c_2, \end{cases}$$

where c_1 and c_2 are arbitrary constants. Comparing, $f(\theta) = 0$ and the constant of integration may be set equal to zero. A parallel development yields the velocity potential.

Thus, in polar coordinates for a point vortex at the origin:

$$\psi_V = \frac{\Gamma}{2\pi} \ln r$$

$$\phi_V = -\frac{\Gamma}{2\pi} \theta,$$

(4.34)

whereas in Cartesian coordinates:

$$\psi_v = \frac{\Gamma}{2\pi} \ln\left(\sqrt{x^2 + y^2}\right)$$

$$\phi_v = -\frac{\Gamma}{2\pi} \tan^{-1}\left(\frac{y}{x}\right),$$

(4.35)

where the subscript v denotes a vortex.

The (constant) circulation, Γ, around the point vortex is termed the *strength* of the vortex because the tangential velocity is directly proportional to this arbitrary constant. Again, substitution demonstrates that Eqs. 4.34 and 4.35 are solutions to the Laplace's Equation.

Source–Sink Pair

The doublet solution is generated by a superposition of a source and a sink. It is such a basic element in the superposition of complex flows that here, it is considered a fourth-elementary flow.

Consider a source–sink pair of equal strength placed a distance h apart along the x-axis, as shown in Fig. 4.11. For convenience, assume that the left singularity (i.e., the source) is located at the origin of coordinates.

At a Point P, which is arbitrarily located, $\psi_1 = \frac{\Lambda}{2\pi}\theta_1$ and $\psi_2 = -\frac{\Lambda}{2\pi}\theta_2$ by applying Eq. 4.31. Superposing (adding) these two stream functions results in a new stream function that also satisfies the Laplace's Equation and represents a new, more complicated flow field. Thus,

$$\psi = \psi_1 + \psi_2 = \frac{\Lambda}{2\pi}(\theta_1 - \theta_2) = \frac{\Lambda}{2\pi}(-\Delta\theta)$$

(4.36)

because $\theta_1 = \pi - \Delta\theta$ and $\theta_2 = \pi - (\pi - \theta_1 - \Delta\theta)$.

Doublet (or Dipole)

A *doublet* (or *dipole*) is defined as a source–sink pair for which the separation distance $h \rightarrow 0$, whereas the absolute value of the strength, Λ, increases such that the

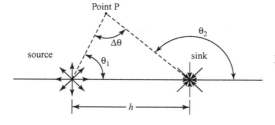

Figure 4.11. Source–sink pair.

product $\Omega = h\Lambda$ remains constant. What this means is that the strength becomes large and the separation distance becomes infinitesimal. Thus, by definition:

$$\psi_D = \lim_{\substack{h \to 0 \\ \Omega = \text{const}}} \left\{ -\frac{\Lambda}{2\pi} d\theta \right\}.$$

Now, referring to Fig. 4.12, consider the Point P to be described by polar coordinates (r,θ) with the origin at the source. Drop a perpendicular A-B of length ℓ onto r from the sink at Point A, as shown in Fig. 4.12. From the geometry, $\ell = h \sin\theta$ and $m = r - h \cos\theta$. Now, in the limit as $h \to 0$, $d\theta \to 0$ also. Then, for small angles:

$$\tan d\theta \cong d\theta = \frac{\ell}{m} = \frac{h\sin\theta}{r - h\cos\theta}.$$

Using this in the expression for the doublet stream function defined previously,

$$\psi_D = \lim_{\substack{h \to 0 \\ \Omega = \text{const}}} \left[-\frac{\Lambda}{2\pi} \frac{h\sin\theta}{(r - h\cos\theta)} \right].$$

However, (Λh) is a constant and, as h goes to zero, the term $h\cos\theta$ in the denominator becomes negligible compared to r; that is, the bracket in the denominator has the form $(1 - \varepsilon)$, where $\varepsilon << 1$. Thus, the stream function for the doublet becomes:

$$\psi_D = -\frac{\Omega\sin\theta}{2\pi r} \quad \text{or}$$

$$\psi_D = -\frac{\Omega y}{2\pi(x^2 + y^2)}, \tag{4.37}$$

where $\Omega = \Lambda h$. To determine the velocity potential for a doublet, we use the definitions and Eq. 4.37; that is, we write:

$$u_\theta = -\frac{\partial\psi}{\partial r} = -\frac{\Omega}{2\pi}\frac{\sin\theta}{r^2} = \frac{1}{r}\frac{\partial\phi}{\partial r} \Rightarrow \phi = \int -\frac{\Omega}{2\pi}\frac{\sin\theta}{r}d\theta = \frac{\Omega}{2\pi}\frac{\cos\theta}{r} + f(r) + \text{constant}$$

and

$$u_r = \frac{1}{r}\frac{\partial\psi}{\partial\theta} = \frac{-\Omega}{2\pi}\frac{\cos\theta}{r^2} = \frac{\partial\phi}{\partial r} \Rightarrow \phi = \frac{\Omega}{2\pi}\frac{\cos\theta}{r} + g(\theta) + \text{constant}.$$

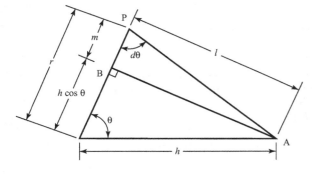

Figure 4.12. Source–sink pair geometry.

Comparing the two expressions for the velocity potential, $f(r) = g(\theta) = $ constant. Setting the potential function to be zero at $\theta = \pi/2$ makes the constant of integration zero, so that:

$$\phi_D = \frac{\Omega}{2\pi} \frac{\cos\theta}{r} \tag{4.38}$$

$$\text{and} \quad \phi_D = \frac{\Omega}{2\pi} \frac{x}{x^2 + y^2}$$

for polar and rectangular coordinates, respectively. The subscript D denotes a doublet.

Again, Eqs. 4.37 and 4.38 can be shown to be solutions of Laplace's Equation by direct substitution. The flow pattern generated by the doublet at the origin, with the source and sink placed on the x-axis, is shown in Fig. 4.13. The flow direction is established by the source being placed on the left side and the sink on the right side.

If the source and sink were located initially on the y-axis instead of the x-axis, the circular streamlines would have been symmetric about the horizontal axis instead of the vertical axis, as depicted in Fig. 4.13, and the axis of the doublet then would be vertical. The doublet shown in the figure is said to have a horizontal axis, meaning that the original source and sink were located on the x-axis.

Four elementary solutions now have been developed and they are collected and tabulated in Table 4.1. It is assumed in this discussion that the source, vortex, and doublet each are located at the origin of coordinates. If it is desired to locate any of them away from the origin, the appropriate expression for the stream function or the velocity potential can be obtained readily from the equations given herein by a simple shifting (i.e., linear transformation) of the appropriate coordinates. It remains to superimpose combinations of these simple Laplace's Equation solutions to produce solutions for more complicated and interesting flow fields. We also are looking for a way to address the singular points that arise in the elementary solutions. Clearly, these cannot correspond to physically real conditions because, for example, there cannot be point sources with infinite velocity or vortex points with infinite strength.

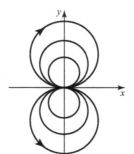

Figure 4.13. The doublet (dipole) with vertical axis of symmetry.

Table 4.1. *Elementary solutions of Laplace's Equation*

	φ		ψ		u	v	V_r	V_θ
	Cartesian	Polar	Cartesian	Polar				
Freestream at α	$V_\infty(x\cos\alpha + y\sin\alpha)$	$V_\infty r\cos(\theta-\alpha)$	$V_\infty(y\cos\alpha + x\sin\alpha)$	$V_\infty r\sin(\theta-\alpha)$	$V_\infty\cos\alpha$	$V_\infty\sin\alpha$	$V_\infty\cos(\theta-\alpha)$	$V_\infty\sin(\alpha-\theta)$
Source–sink	$\dfrac{\pm\Lambda}{4\pi}\ln(x^2+y^2)$	$\dfrac{\pm\Lambda}{2\pi}\ln(r)$	$\dfrac{\pm\Lambda}{2\pi}\tan^{-1}\left(\dfrac{y}{x}\right)$	$\dfrac{\pm\Lambda}{2\pi}\theta$	$\dfrac{\pm\Lambda}{2\pi}\dfrac{x}{(x^2+y^2)}$	$\dfrac{\pm\Lambda}{2\pi}\dfrac{y}{(x^2+y^2)}$	$\dfrac{\pm\Lambda}{2\pi r}$	0
Doublet	$\dfrac{\kappa}{2\pi}\dfrac{x}{(x^2+y^2)}$	$\dfrac{\kappa}{2\pi}\dfrac{\cos\theta}{r}$	$\dfrac{\kappa}{2\pi}\dfrac{y}{(x^2+y^2)}$	$\dfrac{\kappa}{2\pi}\dfrac{\sin\theta}{r}$	$-\dfrac{\kappa}{2\pi}\dfrac{y^2-x^2}{(x^2+y^2)^2}$	$-\dfrac{\kappa}{2\pi}\dfrac{2xy}{(x^2+y^2)^2}$	$-\dfrac{\kappa}{2\pi}\dfrac{\cos\theta}{r^2}$	$-\dfrac{\kappa}{2\pi}\dfrac{\sin\theta}{r^2}$
Vortex	$-\dfrac{\Gamma}{2\pi}\tan^{-1}\left(\dfrac{y}{x}\right)$	$-\dfrac{\Gamma}{2\pi}\theta$	$\dfrac{\Gamma}{4\pi}\ln(x^2+y^2)$	$\dfrac{\Gamma}{2\pi}\ln(r)$	$\dfrac{\Gamma}{2\pi}\dfrac{y}{(x^2+y^2)}$	$-\dfrac{\Gamma}{2\pi}\dfrac{x}{(x^2+y^2)}$	0	$-\dfrac{\Gamma}{2\pi r}$

4.6 Superposition of Elementary Solutions

Although a superposition can be carried out using either the elementary velocity potentials or stream functions, the superpositions discussed here are in terms of the stream function. The advantage of using the stream function is that the streamlines of the complicated flow then are generated automatically; recall that lines of ψ = constant are streamlines. In principle, for any flow generated by superposition, we can choose different constant values for the stream function; find pairs of values of (x,y) or (r,θ), which make the superposed stream function equal to this constant; plot these coordinate pairs; and then find the streamlines by "joining the dots" (e.g., with a contour-plotting software package). In some cases, the streamline pattern may be found analytically from the superposition expression.

Uniform Flow Plus Source at Origin

This superposition uses the two solutions, Eqs. 4.29 and 4.32. Namely:

$$\psi = \psi_{UF} + \psi_S = V_\infty y + \frac{\Lambda}{2\pi}\tan^{-1}\frac{y}{x}. \tag{4.39}$$

For an illustrative flow problem, V_∞ and Λ are constants that may be chosen arbitrarily. The streamlines generated by joining the dots (as previously explained) can be examined by running the software Program **PSI**.

The following comments are pertinent to running Program **PSI**:

1. The flow is symmetrical about the x-axis. One streamline with a value $\psi = \frac{\Lambda}{2}$ passes through a stagnation point located on the x-axis. This may be seen by differentiating Eq. 4.39 to find u, v, and then setting these two velocity components equal to zero (because at a stagnation point, $V = 0$) and solving for (x,y). The result is:

$$x = -\frac{\Lambda}{2\pi V_\infty}, \; y = 0.$$

   (The student should verify this result.) The value of this x,y pair, when substituted into Eq. 4.39, yields the value of ψ on that stagnation streamline.

2. Recall that any streamline can be "cross-hatched" mentally to represent the surface of a solid body. The streamline passing through the stagnation point thus may be thought of as the surface of an open-ended (i.e., semi-infinite) body opening to the right. Vary the source strength, Λ, and the freestream velocity values in the program to verify that the body shape changes as anticipated. Note that the location of the stagnation point along the x-axis changes as these parameters are varied. This is to be anticipated because the stagnation point occurs when the oncoming freestream flow is just balanced by the opposing source-flow streamline directed upstream along the x-axis.

3. Notice that the point source is located at the origin and, hence, inside the open-ended body. Thus, the fact that the velocity is infinite at the source is of no concern because the source is not within the external flow field around the body.

4. One of the other streamlines approaching the body in the second or third quadrants may be "cross-hatched." This streamline describes the flow along a plain and over a hill of continually increasing elevation.
5. Note that although the body shape (or hill) may be varied by varying the flow parameters, as in comment (2), the body shape comes out of the solution. Thus, we cannot easily specify a certain body shape in advance—"You get what you get."

Uniform Flow Plus Source–Sink Pair

As surmised from the preceding section, the superposition of a sink located downstream of the source discussed yields the flow around a closed body—that is, providing that the source and sink are of equal strength (i.e., net strength of zero indicating that the mass emanating from the source is absorbed by the sink). Thus, using Eqs. 4.30 and 4.31:

$$\psi = \psi_{UF} + \psi_{S1} - \psi_{S2} = V_\infty r \sin\theta + \frac{\Lambda}{2\pi}\theta_1 - \frac{\Lambda}{2\pi}\theta_2, \qquad (4.40)$$

where θ is the polar angle to the point in question and θ_1 and θ_2 are the angles between the positive x axis and straight lines joining the source–sink (respectively) and the point in question. By holding V_∞ and Λ fixed and assigning different constant values to the stream function in Eq. 4.40, we can find different combinations of r, θ, θ_1, and θ_2 that satisfy this equation (i.e., θ_1 and θ_2 can be expressed in terms of r and θ for a specified spacing between the source and sink). Then, the streamlines may be found by plotting curves, which are generated by varying the angle θ and using Eq. 4.40 to solve for the radius r for a particular streamline corresponding to a constant value of ψ. Each value of ψ provides another streamline. The result of joining the dots is demonstrated in Program **PSI**.

RUNNING PROGRAM PSI

1. One streamline represents a closed body, as expected. Confirm that the body shape and size depends on the values of freestream velocity, source–sink strength, and source–sink spacing.
2. The body has two stagnation points, one at the front and the other at the rear. Their locations may be found by solving for u and v and setting both velocity components equal to zero. From this, the value of the stream function on this stagnation streamline follows.
3. The velocity singularities are located within the solid body.
4. The body shape looks like an ellipse but it is not; it is called a Rankine Ovoid.*
5. The student is encouraged to run Program PSI to obtain the streamline patterns when the source and sink strengths do not sum to zero.

---

* Named after W. Rankine, a Scottish engineer who first solved this problem in the 1800s.

6. To modify the shape of the oval, additional sources and sinks may be distributed
   along the x axis with the condition that the net strength must remain zero. This
   suggests a way to obtain the flow field for an ellipse by adding more singularities. The difficulty with this approach is that there is no way to know in advance
   where to put the singularities and what their strengths should be. An alternative
   is to distribute source–sink pairs along the x-axis and determine their strengths
   by demanding that the stagnation streamline (i.e., closed body) be an ellipse. A
   better approach to this problem of determining the flow around a specified (and
   general) body shape is discussed later.

Uniform Flow Plus Doublet Flow Around a Cylinder

Both the Cartesian- and polar-coordinate expressions for the superposition are convenient here, so we use Eqs. 4.29, 4.30, and 4.38 to form:

$$\psi = \psi_{UF} + \psi_D = V_\infty y - \frac{\Omega}{2\pi}\left(\frac{y}{x^2+y^2}\right) = V_\infty r\sin\theta - \frac{\Omega}{2\pi}\frac{\sin\theta}{r}. \tag{4.41}$$

The Cartesian-coordinate expression may be written as:

$$\psi = \psi_{UF} + \psi_D = V_\infty y\left[1 - \frac{\Omega/2\pi V_\infty}{x^2+y^2}\right]. \tag{4.42}$$

As before, we try to find the shape of the streamlines ψ = constant. Here, the body
shape may be found analytically. We ask the questions, "What is the shape of the
streamline $\psi = 0$?" From Eq. 4.42, the stream function is zero when:

$y = 0$; that is, along the x-axis
or when

$$\left[1 - \frac{\Omega/2\pi V_\infty}{x^2+y^2}\right] = 0 \Rightarrow x^2 + y^2 = \frac{\Omega}{2\pi V_\infty} = R^2,$$

which is a circle of radius R having a center at the origin.

Recall that when the stream function representing the freestream was determined, the constant of integration was made zero by setting the stream function zero
along the x-axis. Here, the x-axis is the streamline $y = 0$, but there is a singularity at
$x = 0, y = 0$. That is, the stream function at the origin is indeterminate. Notice that
this singularity presents no difficulty because the origin is *inside* the zero-streamline
body. In particular, the zero-streamline body is a circle (i.e., a right circular cylinder
in cross section) of radius R. The other streamlines outside the cylinder may be
found by joining the dots as before, and the results are shown in Program **PSI**. Note
that when running this program:

1. The radius of the cylinder and the resulting flow field may be changed by varying
   the freestream velocity and the strength of the doublet.
2. The flow field is symmetrical with respect to both the x- and y-axes. This indicates that there is no net force on the cylinder in either the lift (i.e., vertical)

or the drag (i.e., streamwise) direction. The fact that there is no lift should not be surprising because there is no asymmetry in either of the superposed stream functions. In fact, introducing asymmetry is the next step. The result that there is no drag contradicts experience until it is realized that the flow model is inviscid, so that there are no boundary-layer effects present and, in particular, there is no large separated region (i.e., wake) behind the cylinder. In fact, as shown next, there are two stagnation points on the cylinder, both on the x-axis at +/- R. Clearly, this solution is of no value in describing the flow over the downstream side of a cylinder in a practical problem. However, the theory satisfactorily predicts the pressure distribution on the upstream side of a cylinder (or sphere), where the boundary layer is thin and remains attached. The solution for a cylinder also is important in the development of inviscid-airfoil theory (see Chapter 5).

3. Note the distribution of static pressure around the cylinder. The pressure distribution on the surface is symmetrical with respect to both the x and y axes; hence, there is no unbalanced force on the cylinder. The distribution-of-pressure coefficient is shown as well, which does not provide additional information but rather is intended to familiarize the student with such data presentations.

The polar-coordinate notation is reviewed in Fig. 4.14 to avoid confusion about signs. Note that the arrows on the velocity components indicate the positive direction of the component and that the polar angle is measured from the positive x-axis, which here is the downstream direction.

The velocity components in polar coordinates are obtained by differentiating Eq. 4.41:

$$u_r = \frac{1}{r}\frac{\partial \psi}{\partial \theta} = V_\infty \cos\theta \left[1 - \frac{R^2}{r^2} \right]$$

$$u_\theta = -\frac{\partial \psi}{\partial r} = -V_\infty \sin\theta \left[1 + \frac{R^2}{r^2} \right].$$

(4.43)

From this, it follows that on the surface of the cylinder $(r = R)$, $u_r = 0$. This is as it should be because there cannot be a flow component normal to a streamline. The fact that $u_r = 0$ at the body surface $r = R$ also indicates that the surface-boundary condition is satisfied. Also, from Eq. 4.43, $u_\theta = -2V_\infty \sin\theta$ at $r = R$. Hence, there is a stagnation point on the body surface at $\theta = \pi$ (i.e., the rear stagnation point on the

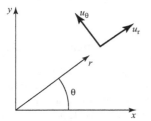

Figure 4.14. Polar-coordinate notation.

x-axis) and at $\theta = 0$ (i.e., the front stagnation point on the x-axis). Upstream of the body along the x-axis, $\theta = \pi$ and $u_\theta = -2V_\infty \sin\theta = 0$ and $u_r < V_\infty$ because $R < r$. By substituting different values for r, Program **PSI** shows how u_r begins to decrease about 10 cylinder radius upstream of the stagnation point and then decreases to zero at an increasing rate as r approaches the cylinder radius. The velocity component u_θ at the top of the cylinder ($\theta = \pi/2$) exhibits similar behavior with r for $r \geq R$. Thus, the disturbance due to even this blunt body dies out rather quickly in this low-speed flow. Finally, Eq. 4.43 shows that far from the body, $r \to \infty$ and $V = V_\infty$, which satisfies the boundary condition.

At $\theta = \pi/2$ on the surface of the cylinder, the tangential velocity has the (maximum) value $u_\theta = -2V_\infty$ (the minus sign indicates that the velocity is directed downstream). Finally, from the definition of the pressure coefficient for an incompressible, inviscid flow, on the surface of the cylinder:

$$Cp = 1 - \frac{V^2}{V_\infty^2} = 1 - 4\sin^2\theta. \tag{4.44}$$

Physically, a fluid particle approaching the cylinder along the x-axis decelerates to zero velocity at the upstream stagnation point. The particle next accelerates along the cylinder surface (recall that the flow is inviscid) until it reaches a velocity twice the freestream value at the top of the cylinder; it then decelerates along the rear half of the cylinder until it reaches a zero velocity at the rear stagnation point. The pressure coefficient has a value of +1.0 at the stagnation points and -3.0 at the top and bottom of the cylinder. Integration of the pressure (or pressure coefficient) distribution on the surface of the cylinder yields zero lift and drag, a conclusion argued previously from symmetry.

EXAMPLE 4.11 *Given*: Consider the steady, incompressible, inviscid flow around a cylinder with zero lift.

Required: Predict the points on the upstream side of the cylinder where the surface static pressure is equal to the static pressure of the oncoming freestream.

Approach: The pressure-coefficient definition and the equation for the variation of pressure coefficient around the cylinder is used.

Solution: From Eqs. 4.27 and 4.44, with p set equal to p_∞ at the point in question:

$$C_p = \frac{(p - p_\infty)}{\frac{1}{2}\rho V_\infty^2} = 0 = 1 - 4\sin^2\theta \Rightarrow \theta = \pm 30°, \pm 150°.$$

Because the angle is the polar angle measured from the positive x-axis, the required points are at $\theta = 150°$ and $\theta = 210°$ on the upstream side of the cylinder.

Appraisal: The flow streamline along the x-axis stagnates on the cylinder at $\theta = 180°$. The flow on the surface of the cylinder (i.e., along that same streamline) then accelerates and, from the Bernoulli Equation, as the velocity (i.e., dynamic pressure) increases, the static pressure must decrease. At a distance along the surface of only 1/12 of the cylinder diameter, the pressure on the surface decreased from a maximum of freestream stagnation pressure to a value equal to the freestream static pressure. Continuing to the top of the cylinder, the velocity increases to a maximum and the static pressure on the cylinder reaches a minimum. The flow field around the cylinder is symmetric, with the freestream static pressure occurring on the surface of the cylinder at mirror-image points in both the x- and y-axes; only the locations on the upstream side of the cylinder were required.

Uniform Flow Plus Doublet Plus Vortex: Flow Around a Lifting Cylinder

The superposition of a uniform flow and a doublet yielded a useful body shape (i.e., a cylinder), but the flow field was symmetrical so that there was no net force on the cylinder. The addition of a vortex in the superposition creates an asymmetry in the flow. In particular, if the vortex is at the origin and has a clockwise sense, then it adds to the local velocity over the top half of the cylinder and subtracts from it over the bottom half. This asymmetry in the velocity field leads to an asymmetry in static pressure. From the Bernoulli Equation, the lower values of static pressure are on top of the cylinder and the higher values of static pressure are on the bottom of the cylinder. When integrated, this pressure asymmetry leads to a net force upward— that is, to a positive lift. Because no asymmetry is introduced about the y-axis by the superposition of the vortex at the origin, there is no unbalanced force in the x-direction. Thus, the drag of the cylinder is still zero.

Recall that in the development of Eq. 4.34 for the vortex, the constant of integration in the stream-function expression arbitrarily was set equal to zero for convenience. However, this constant may have any value because the velocity field is obtained by differentiation and the constant disappears. In this superposition to obtain the flow over a lifting cylinder, we write the constant of integration as the constant:

$$-\frac{\Gamma}{2\pi}\ln R$$

instead of setting the constant equal to zero.

The stream function for the vortex to replace Eq. 4.34 then becomes:

$$\psi_V = \frac{\Gamma}{2\pi}\ln r + \text{const} = \frac{\Gamma}{2\pi}\ln r - \frac{\Gamma}{2\pi}\ln R = \frac{\Gamma}{2\pi}\ell n \frac{r}{R}.$$

The constant of integration was changed so that at $r = R$, $\psi_V = 0$. Thus, the zero streamline for the vortex now is on the circle $r = R$, just as it is for the superposition of a doublet with a uniform flow. Then,

$$\psi = \psi_{UF} + \psi_D + \psi_V \qquad (4.45)$$

$$\psi = V_\infty r\sin\theta\left[1 - \frac{R^2}{r^2}\right] + \frac{\Gamma}{2\pi}\ln\frac{r}{R}.$$

The student should verify by substitution that the stream-function expression given by Eq. 4.45 is indeed a solution to the Laplace's Equation. The streamlines resulting from the superposition in Eq. 4.41 are seen by running Program **PSI**. Notice with respect to this program that:

1. There are two parameters to vary: the freestream velocity and the strength of the vortex. Note what happens when these are varied. In particular, hold the freestream velocity constant and vary the vortex strength.
2. Although the streamline $\psi = 0$ still lies on the circle $r = R$, it no longer lies along the x-axis away from the cylinder. In particular, from Eq. 4.45, along the streamline $\psi = 0$:

$$\sin\theta = -\frac{\frac{\Gamma}{2\pi}\ln\frac{r}{R}}{V_\infty r\left[1-\frac{R^2}{r^2}\right]}.$$

Now, for $\Gamma > 0$ (i.e., for a clockwise vortex according to the sign convention established previously), this expression states that if r is greater than R, then $\sin\theta < 0$, which places the zero streamline in the 3rd and 4th quadrants and not along the x-axis, as was the case for the nonlifting cylinder. Thus, as expected, the flow field is not symmetrical about the x-axis.

Next, we examine the velocity field:

$$u_r = \frac{1}{r}\frac{\partial\psi}{\partial\theta} = V_\infty\cos\theta\left[1-\frac{R^2}{r^2}\right] \qquad (4.46)$$

$$u_\theta = -\frac{\partial\psi}{\partial\theta} = -V_\infty\sin\theta\left[1+\frac{R^2}{r^2}\right]-\frac{\Gamma}{2\pi r}.$$

Eq. 4.46 shows that the radial component of velocity is zero on the surface of the cylinder, $r = R$, so that the "no-flow-through" boundary condition still is satisfied. The boundary condition far from the body (i.e., for very large r) also is satisfied because far away from the body:

$$u_r \to V_\infty\cos\theta \text{ and } u_\theta \to V_\infty\sin\theta \text{ so that } u_r^2 + u_\theta^2 \to V_\infty^2$$

and the disturbance due to the body dies out. Finally, if we go back to the velocity-component equations for the three fundamental solutions that comprise this superposition, it is shown that the superposition represented by the new stream function in Eq. 4.45 also implies an addition of the three constituent velocity components at any point (Eq. 4.46). This fact can be useful.

Next, examine the asymmetry of the flow by locating the stagnation points on the cylinder. Again, refer to the computer solution represented by Program **PSI** and vary the vortex strength. Recall the condition for a stagnation point to exist— namely, that $u_r = u_\theta = 0$. From Eq. 4.46, $u_r = 0$ when $r = R$ (the cylinder surface) or when $\theta = \pi/2$ or $3\pi/2$. Examine each of the following three possibilities:

1. On $r = R, u_r = 0$. If $u_\theta = 0$ as well, then:

$$\sin\theta = -\frac{\Gamma}{4\pi RV_\infty}.$$

Because the sine function is bounded between ± 1.0, this states that $\Gamma \le |4\pi RV|_\infty$. The zero value of circulation corresponds to the stagnation points on the x-axis (i.e., nonlifting case). Confirm this result by running the program.

2. If $u_r = 0$ by virtue of $\theta = \pi/2$ and we demand that $u_\theta = 0$ there as well, then from Eq. 4.46, it follows that for a positive r, the circulation must be negative. Ignore this choice because it was shown that positive (i.e., clockwise) circulation provides positive lift.

3. If $u_r = 0$ because $\theta = 3\pi/2$, then if $u_\theta = 0$ as well, it follows from Eq. 4.46 that:

$$V_\infty\left[1 + \frac{R^2}{r^2}\right] - \frac{\Gamma}{2\pi r} = 0.$$

Solving this quadratic equation for r/R:

$$\frac{r}{R} = \frac{\Gamma}{4\pi RV_\infty} \pm \sqrt{\left(\frac{\Gamma}{4\pi RV_\infty}\right)^2 - 1},$$

which is valid only if $\dfrac{\Gamma}{4\pi RV_\infty} \ge 1.0$.

Now, recall from Condition (1) that if the stagnation point is on the surface of the cylinder, then $\Gamma \le |4\pi RV|_\infty$. Thus, the stagnation point in Condition (3) corresponds to a stagnation point along the negative y-axis (directly below the cylinder) and is either on or away from the surface (i.e., in the external flow field). We observe this by running the program.

Finally, we run Program **PSI** and take note of the pressure (and pressure-coefficient) distribution around the cylinder. The distributions are not symmetrical with respect to the x-axis and there is a resulting unbalanced (lift) force. The magnitude of this force is determined later by integrating the pressure distribution. The pressure-coefficient information in this program is presented with positive values of the coefficient along the positive ordinate to emphasize the physical situation; however, this is *not* the usual format for presenting airfoil pressure-coefficient information.

Does this superposition of a uniform flow, a doublet, and a vortex correspond to a physically realistic flow situation? As in the case of the nonlifting cylinder, the drag of the lifting cylinder is zero by virtue of the symmetry of the flow field about the y-axis. In this respect, the superposition is not realistic. However, it is physically possible to generate an unbalanced force on a symmetrical geometric object such as a cylinder. To understand this, the role of viscosity must be recognized. If we set up an experiment and rotate a cylinder in a clockwise direction in a uniform flow, then the large viscous-shearing action at the surface of this cylinder diffuses into the flow and tends to "pull" the flow along on the upper side and "retard" the flow on the lower side. (To visualize this, imagine rotating a cylinder in a flow of oil.) This viscous action creates a flow asymmetry that manifests as an unbalanced force on the cylinder. There is a circulation set up around the body by virtue of the viscous shear, or vorticity, at the surface; this circulation is represented by the vortex in the inviscid model. Notice that the presence of circulation does not imply that there are any circular streamlines around the cylinder.

As we would expect, the magnitude of the unbalanced force on the spinning cylinder is proportional to the rate of spin, which is an external input. There is no physical mechanism here to specify the magnitude of the spin and, hence, of the circulation. It is shown later that in the case of a sharp-edged body such as an airfoil, there is a physical mechanism present that selects a unique value of circulation for each angle of attack. Again, to explain the presence of this mechanism, it is necessary to appeal to experiment and to the role of viscosity.

Next, the force on a lifting cylinder is evaluated by integrating the surface pressure over the cylinder surface. The force in the lift (y) direction is sought. The drag (x)-direction force can be argued to be zero by symmetry or proven to be zero by integrating the net surface-pressure force in the x-direction.

From Fig. 4.15, dF_n is the normal force due to a pressure p (force per unit area) acting on an element of surface area $Rd\theta$ of the two-dimensional circular cylinder. Notice that the length of the element normal to the x-y plane is taken to be unity; the calculation then gives the force per unit length of the cylinder. Resolving this force into components in the coordinate directions:

$$dF_x = - dF_n\cos\theta = -pR\cos\theta d\theta.$$
$$dF_y = - dF_n\sin\theta = -pR\sin\theta d\theta.$$

Now, we sum these differential component forces by integration over the entire cylinder, recalling that $V = V(\theta)$ is known so that the pressure can be expressed from the Bernoulli Equation as $p = p(\theta)$. The evaluation of F_x is left to the student as an exercise. F_y is found to be:

$$F_y = \int_0^{2\pi} dF_y = -\int_0^{2\pi} pR \sin\theta d\theta = -2R \int_{-\pi/2}^{\pi/2} p \sin\theta d\theta,$$

where the limits of integration are replaced by observing the symmetry about the y-axis.

From the Bernoulli Equation, $p = p_0 - 1/2\rho V^2$. Thus,

$$F_y = -2R \int_{\pi/2}^{\pi/2}\left[p_0 - \frac{1}{2}\rho V^2 \right] \sin\theta d\theta = 2Rp_0 \int_{-\pi/2}^{\pi/2} \sin\theta d\theta + \rho R \int_{-\pi/2}^{\pi/2} V^2 \sin\theta d\theta.$$

The first integral is zero by virtue of the integration of the sine function between the limits. Now, on the surface of a cylinder with circulation:

$$V = u_\theta = -2V_\infty \sin\theta - \frac{\Gamma}{2\pi R},$$

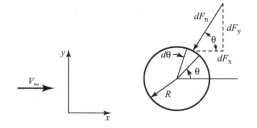

Figure 4.15. Force on an element of a circular cylinder due to a pressure force p (per unit area).

because $u_r = 0$. Thus,

$$F_y = \rho R \int_{-\pi/2}^{\pi/2} \left[-2V_\infty \sin\theta - \frac{\Gamma}{2\pi R} \right]^2 \sin\theta\, d\theta$$

$$F_y = \rho R \int_{-\pi/2}^{\pi/2} \left[4V_\infty \sin^3\theta + 4\frac{\Gamma}{2\pi r}V_\infty \sin^2\theta + \left(\frac{\Gamma}{2\pi R}\right)^2 \sin\theta \right] d\theta.$$

$$\text{(A)} \qquad\qquad \text{(B)} \qquad\qquad \text{(C)}$$

The circulation and the freestream velocity are constants independent of θ. Hence, terms (A) and (C) integrate to zero between the limits, and the integral of $\sin^2\theta$ in term (B) is $\pi/2$. Therefore:

$$F_y = \rho R\left(\frac{4\Gamma}{2\pi R}\right)V_\infty\left(\frac{\pi}{2}\right)$$

or

$$F_y = L' = \rho V_\infty \Gamma, \qquad\qquad (4.47)$$

where L' is the lift on the cylinder (i.e., the force acting perpendicular to the freestream direction) per unit length of cylinder. The "prime" here denotes *per unit length* or *per unit span*.

Evaluating F_y in Eq. 4.47 entails using superposition to find a stream function, differentiating this function to determine velocity components, using the Bernoulli Equation to determine the pressure distribution on the cylinder, and integrating the pressure distribution to find the lift force. The final result is interesting; it shows that the lift per unit span is proportional to the circulation about the cylinder. If this were a general result, Eq. 4.47 indicates that we can evaluate the lift on a body by finding the circulation about the body; it is not necessary first to find the pressure distribution and then to integrate. This is an attractive idea because it is often easier to find the circulation about a body than it is to find the pressure distribution on the body surface and then integrate (e.g., in the thin-airfoil theory in Chapter 5). As shown in the next section, this is indeed a general result and Eq. 4.47 holds for any right cylinder (e.g., an airfoil), not only a right-circular cylinder.

Note that in Eq. 4.47, the magnitude of the circulation is arbitrary. There is nothing inherent in the flow field that determines it. Thus, the lift on the cylinder can have any value depending on the magnitude of the circulation (i.e., the magnitude of the spin that is given to the cylinder in an experiment). As discussed later, if a body in a uniform flow has a sharp edge (e.g., an airfoil with a sharp trailing edge), then the resulting flow field specifies the magnitude of the circulation about the body for any body attitude. The resulting lift force that is generated has a unique value.

EXAMPLE 4.12 *Given:* A lifting cylinder of radius 2 feet is experiencing a lift force of 8 pounds per foot in a freestream with a velocity of 20 ft/s. Assume steady, incompressible, inviscid flow at standard conditions.

Required: (a) What is the circulation about this cylinder? (b) Where are the stagnation points located on the cylinder? (c) What is the maximum velocity on the surface of the cylinder? (d) What is the value of the pressure coefficient on the bottom of the cylinder?

Approach: Eqs. 4.46 and 4.47 yield the circulation strength and give velocity information. Eqs. 4.46 and 4.28 are needed for the pressure-coefficient part of the question.

Solution:

(a) From Eq. 4.47:

$$L' = 8.0 = \rho V_\infty \Gamma = (0.002378)(20)\Gamma \Rightarrow \Gamma = 168 \, \text{ft}^2/\text{s}$$

(b) On the cylinder surface, $u_r = 0$ and, at the stagnation points, u_θ also is zero. Then, from Eq. 4.46:

$$u_\theta = 0 = -V_\infty \sin\theta \left[1 + \frac{R^2}{R^2}\right] - \frac{\Gamma}{2\pi R} = -(20)\sin\theta[2] - \frac{168}{2\pi(2)}.$$

Solving for θ and recognizing that the stagnation points are in the third and fourth quadrants, $\theta = 199.5°$ and $\theta = 340.5°$.

(c) The maximum velocity is at the top of the cylinder. There, $\theta = \pi/2$ and Eq. 4.46 yields a value for the tangential velocity of -53.4 ft/s; the negative sign indicates flow in the downstream direction.

(d) At the bottom of the cylinder, $\theta = 3\pi/2$. From Eq. 4.46, $u_\theta = V = 26.6$ ft/s (the positive sign now indicates the downstream direction), and from the pressure coefficient expression, Eq. 4.28:

$$C_p = 1 - \frac{V^2}{V_\infty^2} = -0.77.$$

Appraisal: (a) Check the units of Eq. 4.47 and confirm that circulation has the units ft²/s. Verify this by realizing that circulation is the line integral (summation) of the product of a velocity and a line segment; hence, it should have the units [ft/s][ft].

(b) The locations of the stagnation points are as expected: They are symmetrical about the *y* axis.

(c) The velocity at the top of a nonlifting cylinder is twice the freestream value. Here, the velocity at that location is about 2.5 times the freestream value. This indicates the additive effect of the clockwise vortex and, from the Bernoulli Equation, implies that the static pressure at that point is less than it would be for the nonlifting case.

(d) The velocity at the bottom of the lifting cylinder (26.6 ft/s) is less than the value for the nonlifting cylinder (40 ft/s), indicating that at that point, the clockwise vortex is opposing the oncoming stream. However, note that this velocity is still greater than the oncoming freestream value (20 ft/s) so that the flow is retarded but not reversed. The same conclusion can be drawn from the negative sign on the value of the pressure coefficient at the bottom of the lifting cylinder. There is a local acceleration of the freestream flow at that point, but it is less of an acceleration than in the nonlifting case. When a stagnation point occurs on the bottom of the cylinder, the influence of the circulation is just sufficient to oppose the local flow in the downstream direction and bring it to rest.

4.7 The Kutta–Joukouski Theorem

It is shown in Section 4.7 that an integration of the pressure distribution on a lifting cylinder leads to the result that the lift per unit span is proportional to the circulation around that cylinder. The Kutta–Joukouski Theorem* states that for any body of arbitrary cross section (e.g., an airfoil), the lift per unit span $L' = \rho V_\infty \Gamma$, where Γ is taken around any closed path enclosing the body. The proof of the theorem is beyond the scope of this book. However, the theorem was demonstrated to be true for a right circular cylinder by integration of the static pressure acting on the surface of a body, and it is shown later to be true for an airfoil shape.

The importance of this theorem is that it provides an alternative way to calculate the lift force on a lifting body. Instead of calculating the velocity magnitude at a point on the body surface, then using the Bernoulli Equation to evaluate the pressure there, and finally integrating to determine the force, the theorem states that the lift force can be found simply by calculating the circulation around the body. As previously mentioned, it often is easier to calculate the circulation than it is to determine the pressure distribution. Of course, if the pressure magnitude at a point or the pressure distribution on the body surface is required, the theorem is of no help because it speaks only of the net force.

This theorem is the basis for the so-called circulation theory of lift. This is a mathematical method of calculating lift that convenient for many inviscid-flow problems. However, remember that the lift (and drag) on a body is *physically* generated by the pressure (and shear-stress) distribution over the surface. The oncoming flow adjusts to accommodate the presence of the lifting body and, in so doing, sets up a velocity and pressure field such that the circulation around the lifting body is nonzero.

To fix ideas regarding lift and circulation, imagine a two-dimensional wing installed at an angle of attack in a wind tunnel. Also imagine that there is a suitable instrument available that measures the flow velocity (i.e., magnitude and direction) at numerous points around the wing. The particular points of interest are located along a closed path in a vertical plane aligned with the oncoming stream. Make the measurement, form the vector-dot product $\mathbf{V} \cdot \mathbf{ds}$ at each measurement station, and sum around the closed path. This calculation of circulation yields a positive quantity that is equal to $L'/\rho V_\infty$. Thus, the lift (i.e., physically, the net pressure force acting upward on the wing) is exhibited as a circulation around the wing. Recall an analogy in Chapter 3 in which a measurement of drag was carried out by evaluating the momentum loss in the wake. There, the drag was due physically to the pressure and

---

* Named for M. Kutta, a German mathematician, and E. Joukouski, a Russian physicist, who independently established the theorem during the early 1900s.

shear forces acting on the body surface and the drag was exhibited as a momentum loss.

Remember that the presence of a circulation around a body does not imply any fluid particles rotating about it. It simply means that the flow above and below the lifting body is higher and lower average velocities than the zero-lift value, respectively.

4.8 The Kutta Condition

In the lifting-cylinder case discussed in Section 4.7, the value of the circulation generated in the spinning-cylinder experiment (and, hence, the lift on the cylinder) is arbitrary. The circulation depends on an external input: the rate of spin of the cylinder. Because the Kutta–Joukouski Theorem states that the lift on an airfoil is proportional to the circulation and because from experience the lift on a particular airfoil is unique for a given orientation, there must be a physical mechanism (not an external input) that specifies a unique value of the circulation about an airfoil. The flow condition to be satisfied that results in the generation of a unique value of the circulation about a lifting airfoil is called the Kutta condition. It is stated formally herein, but first the physical phenomena that establish this condition are explained by introducing viscosity into the flow model. When the flow phenomena are discussed, the viscosity switch is turned "off" again and the physical behavior observed in nature is suitably incorporated into the inviscid-flow model.

To visualize the following discussion, consider a thought experiment. A two-dimensional airfoil is set at a positive angle of attack in a viscous (real) medium at rest, as illustrated in Fig. 4.16.

The flow is started in motion to the right and rapidly brought up to a constant velocity. At the first instant of time, the flow next to the lower surface of the airfoil proceeds toward the sharp trailing edge and then around the trailing edge to the upper surface, as if there were no viscosity present. That is, the flow is essentially a potential

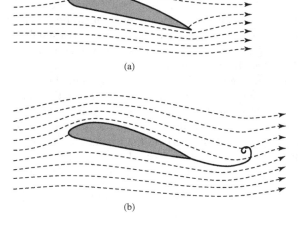

(a)

Figure 4.16. Airfoil started impulsively.

(b)

flow at the initial instant (Fig. 4.16a). As a consequence, this stream tube exhibits a sharp reversal of direction with zero radius of curvature right at the sharp trailing edge. This means that the velocity at the trailing edge at that instant is infinite. An infinite velocity is a physical impossibility. Because the viscous fluid cannot accommodate this zero-radius turn, it instantaneously separates at the trailing edge and rolls up into a vortex with a counterclockwise sense (Fig. 4.16b). As the freestream velocity continues to increase rapidly, approaching a constant value, the fluid at the trailing edge continues to separate and roll up so that the instantaneous vortex grows and becomes stronger. The flow field responds to this strengthening vortex (increasing vorticity) at the trailing edge by setting up a reaction effect (circulation) about the airfoil. As a consequence, the upper-surface stagnation point, which was instantaneously located well upstream of the trailing edge, moves rapidly toward the trailing edge. Finally, when the freestream velocity reaches a constant value, the flow along the airfoil upper surface smoothly leaves the trailing edge and the fluid particles comprising the so-called starting vortex roll up tightly and are swept downstream. This starting vortex need not be included in a steady-flow problem, where the flow is considered to have been going on for a long time. The reason is that the starting vortex is located far downstream of the airfoil in such a problem and its influence at the airfoil is negligible because the velocity field associated with a vortex varies inversely with the distance.

If the airfoil angle of attack is sharply increased in a constant-velocity flow, another vortex of the same sense described previously is formed and sheds downstream. If the angle of attack of the airfoil is suddenly decreased, a vortex of the opposite sense rolls up and sheds. The strength of the circulation around the airfoil likewise changes to a unique value as the airfoil angle of attack is changed. The experimentally observed fact, then, is that the viscous shear layer on the airfoil surface first rolls up at the trailing edge and then the flow field adjusts until this rollup no longer occurs. Each flow adjustment generates a unique value of circulation about the airfoil (see Section 4.10).

Formally, the Kutta condition states that a body with a sharp trailing edge generates a circulation of just sufficient strength that the flow smoothly leaves the trailing edge. The Kutta condition has a vital role in the inviscid thin-airfoil theory developed in Chapter 5 because it serves as a boundary condition.

Now, the statement that the flow smoothly leaves the trailing edge must be stated more carefully because there are two possibilities, as follows:

1. *Finite trailing-edge angle.*

In this case, the two velocity vectors at the trailing edge are parallel to the upper and lower surfaces. In the limit, right at the trailing edge, this implies that there are two flows going in two different directions at the same point, which is physically impossible. The situation is resolved if both velocities at the trailing edge are zero (i.e., then, they have no flow direction). Accordingly, the trailing edge must be a stagnation point and the circulation must be of precisely the right magnitude so as to move the stagnation point, which was initially on the upper surface of the lifting airfoil, downstream to the trailing edge.

2. *Cusped trailing edge.*

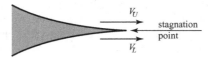

Here, there is no conflict in flow direction. However, the velocity magnitudes (upper and lower) must be identical because, otherwise, the Bernoulli Equation demands that there be a static-pressure difference between the upper and lower surface flows as they leave the trailing edge. However, between the two flows right at the trailing edge, there is a free surface (i.e., interface) that cannot support a pressure difference. The conclusion is that the two velocities at the trailing edge must be the same, and a circulation is set up around the airfoil so as to make this happen. A cusped trailing edge on an airfoil is not a practical configuration because it requires an infinitesimal thickness. However, airfoil designers strive to come as close to this ideal configuration as possible while recalling the practical limitations imposed by structural considerations.

4.9 The Starting Vortex: Kelvin's Theorem

The existence of circulation around an airfoil may be confirmed by using an inviscid-flow model and Kelvin's Theorem.* The argument is based on the experimental evidence of a starting vortex at the trailing edge of an airfoil.

Consider a large closed path, A-B-C-D, enclosing a fixed airfoil in a flow at rest. The circulation around this closed path is zero because the velocity is zero everywhere. Assume that the fluid particles comprising this path were marked in some way and then impulsively set the flow in motion from rest. Kelvin's Theorem states that for an inviscid flow, the time rate of change of circulation around a closed path comprised of the same fluid particles is zero. That is:

$$\frac{D\Gamma}{Dt} = 0$$

in Eulerian derivative notation. This means that the circulation that was initially zero around the original closed path, A-B-C-D, must continue to be zero even when the closed path is swept downstream because A-B-C-D always is assumed to be described by the same fluid particles.

Now, assume that the flow continues at a constant velocity and that a starting vortex was formed and shed. Let the time interval dt be short enough and the closed path be large enough so that both the airfoil and the shed vortex are contained within A-B-C-D, as shown in the snapshot taken at $t = dt$ (Fig. 4.17). Remember that the circulation around this closed path is still zero.

---

* Named after a British physicist who established the theorem in the late 1800s.

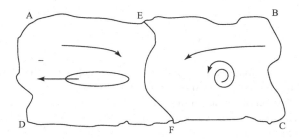

Figure 4.17. The starting vortex.

Now, subdivide the original closed path, A-B-C-D, into two parts by adding an arbitrary line, E-F, located somewhere between the airfoil and the shed vortex. Around the closed path E-B-C-F, there is a nonzero value of circulation because the path encloses a vortex. However, this means that an equal and opposite circulation must exist around A-E-F-D because the sum of the two must cancel to satisfy Kelvin's Theorem. Because the starting vortex was formed when the flow along the lower surface of the airfoil tried to go around the trailing edge and then separated, the sense of the starting vortex is counterclockwise for positive lift. The conclusions, then, are that a circulation must be present around the airfoil, it must have a unique value, and it must have a clockwise sense.

4.10 Summary

In this chapter, we introduce techniques for solving aerodynamics problems involving incompressible, inviscid flow fields. We demonstrate that under these conditions, the governing equations reduce to the Laplace's Equation, which has been studied extensively in many fields. Because it is a linear differential equation, its simplest solutions can be superimposed to produce more complex flows. We carry out solutions for four cases that we identify as the uniform flow, the simple source (or sink), the vortex, and the doublet. Superposition of a uniform freestream with a doublet and a vortex yields a solution that we identify as the flow around a spinning cylinder. Of great significance in this example is the creation of lift. This enables us to search for the fluid-dynamics origin of the lifting force, which we identify as the generation of a circulatory flow around the moving body. This idea is extended in the form of the Kutta–Joukouski Theorem to represent the lift on bodies of arbitrary shape. The need for a sharp trailing edge in creating lift on an airfoil is discussed, and the results are summarized in the so-called Kutta condition.

PROBLEMS

4.1 Find the equation of the streamline passing through the point $(x = 2, y = 3)$ in a two-dimensional flow field where the velocity components are given by $u = 3y^2$ and $v = -2x$.

4.2 A two-dimensional, steady flow field is described by:
(a) the equation of the streamlines, $xy^2/2 = $ constant

(b) the magnitude of the velocity vector at point, as given by:

$$|\mathbf{V}| = \frac{y}{2}[4x^2 + y^2]^{1/2}.$$

Find the expressions for the velocity components (u,v) of this flow field.

4.3 A two-dimensional, steady flow field is described by:
(a) the equation of the streamlines, $xy^2 - x^3/3 = $ constant
(b) the magnitude of the velocity vector as given by $|\mathbf{V}| = [x^2 + 4x^2y^2 + y^2]^{1/2}$.
Find the expressions for the velocity components (u,v) of this flow field.

4.4 Consider a two-dimensional, perfect fluid flow described by:
$$\psi = 2y^2 - x^2 + 9.$$

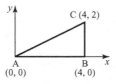

Problem 4.4

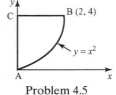

Problem 4.5

Calculate the circulation, Γ, around the closed path, A-B-C, as shown in the figure, by using two methods:
(a) evaluating the line integral
(b) using Stokes' Theorem

4.5 A two-dimensional flow field has velocity components given by $u = 2y$ and $v = 4x$. Find the circulation around the closed path, A-B-C, as shown in the figure, by using two methods:
(a) evaluating the line integral
(b) using Stokes' Theorem

4.6 The velocity potential for a steady, incompressible, two-dimensional inviscid flow is given by $\phi = x^2y - \dfrac{y^3}{3}$. Prove that a stream function exists for this flow, and then find it.

4.7 A two-dimensional, inviscid flow can be described by the velocity vector:

$$\mathbf{V} = u_r\mathbf{e}_r + u_\theta\mathbf{e}_\theta = (\theta^2 + 2r\theta)\mathbf{e}_r + (2\theta + r)\mathbf{e}_\theta.$$

Does a velocity potential exist for this flow? If so, find it. Does the result represent a physically possible flow field? Explain carefully.

4.8 The velocity field for a two-dimensional, perfect fluid is given by:

$$\mathbf{V}(x, y) = xy\mathbf{i} + \frac{y^2}{2}\mathbf{j}.$$

(a) Does a stream function exist for this flow? If so, find it.
(b) Does a velocity potential exist for this flow? If so, find it.

4.9 In a two-dimensional flow of an ideal fluid, the velocity components are given in polar coordinates as $u_\theta = 10r$ and $u_r = 0$.
(a) Does a stream function exist for this flow? If so, find it.

(b) Does a velocity potential exist for this flow? If so, find it.

4.10 A pitot tube is mounted on the nose of a low-speed airplane flying a constant speed at standard sea-level conditions. The pitot tube reads 1.5×10^3 N/m². What is the flight velocity of this airplane?

4.11 An airplane is in steady flight at 150 mph at an altitude of 15,000 feet. Assume inviscid, incompressible flow.
   (a) If the static pressure at a certain point on the wing is measured at 8.0 psia, what is the flow velocity at this point?
   (b) What is the pressure at a stagnation point on the airplane?
   (c) If a pitot probe were mounted on the nose of the airplane, what would it read (give the pressure units)?

4.12 A vortex and a uniform flow are superposed. These elements are described by:

   vortex: $u_r = 0$ $u_\theta = -40/r$
   uniform flow: $u = 15$ $v = 40$

   What is the x-component of the resulting velocity $\mathbf{V}$ at the point $(r,\theta) = (2,30°)$?

4.13 Consider the velocity potential, ϕ, of a flow field generated by a superposition, where $\phi = \phi_1 + \phi_2$ with $\phi_1 = x^2 + x - y^2 + c_1$ and $\phi_2 = 4xy + c_2$, where c_1 and c_2 are constants.
   (a) Find the velocity components (u,v) of the superposed flows at point $(1,2)$.
   (b) Find the magnitude of the velocity of the superposed flows at point $(1,1)$.

4.14 A cylinder of radius 2 feet is in a uniform incompressible flow with a freestream velocity of 40 ft/s. The circulation about the cylinder is zero. Take $\rho = 0.002$ slug/ft³ and assume two-dimensional, steady, inviscid flow. Referring to the figure,

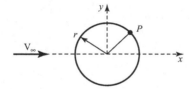

Problem 4.14

   (a) Where is Point P on the surface of the cylinder if at this point $Cp = -2.0$? How many such Points P are there on the cylinder?
   (b) How many body radii upstream of the center of the cylinder is the point on the x-axis where Cp equals half of the maximum value of Cp on the cylinder?

4.15 A static-pressure rake is to be made by mounting a series of static-pressure probes on a long cylinder 2 inches in diameter, as shown in the figure.

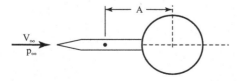

Problem 4.15

A measurement of freestream static pressure in the flow is made by using this rake and measuring the pressure p_A at Point A. The test engineer requires that the static-pressure tap be far enough forward of the supporting cylinder that the measurement error as defined by

$$\frac{(p_A - p_\infty)}{q_\infty}$$

is less than 0.1 percent. Specify the dimension "A" in inches. Assume steady, two-dimensional, perfect fluid flow around the rake.

4.16 A two-dimensional cylinder of radius 2 feet is in a uniform inviscid flow with a velocity of 25 ft/s. The stagnation points on the cylinder are known to be at Points A and B (see the figure) due to a circulation of unknown strength.
   (a) Find the magnitude of the circulation. Give the units of your answer.
   (b) Find the value of the maximum velocity on the surface of the cylinder.
   (c) Find the velocity on the surface of the cylinder at Point P shown on the figure.

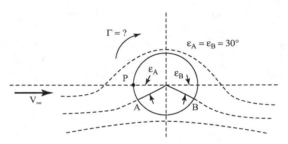

Problem 4.16

REFERENCES AND SUGGESTED READING

Anderson, John D., *Fundamental Aerodynamics*, McGraw-Hill Book Company, New York, 1984.
Kuethe, A. M., and Schetzer, J. D., *Foundations of Aerodynamics,* 2nd ed., John Wiley & Sons, New York, 1961.
Prandtl, L., and Tietjens, O. G., *Fundamentals of Hydro- and Aeromechanics,* Dover Publications, Inc., New York, 1957.
Schlichting, Hermann, *Boundary Layer Theory*, 7th ed., McGraw-Hill Book Company, New York, 1979.
Wilcox, David C., *Basic Fluid Mechanics,* 3rd ed., DCW Industries, Inc., 2007.

5 Two-Dimensional Airfoils

5.1 Introduction

A flight vehicle is sustained in the air by the lift that is generated on the wings. If the vehicle design is to be successful, this lift must be generated as efficiently as possible—that is, with minimum drag and structural weight. Because of their enormous importance, wings (and the airfoil sections that comprise them) have been studied for years both experimentally and analytically. The emphasis in this chapter is on flow about airfoil shapes in two dimensions. Two-dimensional flow implies that the flow field and the body shape are identical in any vertical plane aligned with the flow. Thus, an airfoil section at any spanwise station of an infinite wing of constant section (Fig. 5.1a) behaves as if it were in two-dimensional flow (Fig. 5.1b).

If a lifting wing has a finite span band, hence, wing tips at values of,

$$y_{tip} = \pm\frac{b}{2},$$

then there is a flow around the wing tips from the lower surface (i.e., higher pressure) to the upper surface (i.e., lower pressure). The intensity of the spanwise flow due to this effect varies across the wing span, so that the flow field is no longer the same at every spanwise (y) station. The finite wing, therefore, constitutes a three-dimensional flow problem (see Chapter 6).

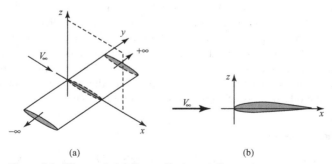

(a)   (b)

Figure 5.1. Wing and airfoil-coordinate notation.

The focus of this chapter is the prediction of the pressure distribution, lift forces, and moments on various airfoil shapes as well as the dependence of these quantities on airfoil-shape parameters. Initially, no viscous forces are accounted for explicitly; therefore, no drag occurs on a two-dimensional shape. This result is known as D'Alembert's Paradox, and its origin can be identified readily in the calculations in Chapter 4 that describe the lift generation on a cylinder with circulation. Modification of the pressure distribution by three-dimensional flow effects and by viscous forces leads inevitably to drag.

Airfoil shapes are defined by mathematical techniques or by prescribing the values of the shape parameters. The emphasis then is on the so-called direct problem. That is, given the airfoil shape, we find the pressure distribution and the force and moment behavior of the airfoil with angle of attack. The direct problem is investigated by experiment, by exact or approximate theories, and—in recent years—by numerical analysis.

Recently, the *inverse problem* has received considerable attention. That is, given the desired chordwise pressure distribution, we find the airfoil shape that leads to this distribution. The motivation for this approach is control of the behavior of the boundary layer on the airfoil surface relative to transition and separation. Streamwise pressure gradients have a major effect on the growth and stability of a viscous boundary layer. This suggests that it is desirable to specify the chordwise pressure distribution on the airfoil and then find the airfoil geometry that would generate this pressure distribution. Thus, the airfoil shape is compatible with the desired boundary-layer development. This method allows significant decreases in drag and optimization of other airfoil characteristics. Numerical analysis provides a powerful technique for solving this type of problem; hence, computational methods now have a major role in modern airfoil design.

It is not possible to cover here, in detail, all of the many approaches used in the analysis of airfoils; rather, the objective is to provide a strong framework within which the student readily can access particular techniques in later study. Therefore, the chapter is organized as follows: First, airfoil shape and behavior parameters are defined. Then, the primary focus is to solve the direct problem of determining pressure distribution, forces, and moments on a specified airfoil shape. This development begins with an example of a classical technique for determining the shape and performance of an airfoil of arbitrary thickness ratio by an analytical method. Following this, the experimental performance assessments carried out by the National Advisory Committee for Aeronautics (NACA; now NASA) using systematic shape variations are explained. Next, thin-airfoil theory is developed for the prediction of forces and moments on airfoils of arbitrary (but thin) shapes. Thin-airfoil theory is considered in detail because this analysis provides excellent insight into the role of airfoil parameters and geometry on their aerodynamic performance. The chapter concludes with a brief discussion of numerical methods as applied to the direct problem, as well as comments on modern approaches to the inverse problem with sample results.

Airfoil-Shape Parameters

An airfoil shape is usually perceived as a symmetric *thickness envelope* distributed above and below a *mean camber line* (*camber* is a measure of the curvature of an

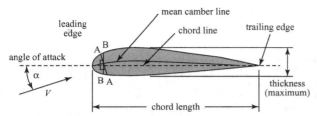

Figure 5.2. Airfoil terminology.

airfoil). The *chord line* is a straight line joining the two ends of the mean camber line, which are termed the *leading edge* and the *trailing edge* and are depicted in Fig. 5.2.

Strictly speaking, the thickness envelope is defined as measured along a line perpendicular to the mean camber line (e.g., A-A). However, if the maximum camber is small, then the thickness distribution may be defined as symmetrical above and below the mean camber line as measured in a direction perpendicular to the chord (e.g., B-B). For small camber, the difference between A-A and B-B is minor.

Following are the major parameters that define an airfoil shape:

1. *Camber (curvature)*: The shape of the mean line of the airfoil. This shape is usually expressed in terms of camber distribution (i.e., distance to the chord line as a function of chord length) and camber ratio (i.e., maximum camber as a percentage of chord).
2. *Thickness*: Usually expressed in terms of thickness distribution (i.e., the height of the airfoil relative to the mean camber line approximately the chord line, as a function of chord length) and thickness ratio (e.g., the maximum thickness of the airfoil expressed as a percentage of chord).
3. *Nose radius*: (radius of a circle fitted to the nose).
4. *Slope of the airfoil at the trailing edge*: (angle between the camber line and chord line at the trailing edge).

Expressing maximum camber and maximum thickness as a ratio with the chord is convenient because an airfoil of given camber and thickness can appear to the flow as a large or a small obstacle, depending on the relative magnitude of the chord. It follows from these definitions that a symmetrical airfoil has zero camber, with the thickness distributed symmetrically about the chord line.

Two-Dimensional Airfoil Behavior

Figure 5.3 shows the behavior of a typical cambered, two-dimensional airfoil as a function of geometric angle of attack (i.e., the angle between the freestream and the chord line). The data presented in this figure, and other similar figures, are from experiments. The nomenclature that specifies the shape of the airfoil illustrated in this figure is defined later. In Fig. 5.3(a), notice that up to an angle of attack of about 10 degrees, the lift increases linearly with angle of attack. Above 10 degrees, the lift begins to drop off and, at about 16 degrees (depending on the Reynolds number), the loss in lift is catastrophic; that is, the airfoil stalls. The upper lift curve shows the

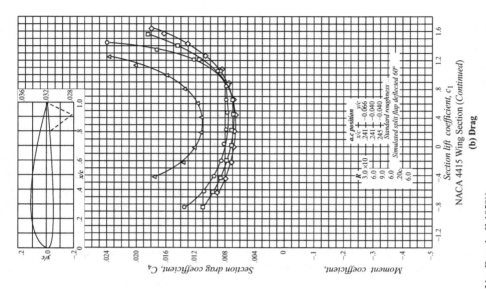

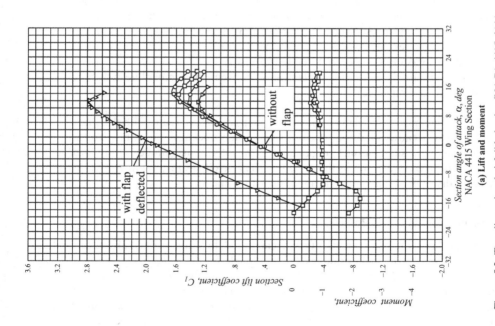

Figure 5.3. Two-dimensional airfoil behavior (NACA 4415, (Abbott and Van Doenhoff, 1959)).

effect of a deflected flap. The lift curve does not maintain a constant slope because viscous effects begin to dominate. As the angle of attack increases, the adverse pressure gradient over the aft-upper surface of the airfoil becomes larger and the boundary-layer separation point begins to move upstream from near the trailing edge. As a result, a loss in lift occurs over the rear portion of the airfoil. Suddenly, the separation point jumps to near the leading edge, the aerodynamic shape as "seen" by the flow changes dramatically, and the airfoil stalls. The maximum lift coefficient, $C_{\ell max}$, is obtained immediately before the stall. Notice how $C_{\ell \, max}$ for the airfoil depends on the Reynolds number (Re), which indicates that the airfoil boundary-layer behavior is varying with the Reynolds number. Of course, the inviscid theory considered here predicts neither stall nor $C_{\ell \, max}$; rather, it predicts the lift curve slope for moderate angles of attack.

Figure 5.3(a) shows that at zero lift, the angle of attack for this airfoil is negative because the airfoil has positive camber. The negative angle of attack required for zero lift is called the *angle of zero lift*. If the airfoil is symmetric, the angle of zero lift is zero.

The pitching-moment coefficient also is shown in Fig. 5.3(a). Here, it is taken about the *aerodynamic center*, which is that point on the airfoil about which the pitching moment is *independent of angle of attack*. Do not confuse this with the *center of pressure*, which is the point about which the *total moment is zero*. The location of the aerodynamic center and the moment about the aerodynamic center can be found from testing and also predicted from inviscid theory (for moderate angles of attack). Because the pitching moment is defined as positive clockwise (i.e., nose up), the negative pitching-moment coefficient shown in Fig. 5.3(a) implies a nose-down (i.e., counterclockwise) sense and, hence, a restoring (i.e., stable) moment.

Fig. 5.3(b) shows experimental data expressing the drag behavior of the same airfoil. This curve, called the *drag polar,* presents drag coefficient versus lift coefficient rather than drag coefficient versus angle of attack. The drag in these two-dimensional tests is called the *profile drag*. This is the part of the drag due to the action of viscosity. It is the sum of two drag contributions—namely, skin-friction drag and pressure drag due to flow separation. Notice how the drag increases dramatically as boundary-layer separation begins to dominate the flow field. Airfoil drag is predicted to be zero for two-dimensional inviscid flow. The subject of drag appears in many places in the text because it originates in several ways. Detailed discussions are in Chapter 9, in which the subject of compressible wave drag is set forth.

5.2 The Joukowski Airfoil

To illustrate what can be accomplished with an analytical technique, we consider representing an airfoil as a *mapping* of a known flow solution. In particular, we consider the circular cylinder with a superposed vortex flow studied in Chapter 4. This model was used to demonstrate the connection between lift and the creation of circulation. If we could distort the coordinates correctly, perhaps we could use this known solution to understand the flow around actual airfoils.

Mathematical Approach

The shape and behavior of the Joukowski airfoil are determined by applying an exact mathematical method. The analysis uses the theory of complex variables—in particular, conformal mapping—to map (i.e., transform) a right-circular-cylinder shape (for which the flow with circulation already was determined) in one complex plane into an airfoil shape in a second complex plane. Conformal mapping means that intersecting lines are mapped such that the angles between the lines in the two planes are preserved. To accomplish this mapping, the Joukowski transformation uses a simple transformation function—namely, the first two terms of an infinite series that is the most general transformation formula. This simplicity leads to constraints on the resulting airfoil geometry.

If it is to look like an airfoil, the shape resulting from the transformation must have a rounded leading edge and a sharp trailing edge. To generate a sharp trailing edge by means of the transformation, the circular cylinder (i.e., a circle in the cylinder plane) must pass through one of the *critical points* of the transformation—that is, through a point where the transformation is not conformal. This critical point is located on the x-axis in the circle plane and the center of the cylinder is offset from the origin, being located in the second quadrant of the plane. Figure 5.4 illustrates the geometry of the mapping for several cases of interest. The distance that the center of the circle lies above the x-axis determines the thickness ratio of the airfoil, whereas the distance it lies to the left of the y-axis determines the magnitude of the camber ratio. The cylinder radius is simply a scale factor.

The z (i.e., cylinder) plane is a complex plane in which the cylinder appears as a circle in cross section and a point on the plane is given by the complex variable $z = x + iy$. Similarly, a point in the ς (i.e., circle) plane is described by $\varsigma = \xi + i\eta$. The Joukowski transformation is given by $\varsigma = z + C^2/z = \varsigma_1 + \varsigma_2$, where C is a constant.

Referring to the z (i.e., cylinder) plane in Fig. 5.4, the first term of the Joukowski transformation ($\varsigma_1 = z$) maps a point-by-point reproduction of the cylinder in the z plane into a "major" circle in the ς (i.e., circle) plane. The second term ($\varsigma_2 = C^2/z$) maps the cylinder in the z-plane into a second ("minor") circle shown as a dashed line in the circle (ς) plane. However, compared to the major circle, the minor circle has a reduced radius, the center of the circle is transformed across the imaginary (i.e., vertical) axis, and a point on the major circle described by the polar angle θ is reflected in the real (i.e., horizontal) axis as ($-\theta$).

Finally, the complete transformation of the cylinder into the ς_J (i.e., airfoil) plane is accomplished by the vector addition of two complex quantities, $\varsigma_J = \varsigma_1 + \varsigma_2$. One point, z, on the cylinder and its image on the major and minor circles and then on the airfoil surface is illustrated in Fig. 5.4. A complete illustration is found in Program **JOUK**.

Run Program **JOUK** to see how the points on the cylinder map into points on the airfoil in the second complex plane and how the shape of the airfoil is developed. The program also allows the user to compare the shape of a Joukowski airfoil with that of a more familiar airfoil.

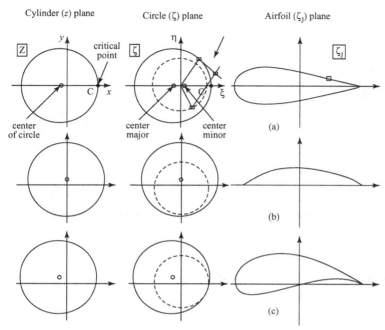

Figure 5.4. Mapping of circle into an airfoil: (a) symmetric Joukowski airfoil; (b) circular-arc airfoil; and (c) Joukowski airfoil with camber.

With the geometry established, a function of a complex variable that relates the velocity components on the airfoil to the velocity components on a cylinder at corresponding mapping points can be determined and then transformed. Because the flow field around a cylinder is known, the velocity (and, hence, the static pressure) on the airfoil can be found at corresponding points. If only the lift is required, finding the circulation around the airfoil provides the answer through the vortex theory of lift.

It may be shown that the Joukowski transformation preserves the value of the circulation in the two planes. That is, the circulation around the airfoil is the same as the circulation around the related cylinder, and the lift force on the airfoil follows from the Kutta–Joukowski Theorem once the circulation is known. How is the unique value of the circulation around the cylinder established? Recall that the critical point on the cylinder transforms into the airfoil trailing edge. Now, to satisfy the Kutta condition, the trailing edge of the airfoil must be a stagnation point. This means that the critical point in the circle plane also must be a stagnation point, as shown in Fig. 5.5.

Once the freestream flow direction, α, is selected (which is also the airfoil angle of attack) and the location of the center of the cylinder is chosen, the magnitude of the circulation required to hold the rear stagnation point on the cylinder at the critical point can be calculated. The lift per unit span, L', on the corresponding Joukowski airfoil follows directly. Inviscid theory predicts that the drag of the Joukowski airfoil is zero in accordance with the D'Alembert's Paradox discussed previously.

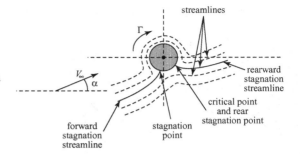

Figure 5.5. Flow around a
cylinder linked to flow around a
Joukowski airfoil.

Mathematically, this is because the flow is symmetrical about this cylinder relative to a line through the center and perpendicular to the freestream flow direction.

Results

Recall the definition of the two-dimensional lift coefficient:

$$C_\ell = \frac{L'}{\frac{1}{2}\rho V^2 c}.$$

The result for a symmetrical Joukowski airfoil at moderate angle of attack is:

$$C_\ell = 2\pi(1+\varepsilon)\alpha,$$

where α is the geometric angle of attack of the airfoil (i.e., the angle between the freestream direction and the airfoil chord line) and ε is a small number proportional to the thickness ratio. For a vanishingly thin ($\varepsilon \to 0$) symmetric Joukowski airfoil, the lift curve slope is:

$$\frac{dC_\ell}{d\alpha} = 2\pi. \tag{5.1}$$

For a 10 percent–thick Joukowski airfoil, the theoretical lift-curve slope is only about 7 percent higher. Remember that these results are exact; they suggest that if an airfoil is thin (i.e., less than 10 percent), then the effect of the thickness ratio on the lift-curve slope can be ignored with minor error. Keep this in mind when the assumptions involved in the thin-airfoil theory are presented herein.

Unfortunately, the Joukowski transformation leads to an airfoil shape that is not practical based on experience. The mean camber line is always a circular arc regardless of which airfoil camber and thickness ratios are chosen, and the maximum thickness of the Joukowski airfoil is always at about the quarter-chord—which, in practice, is too far forward. Finally, the trailing-edge angle is always a cusp (i.e., has a zero-included angle). It is possible to generalize the Joukowski airfoil shape to incorporate a finite trailing-edge angle by taking more terms in the general transformation formula; however, the method becomes complicated even for symmetrical airfoils.

The Joukowski airfoils were used in the 1930s in Europe in several sailplane designs. They formed the basis of the successful Göttingen series of airfoils that are used in the present day in low-speed designs. Preceding World War II, they were considered state-of-the-art for high-performance sailplane applications. Even if the Joukowski airfoil is not currently considered to represent a practical airfoil-design approach, it has a role in modern analysis because the pressure distribution and lift results are based on a rigorous theory with no approximations. The complex variable analysis thus provides useful benchmark results against which to compare numerical or approximate solutions.

Numerical-analysis methods for inviscid flow were developed that are, in a sense, the inverse of the approach described previously. In these methods, an arbitrary airfoil shape is selected (including a finite trailing-edge angle) and then is transformed into a circle. The flow field for the circle (cylinder) is solved and this solution is transformed back to the airfoil plane, yielding the velocity distribution (and, hence, the pressure distribution) on the airfoil. However, numerical-panel methods (discussed herein) are preferred for solving this problem.

Finally, although the conformal-mapping approach and the underlying complex-variable theory now are not used directly in the solution of practical aerodynamics problems, the ideas involved are of interest in the sense described. If students are interested in more details on the application of conformal-mapping methods, they should consult the extensive literature on the subject published previously in the 1930s (cf. Glauert, 1926/1947).

5.3 The NACA Series of Airfoils

Beginning circa 1930, it was realized that there was a need to put the subject of airfoils on a rational basis. Accordingly, NACA designed and tested numerous airfoil shapes with systematic variations in camber and thickness. Airfoil shapes were generated by expressing various mean camber lines in mathematical-equation form and then wrapping different families of symmetrical thickness distributions around them. The resulting airfoils were tested experimentally in low-speed wind tunnels at different Re values, yielding force and moment data such as those shown in Fig. 5.3. Computer codes now are available that accurately predict pressure distribution, lift, and moment for these airfoil shapes. In a sense, it is now possible to perform the experiments on a computer. The benefit is lower cost and the capability to numerically optimize certain desirable characteristics.

Many of the NACA airfoils are still in use today, which is the result of the great care exerted in securing the airfoil data. We always can be sure that performance predictions based on these data are reliable. Each "family" of the NACA airfoils has a numbering system (explained herein).

Run Program **AIRFOILS** to see illustrative airfoil shapes and performance. Abbott and Van Doenhoff, 1959, has an excellent summary of NACA airfoil data.

NACA Four-Digit Airfoils

This series has one basic thickness form, with the maximum thickness at 30 percent chord. Camber lines were chosen so that the maximum camber occurred from 20 to 70 percent of chord. For any NACA *xxxx* section:

> *First integer:* maximum value of the mean camber line, percent chord
> *Second integer:* distance from the leading edge to point of maximum camber, tenths of a chord
> *Third and Fourth integers:* thickness of airfoil, percent chord

Thus, the NACA 4415 airfoil section shown in Fig. 5.3(b) has a maximum camber of 4 percent chord located at 40 percent chord with a thickness ratio of 15 percent.

NACA Five-Digit Airfoils

Five-digit airfoils were designed in the mid-1930s to improve the pitching-moment characteristics of four-digit airfoils. Accordingly, they have the maximum camber farther forward than the four-digit series. Both the four- and five-digit series have the same basic thickness form (i.e., maximum thickness at 30 percent chord) as the four-digit sections.

For any NACA *xxxxx* section:

> *First integer:* the section-lift coefficient in tenths is 1.5 times the first integer; this first integer is indicative of the amount of camber
> *Second and Third integers together:* twice the value of the maximum camber, percent chord
> *Fourth and Fifthe integers together:* thickness ratio, percent chord

Thus, the NACA 23012 section has a design lift coefficient of 0.3, a camber ratio of 15 percent, and a thickness ratio of 12 percent. The design-lift coefficient is defined as the theoretical value of the section-lift coefficient at an angle of attack such that the oncoming flow is tangent to the mean camber line at the leading edge.

NACA Six-Digit Airfoils

Six-digit airfoils were designed in the early 1940s with the objective of encouraging a longer run of laminar-boundary-layer flow along the surface to reduce the skin-friction drag. As described in detail herein, early transition from a smooth laminar flow to a turbulent flow leads to increased drag. In this airfoil configuration, the mean camber line and the thickness distribution are designed to generate a specific pressure distribution at a certain value of lift coefficient. These sections were the first to reflect the importance of delaying boundary-layer transition on the airfoil surface. The value of this design approach was demonstrated in the highly successful North American P-51 fighter of World War II fame; the high speed and impressive range were directly attributable to such drag-reducing design features.

For any NACA *xxxxx* section:

First integer: identifies series (i.e., 6)
Second integer: chordwise location of minimum pressure in tenths of a chord for a basic symmetrical thickness distribution at zero lift
Third integer: range of C_ℓ (tenths) above and below the design-lift coefficient for which favorable pressure gradients exist on both airfoil surfaces
Fourth integer: design-lift coefficient in tenths (reflects the amount of camber)
Fifth and Sixth integers together: thickness ratio of section, percent chord

Thus, the NACA 64,3-212 airfoil is a six-series section with minimum pressure at 40 percent chord, a favorable pressure gradient for a range in lift coefficient of 0.3 above and below design, a design-lift coefficient of 0.2, and a thickness ratio of 12 percent.

There are variations on many of these airfoil shapes, and the nomenclature becomes more complicated. For example, the letter A appearing in the six-series designation means that the contour was altered near the trailing edge. Abbott and Van Doenhoff (1959) is a useful collection of results for NACA airfoils and also provides details of the airfoil-numbering system. The Riegels (1961) book is a comprehensive catalog of airfoils developed at NACA and elsewhere, as well as a review of airfoil theory.

It often is useful to have a computer program that produces the coordinates of the NACA series. Program **NACAFOIL** is provided with the software package accompanying this text for that purpose. This program generates airfoil coordinates for the NACA four- and five-digit-series airfoils and stores them in a user-defined text file. This file is readable by other programs in the software package that can use airfoil data.

5.4 Thin-Airfoil Theory

Catalogs of experimental airfoil data can be useful in airplane design. It is difficult, however, to understand the physical behavior of airfoils and the relationships between airfoil aerodynamic performance and airfoil geometry simply by studying such data. There clearly is a need for an analytical method that allows the straightforward prediction of airfoil behavior with satisfactory accuracy. Such a theory would allow the role of the airfoil shape parameters to be studied. For example, for a given maximum camber, is it better to place it near the leading or trailing edge of an airfoil to achieve the largest increase in lift coefficient compared to a symmetrical airfoil at the same angle of attack? The answers to this and related questions are derived readily from thin-airfoil theory.

Thin-airfoil theory is an approximate inviscid-flow theory that relies on an assumption of small thickness ratio (i.e., 10 to 12 percent or less) at moderate angle of attack (i.e., several degrees or less). Within this framework, the theory adequately predicts lift and moment for arbitrary thin airfoils. It does not yield information on drag because the D'Alembert's Paradox interferes, which is a consequence of neglecting viscous-flow effects in constructing a simple theory.

Distribution of Singularities on the Surface of a Body

A distribution of vortex singularities is used to represent a thin airfoil in the development of an analytical thin-airfoil solution. A distribution of vortex singularities also is used in the treatment of three-dimensional lifting wings in Chapter 6. Distributions of other types of singularities are useful for certain applications, which now are introduced. For convenience, a distribution of singularities running from Point 1 to Point 2 along a Cartesian coordinate axis is considered for simplicity. In this interval, 1-2, there is an infinite number of singularities of infinitesimal strength; such a distribution is called a *vortex sheet*.

It is demonstrated in Chapter 4 that there is a jump in tangential velocity across a vortex sheet—that is, $\Delta u \neq 0$. As a result, the circulation, Γ, around the sheet between x_1 and x_2 (Fig. 5.6) is nonzero.

It follows that a vortex sheet is a useful representation for a lifting body. The vortex is the singularity chosen in this chapter to analyze the behavior of two-dimensional airfoils and in later chapters for three-dimensional wings and bodies of revolution at angle of attack.

A distribution of sources (and sinks) is useful for flows that are symmetrical about an axis as depicted in Fig. 5.7. This type of distribution of singularities is used to treat bodies of revolution at zero angle of attack in Chapter 7.

Across the source sheet, u is continuous; however, there is a jump in w across the x-axis. Thus, the source sheet splits the streamlines and represents body thickness (recall the superposition of sources and sinks described in Chapter 4). The circulation around the source sheet is zero so that it is not useful for representing a lifting body. However, the source singularity may be used in conjunction with a vortex distribution to represent the thickness effect on a lifting body with finite thickness.

A doublet distribution also may be used to represent a lifting body. This statement must be considered carefully because a doublet was superposed with a uniform stream in Chapter 4 to represent the flow around a nonlifting cylinder and a vortex singularity was added to produce asymmetric flow and lift. The distinction is that the doublet used to generate the flow around the cylinder was developed by placing the source–sink pair along the x-axis (i.e., streamwise direction) and then considering the limiting case with the two singularities meeting at the origin. The resulting doublet then was said to have its axis in the x-direction.

Figure 5.6. Vortex-singularity distribution.

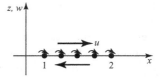

Figure 5.7. Source-singularity distribution.

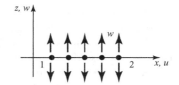

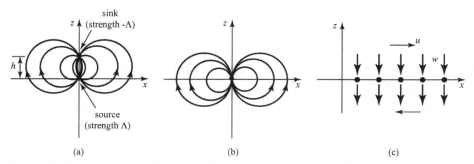

Figure 5.8. Source-singularity distribution for source–sinkpair and doublet.

Now consider a source–sink pair located on the z-axis. Let the source be at the origin of the coordinate system and the sink be located on the positive z-axis, as shown in Fig. 5.8(a). This represents the same source–sink pair in Chapter 4 rotated clockwise by 90 degrees. The streamlines are as shown. Now, we generate a doublet at the origin of the coordinates by letting $h \to 0$ while keeping the product (Λh) a constant, where Λ is the strength of the source–sink pair (Fig. 5.8(b)). We focus our attention on the flow direction at the origin and let the doublet at the origin represent one of a series of doublets of infinitesimal and variable strength distributed along the x-axis (but with the doublet axes in the z-direction), as shown in Fig. 5.8(c). Notice that the resulting doublet-sheet flow in the z-direction is into the top of the sheet and out of the bottom. Also note that this is not the same behavior found in the source sheet.

Detailed analysis (see Karamcheti, 1980) shows that there is no jump in w across the sheet. However, there is a jump in the velocity potential across the sheet and, hence, a jump in the u component of velocity across the sheet. This means that the circulation around the doublet sheet is nonzero. It follows that a two- or three-dimensional lifting body in a freestream can be represented by superposing a uniform flow and a distribution of doublet singularities on the surface of the body with the axes of the doublets in the z-direction. Physically, the doublet sheet may be thought of as imparting a downward momentum to the oncoming flow similar to a vortex sheet. This change in z-momentum generates a net lift force on the sheet and, therefore, on the body it represents.

Thus, flows involving lifting bodies may be modeled using either a vortex or a doublet distribution. In fact, on analysis, the local vortex-sheet strength per unit length, $\gamma(x)$, can be identified with the derivative of the doublet-strength distribution per unit length, $d\lambda/dx$. In this textbook, the vortex-sheet distribution is used to represent a lifting body rather than the doublet distribution because the interaction of the vortex and the flow is physically more apparent. The concept of lift introduced in Chapter 4 by means of a cylinder with circulation induced by a bound vortex then can be adapted readily to the study of airfoils.

The Vortex Sheet

The concept of the vortex sheet is central to thin-airfoil theory. A vortex sheet is a collection of vortex filaments, or threads, placed side by side. Each filament is

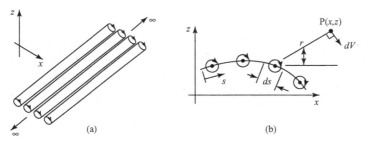

Figure 5.9. The vortex sheet.

infinitesimal in strength and extends to ± infinity in a direction perpendicular to the
x-z plane that contains the two-dimensional airfoil (Fig. 5.9(a)). The vortex filaments
appear in cross section in the x-z plane as a series of point vortices (Fig. 5.9(b)).
Notice that these coordinate axes depart from the simple (x, y) coordinates used in
Chapter 4, which anticipates extension of the theory to the three-dimensional wing
problem, as suggested in Fig. 5.1(a).

Recall that the velocity field associated with a single two-dimensional point
vortex is everywhere perpendicular to a radius taken from the vortex as center. Thus,
an isolated vortex may be thought of as *inducing* at a point of a velocity component
that is at right angles to a line joining the center of the vortex to that point. The word
inducing is emphasized because it is only a conceptual aid. The vortex does not *cause*
a velocity; rather, a real vortical flow configures itself physically such that there is a
viscous-dominated center and an associated velocity field with circular streamlines.

Referring to Fig. 5.9(b), let $\gamma = \gamma(s)$ represent the strength of the vortex sheet
per unit length along s. Thus, the strength of an infinitesimal segment of the sheet
is the strength per unit length multiplied by the lengths—namely, $\gamma\,ds$. The vortex
filaments contained within this small segment are combined and treated as a point
vortex. These point vortices are assumed to rotate in a clockwise sense, as shown
in Fig. 5.9. Recall from the definition of circulation, Eq. 4.8, and from Section 4.6
regarding a point vortex that for a vortex of finite strength with a clockwise sense,
the vortex strength, Γ, is positive and then

$$u_\theta = -\frac{\Gamma}{2\pi r}$$

is negative. The notation for polar coordinates (Fig. 5.9(a)) thus indicates that the
induced velocity u_θ is also in a clockwise sense. It follows that the velocity, dV,
induced by a vortex filament is clockwise, as shown in Fig. 5.9(b), and that

$$dV = -\frac{\gamma\,ds}{2\pi r},$$

where dV is the differential velocity induced by a small segment of the vortex sheet.

Now, we consider a line integral taken around a length ds of vortex sheet. By
definition, Eq. 4.8:

$$\Gamma = -\oint \mathbf{v} \cdot \mathbf{ds}$$

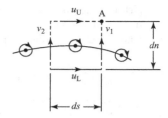

Figure 5.10. Circulation around a vortex-sheet element.

and integrating counterclockwise from Point A in Fig. 5.10:

$$d\Gamma = -(-u_U ds - v_2 dn + u_L ds + v_1\, dn),$$

where all of the velocity components around the path are assumed to be in the positive-coordinate directions. Letting $dn \rightarrow 0$, this reduces to $d\Gamma = (u_U - u_L)ds$.

Now, the strength of the segment of the vortex sheet contained within the closed path is γds. Thus,

$$d\Gamma = \gamma ds = (u_U - u_L)ds \tag{5.2}$$

and $\gamma = (u_U - u_L)$. This result states that the *local jump* in tangential velocity across the vortex sheet at any point is precisely equal to the local *sheet strength*.

Vortex-Sheet Representation of an Airfoil

An airfoil may be represented in an inviscid flow by wrapping a vortex sheet around its surface, as depicted in Fig. 5.11. This vortex sheet is of a variable strength, γds. The strength can be found by applying the boundary condition that there is no flow through the solid surface of the airfoil. The boundary condition at infinity already is satisfied by the vortex-sheet property that at a large distance from a vortex, the induced velocity goes to zero $\lim_{r \to \infty}(1/r) = 0$. If the airfoil is lifting, the Kutta condition also must be imposed.

Once the variable-sheet strength, $\gamma(s)$, is found, then:

$$\Gamma = \sum \gamma \Delta s \rightarrow \int \gamma ds \tag{5.3}$$

and the circulation around the airfoil can be determined. The value of the lift per unit span, $L' = \rho V_\infty \Gamma$, follows immediately from the Kutta–Joukowski theorem.

This representation of an airfoil by a vortex sheet (i.e., a collection of singularities, each of which is a solution to the linear Laplace's Equation) is physically appealing as well as an application of the superposition principle for inviscid flows. Recall that in a real (viscous) flow, there is a thin boundary layer on the surface of

Figure 5.11. Vortex sheet wrapped around an arbitrary airfoil.

the airfoil, which is a region of high vorticity. The vorticity of the vortex sheet at the surface of the inviscid-flow model may be thought of as generated there by the boundary layer that would be present in the actual flow field.

Vortex-Sheet Representation of a Thin Airfoil

An analytical solution for the vortex-sheet strength is not possible for an airfoil of arbitrary thickness ratio. Full numerical solutions were developed, as discussed in Section 5.7. However, an analytical solution can be constructed if the vortex-sheet model is applied to thin airfoils. If an airfoil is thin (i.e., 10 to 12 percent thickness ratio or less), then the vortex sheets on the airfoil upper and lower surfaces are close to the mean camber line. This fact can be approximated by distributing the vortex sheet along the airfoil mean camber line. This entails stripping off and discarding the symmetrical thickness distribution above and below the mean camber line, leaving only the mean camber line to represent the airfoil as shown in Fig. 5.12.

This approach provides results that are in good agreement with experiments for thin airfoils. The exact solution for a Joukowski airfoil discussed previously showed that the influence of thickness on lift and moment coefficient is minor if the thickness ratio is small. It is appealing to replace the mean camber line with a vortex sheet, because there is a jump in velocity across the vortex sheet as there would be across a mean camber line representing the airfoil because the mean camber line then is a streamline. Thin-airfoil theory represents a symmetrical airfoil as a flat plate and a cambered airfoil as a curved plate.

Again, the vortex-sheet strength as a function of chordwise distance is found by imposing the Kutta condition as well as the condition that the mean camber line must be a streamline (i.e., no flow component normal to the camber line). The circulation and the lift per unit span follow.

The uniqueness of the circulation about the airfoil as a function of angle of attack is a result of the Kutta condition. Recall from Section 4.9 that for a finite trailing-edge angle, the Kutta condition demands that the trailing edge be a stagnation point; whereas for a cusped trailing edge, the condition demands that the flow exiting the trailing edge at the upper and lower surfaces has the same velocity. At the trailing edge of the camber line, let the velocity be V. Thus, at the trailing edge (i.e., finite angle), $V_U = V_L = 0$ or (cusp) $V_U = V_L$. Recalling that the local sheet strength was shown to be equal to the local jump in velocity across the sheet, $\gamma = (V_U - V_L)$, it follows that in either case, $\gamma(TE) = 0$. The other condition on $\gamma(s)$ is that the mean camber line must be a streamline.

Consider a thin airfoil at angle of attack and represented by a vortex sheet distributed along the mean camber line. Place the airfoil in a body–axis coordinate

Figure 5.12. Mean camber line represented as a vortex sheet.

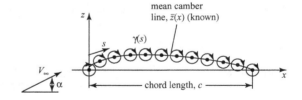

system (Fig. 5.12) with the x-axis aligned with the chord line. The equation of the mean camber line, $\bar{z}(x)$, is known. The vortex-sheet strength, $\gamma(s)$, is determined and then used to find the pressure distribution, $C_p(x)$; the lift coefficient, C_ℓ; and the pitching moment coefficient, C_m.

If the mean camber line is to be a streamline, there can be no flow through the vortex sheet. This means that at every point along the vortex sheet, the velocity component, $V(s)$, is perpendicular to the mean camber line (see Fig. 5.13). $V(s)$ is induced by all other elements of the vortex sheet. It must be equal and opposite to the normal component of the freestream velocity, V_{∞_n}, which is normal to the mean camber line at that point. Thus, $V(s) + V_{\infty_n} = 0$ at every point along the mean camber line, as shown in Fig. 5.13.

Formulate V_{∞_n} by expressing V_{∞_n} in terms of the slope of the mean camber line at any point. The slope of the mean camber line is given by $d\bar{z}/dx$, which is a known. Then, the angle $\delta = \tan^{-1}(-d\bar{z}/dx)$, where the minus sign is added to keep the sense of the angle correct. That is, when the camber-line slope, $d\bar{z}/dx$, is negative, then $(\alpha + \delta)$ should be additive, as shown in Fig. 5.14. Near the front of the airfoil, when the mean camber-line slope is positive, the opposite is true.

From Fig. 5.14:

$$V_{\infty n} = V_\infty \sin(\alpha + \delta) = V_\infty \sin\left[\alpha + \tan^{-1}\left(-\frac{d\bar{z}}{dx}\right)\right].$$

For a thin airfoil at a moderate angle of attack, this may be approximated as:

$$V_{\infty n} = V_\infty\left[\alpha - \frac{d\bar{z}}{dx}\right]. \tag{5.4}$$

It remains to derive an expression for $V(s)$ to complete the boundary-condition statement.

Finding the expression for $V(s)$ is not particularly straightforward because $V(s)$ is the resultant velocity induced at a point on the curved sheet by all of the other elements of the sheet. Thus, referring to Fig. 5.9(b), the Point P now is on the sheet

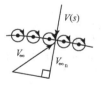

Figure 5.13. The camber-line boundary condition.

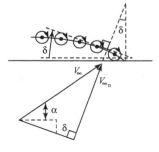

Figure 5.14. The streamline boundary condition in terms of camber-line slope

but the line r from each element of sheet ds to the Point P has a different orientation in the general case of a curved sheet. In addition to determining each slant distance, r, this means that each dV induced by $\gamma \, ds$ at Point P is at a different angle. The complexity is overcome again by taking advantage of the thin-airfoil geometry. We realize that if the thickness ratio is small, then the maximum camber ratio must be even smaller (i.e., a few percentage points), and the mean camber line is not far from the chord line—that is, the x-axis. Thus, only a small error results if the vortex sheet is placed along the chord line and the induced velocity is evaluated there. We assume in effect, that $s \approx x$ and $V(s) \approx V(x)$. This approximation is termed *satisfying the boundary condition in the plane*, a method that is often used in similar geometrical situations. Remember that even though the induced velocity is evaluated along the x-axis, the theory still insists that the mean camber line be a streamline. That is, the sheet strength is evaluated by demanding that the mean camber line be a streamline, *not* that the chord line be a streamline. This approximation greatly simplifies the geometry that enters the induced velocity calculation, as indicated in Fig. 5.15.

Recalling from the previous discussion of the vortex sheet that the induced velocity $dV = -\gamma \, ds / 2\pi r$, it follows from Fig. 5.15 that:

$$V(x) = -\int_0^c \frac{\gamma(\xi) d\xi}{2\pi(x - \xi)}, \tag{5.5}$$

where the integration sums all of the dV at an arbitrary (but fixed) point, x, induced at that point by the element of the vortex sheet at the location ξ. Here, ξ is a running variable that disappears on integration. Notice that the integrand is singular (i.e., blows up) at $x = \xi$. (This is addressed later.) Remember that even though this $V(x)$ is perpendicular to the chord line, it is treated as if it were perpendicular to the mean camber line. Thus, in the boundary-condition expression, $V(x)$ is set equal and opposite to V_{∞_n} at the mean camber line—namely, $V_{\infty_n} + V(x) = 0$.

Assembling Eqs. 5.4 and 5.5, the boundary condition requires that at any arbitrary (but fixed) chordwise station:

$$\frac{1}{2\pi}\int_0^c \frac{\gamma(\xi) d\xi}{(x - \xi)} = V_\infty \left[\alpha - \left(\frac{d\bar{z}}{dx} \right) \right]. \tag{5.6}$$

Eq. 5.6 is the basic equation of thin-airfoil theory. The right side is known because the angle of attack must be given and the mean camber line is specified for a given airfoil so that $d\bar{z}/dx$ can be found. Eq. 5.6 represents an integral equation for the unknown $\gamma(x)$ and must be solved subject to the Kutta condition that $\gamma(c) = 0$, as described in Chapter 4.

Figure 5.15. The chord-line approximation.

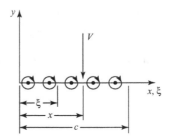

The symmetric thin airfoil and the airfoil with arbitrary camber are examined in the next two sections. The appeal of thin-airfoil theory is that it provides a completely analytical solution; the disadvantage is that it applies only to thin airfoils. Applying the results to airfoils thicker than about 12 percent thickness ratio leads to increasingly larger errors when compared to experimental data.

The Symmetrical Thin Airfoil

In this case, the mean camber line and the chord line coincide. Thus, when the thickness is stripped away, the airfoil is a flat plate and the vortex-sheet representation is along the chord so that the planar-boundary-condition approximation becomes exact. Furthermore, for the symmetric airfoil, $d\bar{z}/dx = 0$, and Eq. (5.6) becomes:

$$\frac{1}{2\pi}\int_0^c \frac{\gamma(\xi)d\xi}{(x-\xi)} = V_\infty\alpha. \tag{5.7}$$

It is convenient to make a change of variable to perform the integration. Let:

$$\xi = \frac{c}{2}(1-\cos\theta), \tag{5.8}$$

where θ is the new running variable. The fixed point in question, x, is denoted by:

$$x = \frac{c}{2}(1-\cos\phi). \tag{5.9}$$

This change of variable is equivalent to specifying chordwise location by angular rather than linear measure. Imagine a circle of radius $c/2$ centered at mid-chord (Fig. 5.16).

At the limits of integration, $x = 0 \to \theta = 0$ and $x = c \to \theta = \pi$. Finally, $d\xi = \frac{c}{2}\sin\theta\, d\theta$.

Then, the defining equation for a symmetric airfoil, Eq. 5.7, may be written as follows:

$$\frac{1}{2\pi}\int_0^\pi \frac{\gamma(\theta)\sin\theta\, d\theta}{(-\cos\phi + \cos\theta)} = V_\infty\alpha. \tag{5.10}$$

This equation may be solved by integral-equation techniques that are beyond the scope of this text. The result is:

$$\gamma(\theta) = 2\alpha V_\infty \frac{(1+\cos\theta)}{\sin\theta}. \tag{5.11}$$

 Figure 5.16. Transformation variable.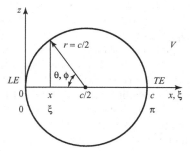

The claim that this is the solution may be verified by substituting Eq. 5.11 into the integrand of Eq. 5.10); namely:

$$\frac{1}{2\pi}\int_0^\pi \left[\frac{2\alpha V_\infty(1+\cos\theta)}{\sin\theta}\right]\left[\frac{\sin\theta d\theta}{(\cos\theta-\cos\phi)}\right]\overset{?}{=}V_\infty\alpha \qquad (5.12a)$$

or

$$\frac{\alpha V_\infty}{\pi}\int_0^\pi\left[\frac{1}{(\cos\theta-\cos\phi)}+\frac{\cos\theta}{(\cos\theta-\cos\phi)}\right]d\theta=V_\infty\alpha. \qquad (5.12b)$$

Evaluating the integral in Eq. 5.12b requires that we address the singular behavior of the integrand at $\theta=\phi$ (i.e., when the running variable is at the point in question, $\xi=x$). This is done by using the "principal value" of the integral (see Appendix E in Abbott and Van Doenhoff (1959)):

$$\int_0^\pi\frac{\cos n\theta d\theta}{(\cos\theta-\cos\phi)}=\frac{\pi\sin n\phi}{\sin\phi}, \qquad (5.13)$$

where n may be 0, 1, 2, ... depending on the form of the integrand. In the case of interest here, $n=0$ and $n=1$ in Eq. 5.12b. Thus, Eq. 5.12b becomes:

$$\frac{\alpha V_\infty}{\pi}[0+\pi]\equiv V_\infty\alpha$$

and the solution, Eq. 5.11, was verified.

Notice also that if Eq. 5.11 is the solution, then it must exhibit the proper behavior at the trailing edge, $\theta=\pi$. At the trailing edge:

$$\gamma(\pi)=\lim_{\theta\to\pi}\left\{\frac{2V_\infty\alpha(1+\cos\theta)}{\sin\theta}\right\}=\frac{0}{0},$$

which is indeterminate. However, using L' Hôpital's Rule (recall that the numerator and denominator are differentiated within the limit):

$$\lim_{\theta\to\pi}\left\{\frac{2V_\infty\alpha(-\sin\theta)}{\cos\theta}\right\}=0.$$

Thus, $\gamma(\pi)=0$ and the Kutta condition is satisfied.

At the leading edge, $\theta\to0$ and Eq. 5.11 indicates that the vortex-sheet strength becomes infinite. This is allowable because the Kutta condition makes no demands on leading-edge behavior. The leading edge of even a thin airfoil is rounded, and the flat or curved plate representation of the airfoil simply breaks down at this single point. This is interpreted here as a limiting representation of what is called *leading-edge suction*, the low pressure (i.e., high velocity) caused by the flow rapidly accelerating around the leading edge of an airfoil from the bottom to the top surface. The vortex-sheet strength then has the general characteristics depicted in Fig. 5.17. Having verified the solution, Eq. 5.11, we now can determine the expressions for the lift and moment coefficients for the symmetrical airfoil.

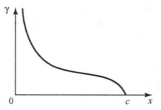

Figure 5.17. Vortex-sheet-strength distribution.

The lift coefficient is found by (1) appealing to the Kutta–Joukowski Theorem, and (2) finding the pressure distribution over the plate and then integrating. The fact that the two results are identical serves as further verification of the vortex theory of lift.

1. The total circulation around the airfoil is found by summing all of the contributions due to the vortex elements—namely:

$$\Gamma = \int_0^c \gamma(\xi)d\xi = \frac{c}{2}\int_0^\pi \gamma(\theta)\sin\theta d\theta$$

after the change of variable. Substituting the solution, Eq. 5.11, into this integrand yields:

$$\Gamma = c\alpha V_\infty \int_0^\pi (1+\cos\theta)d\theta = c\alpha\pi V_\infty. \tag{5.14}$$

The Kutta–Joukowski Theorem then states that the lift per unit span is:

$$L' = \rho V_\infty \Gamma = \rho V_\infty^2 c\pi\alpha$$

and the two-dimensional lift coefficient is:

$$C_\ell \equiv \frac{L'}{q_\infty c(1)} = 2\pi\alpha. \tag{5.15}$$

The quantity unity in the denominator is a reminder that L' is the lift per unit span.

2. The lift coefficient now is found by integrating the pressures acting on the symmetrical thin airfoil (i.e., flat plate). These pressures always act perpendicular to the surface (Fig. 5.18).

Figure 5.18. Pressures acting on the surfaces of the flat plate.

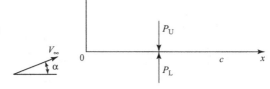

At first glance, Fig. 5.18 is puzzling. It indicates that the net pressure force on the plate acts in the z-direction; that is, it acts as a force normal to the chord. Then, this normal force apparently can be decomposed into two forces: one perpendicular to the freestream direction and the other parallel to the freestream direction. By definition, the force perpendicular to the freestream direction is the lift force. Also by definition, the force component parallel to the freestream direction is the drag force. However, there can be no drag force in a two-dimensional, inviscid flow, as indicated in D'Alembert's Paradox. There can be neither frictional nor pressure drag because no boundary layer is represented in the inviscid model.

The answer to this apparent contradiction is that there is another force acting on the plate—namely, a suction force at the leading edge acting in the chordwise (i.e., x-axis) direction. Recall that there is no boundary condition imposed at the leading edge, such as the Kutta condition at the trailing edge, and that this presents no practical problem because even thin airfoils have rounded leading edges with no anomalous velocity behavior. However, in the analytical model, in the limit of zero thickness, the inviscid flow has an infinite velocity as it accelerates around the sharp leading edge of the flat plate (Fig. 5.19). This infinite velocity produces an infinite negative pressure (i.e., suction) at the leading edge (according to the Bernoulli Equation), and this infinite suction acts on a zero-thickness plate. In the limit, the product of the infinite suction per unit area acting on an infinitesimal plate area is a finite force in the chordwise upstream (i.e., thrust) direction. The forces acting on the flat plate then are as shown in Fig. 5.20, where F'_N is the normal force per unit span and F'_C is the chordwise force per unit span. In the figure, each of these two forces is decomposed into lift and drag components.

From Figs. 5.18 and 5.20, it follows that:

$$F'_N = \int_0^c (p_L - p_U)dx \quad \text{and} \quad p_L = p_0 - \frac{1}{2}pV_L^2; \quad p_U = p_0 - \frac{1}{2}pV_U^2,$$

Figure 5.19. Inviscid flow around lifting flat plate.

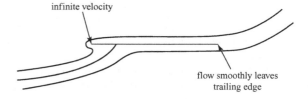

Figure 5.20. Forces acting on a flat plate.

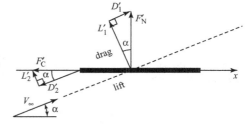

where Bernoulli's Equation was used. Because the stagnation pressure, p_0, is constant throughout the flow, we find that:

$$F_N' = \int_0^c \frac{\rho}{2}\left[V_U^2 - V_L^2\right] = \rho\int_0^c (V_U - V_L)\left(\frac{V_U + V_L}{2}\right)dx.$$

As previously noted, $(V_U - V_L) = \gamma$, so the first term in the integrand is the local vortex-sheet strength. Also, by applying the exact Joukowski-transformation method discussed in the last section to the special case of flow around a flat-plate airfoil, it may be shown that:

$$\left(\frac{V_U + V_L}{2}\right) = V_\infty \cos\alpha.$$

Thus,

$$F_N' = \rho V_\infty \cos\alpha \int_0^c \gamma dx = \rho V_\infty \Gamma \cos\alpha. \tag{5.16a}$$

Notice that Eq. 5.16 does *not* state that exactly $F'_N = \rho V_\infty\Gamma$, nor should it because it is an expression for the normal force, not the lift force, per unit span.

Rather than pondering how to evaluate the product of an infinite pressure and an infinitesimal area, a simpler approach is to argue that in this two-dimensional, inviscid flow, the drag force on the plate must be zero. That is, $D'_1 - D'_2 = 0$, where a positive drag force is in the streamwise direction, so that $D'_2 = D'_1$. From Eq. 5.16a and Fig. 5.20:

$$D_1' = F_N' \sin\alpha = (\rho V_\infty \Gamma \cos\alpha)\sin\alpha.$$

Furthermore, from geometry, $\frac{D_2'}{F_C'} = \cos\alpha$. Solving for F'_C and substituting for $D'_2 = D'_1$:

$$F_C' = \frac{(\rho V_\infty \Gamma \cos\alpha)\sin\alpha}{\cos\alpha} = (\rho V_\infty \Gamma)\sin\alpha. \tag{5.16b}$$

Finally, we recognize that the total lift on the plate is given by $L' = L'_1 + L'_2$, so that:

$$L' = F_N' \cos\alpha + F_C' \sin\alpha.$$

Substituting from Eq. 5.16:

$$L' = \rho V_\infty(\cos^2\alpha + \sin^2\alpha) = \rho V_\infty\Gamma,$$

which agrees with the Kutta–Joukowski Theorem result, and the vortex theory of lift is verified. Finally, using Eq. 5.15, $C_\ell \equiv 2\pi\alpha$.

Thus, according to thin-airfoil theory, for a symmetrical airfoil, the airfoil lift coefficient is given by:

$$C_\ell \equiv 2\pi\alpha \tag{5.17a}$$

and the corresponding lift-curve slope is:

$$\frac{dC_\ell}{d\alpha} = 2\pi. \tag{5.17b}$$

These equations state that the lift coefficient varies linearly with the angle of attack. It is shown later that this prediction is in excellent agreement with experimental data at moderate angles of attack before flow-separation effects begin to dominate and stall is approached. This can be seen for the special case illustrated in Fig. 5.3: The linear behavior applies to the unflapped NACA 4415 airfoil in the angle-of-attack range $-10° < \alpha < 10°$.

Notice in Eq. 5.17 that because the lift coefficient is dimensionless, the angle of attack must be in radians, which also is a dimensionless quantity. Eq. 5.17a is sometimes written as $C_\ell = m_0\alpha$, where m_0 is the lift-curve slope in radians. The coefficient, m_0, may be assigned its theoretical value, $m_0 = 2\pi$, or it may be determined from experimental, two-dimensional test results. Alternately, Eq. 5.17a may be written as $C_\ell = a_0\alpha$, where a_0 is the lift-curve slope per degree (i.e., thin-airfoil-theory result: 0.11 per degree) and the angle of attack then is written in degrees.

The airfoil pitching moment may be found about any point on the chord line and then transferred to any other point. For convenience, the pitching moment per unit, span, M', is taken about the airfoil leading edge as shown is Fig. 5.21. The moment may be evaluated either by summing up the contribution due to each element of the vortex sheet or by evaluating the moment due to the pressure distribution on the airfoil (i.e., plate). The latter approach is more physical and is used here.

Let a clockwise moment about the leading edge be positive. Then, the moment about the leading edge due to a pressure difference over an interval dx, at any distance x from the leading edge is obtained by integration—namely:

$$M'_{LE} = -\int_0^c (P_L - P_U)x\,dx.$$

The minus sign is introduced because if $P_L > P_U$, the moment contribution is counterclockwise and therefore must be negative.

In the development of Eq. 5.16a, it is shown that $(P_L - P_U) = \rho\gamma V_\infty \cos\alpha$. Because the angle of attack is small, this can be approximated by:

$$(P_L - P_U) = \rho\gamma V_\infty$$

and the small angle approximation is used wherever needed. Then, we use Eq. 5.11 for the vortex-sheet strength, γ, and the transformation to angular measure from

Figure 5.21. Pitching moment about the leading edge.

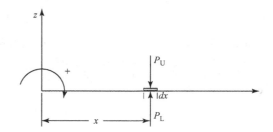

Eq. 5.8. Writing the variable of integration in the moment definition as ξ rather than x to emphasize that it is a dummy integration variable:

$$M'_{LE} = -\int_0^c (\rho V_\infty \gamma) \xi d\xi$$

or

$$M'_{LE} = -\rho V_\infty \int_0^\pi \left[\frac{2V_\infty \alpha (1 + \cos\theta)}{\sin\theta} \right] \left[\frac{c}{2}(1 - \cos\theta) \right] \left[\frac{c}{2}\sin\theta d\theta \right].$$

Carrying out the integration:

$$M'_{LE} = -\frac{\pi \rho V_\infty^2 c^2 \alpha}{4}.$$

Finally,

$$C_{m_{LE}} \equiv \frac{M'_{LE}}{q_\infty c^2} = -\frac{\pi}{2}\alpha = -\frac{C_\ell}{4}. \tag{5.18}$$

Notice that for the moment coefficient to be dimensionless, the definition must contain a length squared in the denominator because the numerator now contains a lever-arm length. Also, notice that at positive angles of attack, the airfoil pitching moment is negative or clockwise; hence, it is a nose-down pitching moment, or a so-called restoring moment.

Two other properties of the symmetrical airfoil can be deduced from Eq. 5.18. Recall from Section 5.1 that the center of pressure is the point on the airfoil about which the moment is zero. We multiply both sides of Eq. 5.18 by $(q_\infty c^2)$, which gives:

$$M'_{LE} = -\frac{c}{4}L'.$$

This result states that lift is the only contributor to moment and that the integrated lift force can be interpreted as a point force with its line of action at the quarter-chord point, as shown in Fig. 5.22. Thus, at the point $c/4$, the lever arm of the lift force is zero and the moment about that point is zero. It follows that the quarter-chord point of the symmetrical airfoil is the center of pressure.

Also recall from Section 5.1 that the aerodynamic center is the point on the airfoil where the moment is independent of the angle of attack. From Eq. 5.17a, the lift varies with the angle of attack. Because lift is the only contributor to moment, then the only point on the airfoil where the moment is independent of the angle of attack is the point where the lift-force lever arm is of zero length. Again, this is the

Figure 5.22. Location of resultant lift force on symmetrical airfoil.

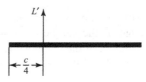

quarter-chord point. At this point, the pitching moment is not only independent of the angle of attack, it also is zero. Thus, for a symmetrical thin airfoil, both the center of pressure and the aerodynamic center are located at the quarter-chord point.

The Cambered Thin Airfoil

The cambered-thin-airfoil problem requires that the vortex-sheet strength be found from Eq. 5.6, given the angle of attack and the equation of the mean camber line. Thus, the following integral equation must be solved for $\gamma(x)$:

$$\frac{1}{2\pi}\int_0^c \frac{\gamma(\xi)d\xi}{(x-\xi)} = V_\infty \left[\alpha - \left(\frac{d\bar{z}}{dx}\right)\right]. \tag{5.6}$$

As in the symmetrical-airfoil case, it is more convenient to cast this equation in angular measure, as in deriving Eq. 5.10. Thus, Eq. 5.6 becomes:

$$\frac{1}{2\pi}\int_0^\pi \frac{\gamma(\theta)\sin\theta d\theta}{(-\cos\phi + \cos\theta)} = V_\infty \left[\alpha - \left(\frac{d\bar{z}}{dx}\right)\right], \tag{5.19}$$

where θ is an integration variable, ϕ designates the fixed point in question, and $d\bar{z}/dx$ is the slope of the mean camber line evaluated at the point in question. The solution for the vortex-sheet strength, $\gamma(\theta)$, in Eq. 5.19 is subject to the Kutta condition, $\gamma(\pi) = 0$.

A solution to Eq. 5.19 is sought by applying a Fourier-series approach. We realize that because the cambered airfoil is represented as an airfoil of zero thickness (i.e., a curved flat plate), the vortex-sheet strength for the cambered airfoil must have the same general character as the distribution shown in Fig. 5.17 for a symmetrical airfoil; namely: $\gamma(x)$ must be zero at the trailing edge and infinite at the leading edge. Now, recall from engineering mathematics that any curve or function may be represented over an interval by an infinite series comprising both sine and cosine terms—that is, a Fourier series.

Run Program **FOURIER** to review features of the Fourier series.

For the current problem, the interval is from the leading edge ($\theta = 0$) to the trailing edge ($\theta = \pi$). In this problem, the cosine terms in the Fourier series are omitted because they would violate the Kutta condition—that is, the cosine terms would make a nonzero contribution to the source strength at the trailing edge, $\theta = \pi$. Thus, we represent the vortex-sheet distribution by the Fourier sine series as follows:

$$\gamma(\theta) = \sum_{n=0}^\infty A_n \sin n\theta. \tag{5.20}$$

Now, the flow velocity and the vortex-sheet strength must be infinite at the leading edge of the cambered airfoil as in the symmetrical airfoil. To ensure the proper behavior at the leading edge, the first term of the series in Eq. 5.20, ($n = 0$), is written as a "flat-plate" term given by the symmetrical airfoil solution in Eq. 5.11. Some generality is retained by writing the coefficient A_0 in place of α. Then, the solution to Eq. 5.19 is represented by the Fourier series as follows:

$$\gamma(\theta) = 2V_\infty A_0 \frac{(1+\cos\theta)}{\sin\theta} + 2V_\infty \sum_{n=1}^{\infty} A_n \sin n\theta. \tag{5.21}$$

The coefficients A_n in the terms of the series $n \geq 1$ appearing in Eq. 5.20 were replaced by $2V_\infty A_n$ in Eq. 5.21 to make a common multiplier $(2V_\infty)$ appear in all of the terms in Eq. 5.21. The constant value $(2V_\infty)$ simply is absorbed in the definition of A_n in the summation part of Eq. 5.21.

Of course, Eq. 5.21 is a solution only if there is a mechanism by which to solve for the coefficients of the series. This mechanism is supplied by imposing the boundary condition that the mean camber line must be a streamline. When the solution is completed, it should contain the symmetrical airfoil as a special case—namely, that A_0 must equal α and all of the other A_n's must be zero.

Substituting Eq. 5.21 into Eq. 5.19 yields:

$$\frac{1}{\pi}\int_0^\pi \frac{A_0(1+\cos\theta)d\theta}{\cos\theta - \cos\phi} + \frac{1}{\pi}\int_0^\pi \sum_{n=1}^{\infty} \frac{A_n \sin n\theta \sin\theta d\theta}{\cos\theta - \cos\phi} = \alpha - \left(\frac{d\bar{z}}{dx}\right). \tag{5.22}$$

The first term can be integrated directly by using the "principal value" as given by Eq. 5.13. The second term can be written in a form suitable for integration by application of the principal value in Eq. 5.13 if the numerator is written in terms of cosines by using the trigonometric identity:

$$\sin n\theta \sin\theta = 1/2[\cos(n-1)\theta - \cos(n+1)\theta].$$

Substituting and integrating Eq. 5.22 leads to:

$$A_0 + \frac{1}{2\pi}\left\{A_1(0) + A_2(\pi) + A_3(2\pi\cos\phi) + \cdots\right\} -$$

$$-\frac{1}{2\pi}\left\{A_1(2\pi\cos\phi) + A_2\left[\pi\left(4\cos^2\phi - 1\right)\right] + \cdots\right\} = \alpha - \frac{d\bar{z}}{dx}.$$

Combining terms and rearranging:

$$A_0 - A_1\cos\phi - A_2\cos 2\phi - A_3\cos 3\phi - \cdots = \alpha - \frac{d\bar{z}}{dx}$$

The student should verify this result. Thus, on integrating, Eq. 5.22 can be written as:

$$\left(\frac{d\bar{z}}{dx}\right) = \alpha - A_0 + \sum_{n=1}^{\infty} A_n \cos n\phi. \tag{5.23}$$

Eq. 5.23 provides the mechanism for evaluating the A_n's. The equation must hold at any chordwise station, x, at which station the left side is known and the angle ϕ is determined by the transformation $x = (c/2)(1-\cos\phi)$. Notice that Eq. 5.23 also could be written in terms of θ as ϕ because both angles correspond to "a chordwise station." The two angles must be distinguished carefully only when they both appear together in an integrand.

The A_n's in Eq. 5.23 are solved in two steps, using standard Fourier-series techniques, as follows:

1. *Solve for A_0*: Integrate Eq. 5.23 over the interval $(0 - \pi)$; namely:

$$\int_0^\pi \frac{d\bar{z}}{dx} d\phi = \pi(\alpha - A_0) + \int_0^\pi \sum_{n=1}^\infty (A_n \cos n\phi) d\phi = \pi(\alpha - A_0)$$

because over this interval:

$$\int_0^\pi (A_n \cos n\phi) d\phi = 0$$

for all n. Rearranging:

$$A_0 = \alpha - \frac{1}{\pi} \int_0^\pi \frac{d\bar{z}}{dx} d\phi. \tag{5.24}$$

In this equation, the camber-line slope can be expressed as a function of the angle ϕ through the change of variable expression (see Example 5.1).

2. *Solve for the remaining A_n's*. Multiply both sides of Eq. 5.23 by $\cos m\phi$, where m is a dummy counter, and again integrate over the interval:

$$\int_0^\pi \frac{d\bar{z}}{dx} \cos m\phi d\phi = \int_0^\pi (\alpha - A_0) \cos m\phi d\phi + \int_0^\pi \sum_{n=1}^\infty (A_n \cos n\phi) \cos m\phi d\phi = I_1 + I_2.$$

From calculus, $I_1 = 0$ for all values of m, whereas $I_2 = 0$ when $n \neq m$ and $I_2 = \frac{\pi}{2} A_n$ when $n = m$.

Using these facts and rearranging:

$$A_n = \frac{2}{\pi} \int_0^\pi \frac{d\bar{z}}{dx} (\cos n\phi) d\phi \qquad (n \geq 1). \tag{5.25}$$

The vortex-sheet-strength distribution for the cambered thin airfoil now is known because Eqs. 5.24 and Eq. 5.25 provide the mechanism for evaluating the coefficients in the Fourier-series expression for $\gamma(\theta) = \gamma(\phi)$ in Eq. 5.21. Notice that the vortex-sheet strength depends on the local slope of the mean camber line because the slope appears in both coefficient equations. Also, note that for a symmetrical airfoil, $A_0 = \alpha$, and all of the other A_n's are zero so that Eq. 5.21 reduces to the prior result for the symmetrical airfoil in this special case, as expected.

If the chordwise pressure distribution on the airfoil is required, all of the A_n's must be found. Evaluating the aerodynamic coefficients is less demanding and the expressions for these coefficients are now determined.

Lift Coefficient

We recall that:

$$L' = \rho V_\infty \Gamma = \rho V_\infty \int_0^c \gamma(\xi) d\xi = \rho V_\infty \int_0^\pi \frac{c}{2} \gamma(\theta) \sin \theta d\theta.$$

Then, we substitute for $\gamma(\theta)$ the Fourier-series expression, Eq. 5.21, so that:

$$L' = \rho V_\infty \frac{c}{2} \left[\int_0^\pi A_0 (1+\cos\theta)d\theta + \int_0^\pi \sum_{n=1}^\infty A_n \sin n\theta \sin\theta d\theta \right].$$

These two integrals may be evaluated using basic trigonometric identities and integral tables. The result is as follows:

$$C_\ell = \frac{L'}{q_\infty c} = \pi(2A_0 + A_1). \tag{5.26}$$

Notice that the value of the lift coefficient depends on both the angle of attack and the camber of the airfoil and that only two Fourier coefficients need to be evaluated. Substituting for A_0 and A_1, Eq. 5.26 may be written as:

$$C_\ell = 2\pi \left[\alpha + \frac{1}{\pi} \int_0^\pi \frac{d\overline{z}}{dx}(\cos\theta - 1)d\theta \right] = 2\pi(\alpha + \beta), \tag{5.27}$$

where β is an angle (in radians) whose magnitude (a constant for a given problem) only depends on the camber of the airfoil, a geometric property of the airfoil shape. From Eq. 5.27, the following facts emerge:

1. The *lift-curve slope for an arbitrary* thin airfoil is:

$$\frac{dC_\ell}{d\alpha} = 2\pi, \tag{5.28}$$

   which is the same expression as for a symmetrical airfoil.
2. When the geometric angle of attack has the value $\alpha = -\beta$, the lift coefficient (and the lift) is zero. This special angle of attack only depends on the character of the airfoil camber and is called the *angle of zero* lift, α_{L_0}. Setting $C_\ell = 0$ in Eq. 5.27, the magnitude of this angle (in radians) is given by:

$$\alpha_{L_0} = -\frac{1}{\pi} \int_0^\pi \frac{d\overline{z}}{dx}(\cos\theta - 1)d\theta. \tag{5.29}$$

   The angle of zero lift is always negative for a positive-camber airfoil.
3. A new angle of attack, the *absolute angle of attack* α_a, now may be defined from Eq. 5.27 with β set equal to $-\alpha_{L_0}$. Thus,

$$C_\ell = 2\pi \left[\alpha - \alpha_{L_0} \right] = 2\pi\alpha_a. \tag{5.30}$$

Examine Fig. 5.23. If an airfoil has positive camber, then when the geometric angle of attack, α, is zero and the chord line is coincident with the flow direction, the lift coefficient is greater than zero. It is necessary to use a negative geometric angle of attack when the zero lift line (ZLL) is aligned with the flow to make the lift coefficient zero. The angle between the chord line and the ZLL is the angle of zero lift, a quantity that usually carries a negative sign. The absolute angle of attack is the angle between the oncoming flow and the ZLL. Notice also that the angle of zero lift is zero for a symmetrical airfoil.

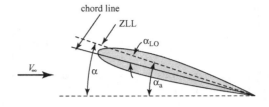

Figure 5.23. Absolute angle of attack.

Moment Coefficient

Because the maximum camber ratio of a thin airfoil is small by definition, the moment expression for a cambered airfoil can be represented adequately by the expression derived for a symmetrical airfoil (i.e., flat plate). Then, referring to Fig. 5.21 and the development for a symmetrical airfoil:

$$M'_{LE} = -\int_0^c (P_L - P_U)(x)dx = -\int_0^c (\rho V_\infty \gamma)(x)dx.$$

Applying the transformation from linear to angular measure and writing this equation in coefficient form results in:

$$C_{m_{LE}} = -\frac{2}{V_\infty}\int_0^\pi \gamma(\phi)[1 - \cos\phi](\sin\phi)d\phi.$$

Substituting for $\gamma(\phi) = \gamma(\theta)$ from Eq. 5.21 and integrating using basic trigonometric identities and integral tables results in:

$$C_{m_{LE}} = -\frac{\pi}{2}\left[A_0 + A_1 - \frac{A_2}{2}\right] = -\frac{C_1}{4} + \frac{\pi}{4}(A_2 - A_1). \tag{5.31}$$

The moment coefficient for a cambered airfoil is seen as depending on only three Fourier-series coefficients and as a function of the angle of attack and camber.

Aerodynamic Center

Other properties of the cambered airfoil may be determined as well. Recall that the aerodynamic center is the point on the airfoil about which the moment is independent of the angle of attack. Now, Eq. 5.31 may be written as:

$$C_{m_{LE}} = -\frac{C_1}{4} + \frac{\pi}{4}(A_2 - A_1) = \left[C_{m_{LE}}\right]_{\alpha_a} + \left[C_{m_{LE}}\right]_{camber} \tag{5.32}$$

because the $C_\ell/4$ term is a function of the absolute angle of attack, α_a (which depends on both the geometric angle of attack and camber) and the second term is a function only of camber (i.e., airfoil geometry) because A_1 and A_2 depend on the slope of the mean camber line, *not* on the angle of attack.

We consider the first term:

$$\left[C_{m_{LE}}\right]_{\alpha_a} = -\frac{C_\ell}{4}$$

and multiply through by $\frac{1}{2}\rho V_\infty^2 c^2$, we find:

$$\left[M'_{LE}\right]_{\alpha_a} = -L'\left(\frac{c}{4}\right).$$

The second equation implies that the integrated pressure force (i.e., the lift) effectively acts at the quarter-chord. Consider what happens at the quarter-chord point when the angle of attack changes and, consequently, Eq. 5.30 lift changes. Because the lift acts through the quarter-chord point, the lift-moment arm is zero. This means that as the lift changes, the moment due to lift at the quarter-chord point does not change. Thus, at the quarter-chord point the moment is independent of L' and, therefore, of the angle of attack.

It follows that for an arbitrary airfoil, the aerodynamic center is located at:

$$x_{ac} = \frac{c}{4}. \qquad (5.33)$$

This is the same result as for the symmetrical airfoil. However, the moment about the quarter chord is *not* zero as before; there now is a camber contribution to moment. Consider the case in which $\alpha_a = 0$ so that $L' = C_\ell = 0$. Equation 5.32 states that for this case, the moment about the leading edge is not zero. This means that although the downward-directed integrated pressure force in the lift direction on the upper surface is equal and opposite to the upward directed integrated pressure force, in the lift direction on the lower surface (i.e., the net lift is zero), these two forces are not collinear. Then a net moment must be present. Recall from statics that two equal parallel forces that are opposite in sense and are not collinear are called a *couple*. Recall also that the moment of a couple is the same about any point in the plane. Thus, evaluate the zero-lift moment at the leading edge and then transfer it unchanged to the quarter-chord point to yield the moment about the aerodynamic center. From Eq. 5.76 for zero-lift coefficient,

$$\left(C_{m_{LE}}\right)_{camber} = C_{m_{ac}} = \frac{\pi}{4}\left(A_2 - A_1\right). \qquad (5.34)$$

Equation 5.34 represents the moment coefficient about the aerodynamic center whether or not the lift is zero, because the resultant lift force acts at the aerodynamic center and, hence, creates no moment about that point.

Center of Pressure

We recall that the center of pressure is the point about which the *total* moment is zero. To find the location of the center of pressure, move the couple to that point, as shown in Fig. 5.24, where ac is the aerodynamic center and the center of pressure, cp, is located a distance Δx downstream. For equilibrium, the total moment about the center of pressure must be zero. Thus,

$$L'\Delta x + M'_{camber} = 0 \quad \rightarrow \quad c_l q_\infty c \Delta x + \left(C_m\right)_{camber}\left(q_\infty c^2\right) = 0.$$

Solving:

$$\frac{\Delta x}{c} = -\frac{\left(C_m\right)_{camber}}{C_\ell} = -\frac{\pi}{4}\left[\frac{A_2 - A_1}{C_\ell}\right].$$

Because the aerodynamic center was already found at the quarter-chord point, the nondimensional distance from the leading edge to the center of pressure is given by:

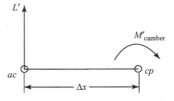

Figure 5.24. Location of the center of pressure.

$$\frac{x_{cp}}{c} = \frac{1}{4} - \frac{\pi}{4}\frac{(A_2 - A_1)}{C_\ell} = \frac{1}{4} - \frac{C_{mac}}{C_\ell}. \tag{5.35}$$

Notice that the location of the center of pressure changes with lift. Also, for small values of lift coefficient, the center of pressure may be downstream of the airfoil. Referring to Fig. 5.24, this is because the moment due to camber is a constant, depending only on geometry, so that as the L' decreases, the lever arm Δx must become increasingly larger to balance the constant moment. The fixed aerodynamic center, then, is a more convenient reference point than the center of pressure, and the load system on an airfoil is described most conveniently by a lift force and a constant moment, both acting at the aerodynamic center.

This discussion concludes the mathematical development of the thin-airfoil theory. Important properties of arbitrary thin airfoils now can be evaluated with relative ease. Also, useful physical insight into the role of camber can be gained.

For example, consider a simple thin airfoil as shown in Fig. 5.25.

We let the maximum camber, H, be fixed and let the chordwise location of the maximum camber, L, vary. Applying the thin-airfoil solutions developed herein, we find that as the location of (fixed) maximum camber ratio H/c moves aft from 25 to 95 percent chord, the angle of zero lift (and, hence, the lift coefficient at an angle of attack) increases dramatically. The magnitude of the zero lift angle increases by about a factor of 5. A similar result is observed regarding the moment coefficient about the aerodynamic center as the maximum camber moves aft. This says, for instance, that flaps and aerodynamic controls that act to change the camber of a wing section should be located near the trailing edge where there is maximum sensitivity to changes in camber.

Agreement of Thin-Airfoil Theory with Experimental Data

The value of any theory lies in its ability to accurately predict physical behavior. Even if a theory is simple, it is worthless if it cannot provide results that satisfactorily agree with experimental data. Thin-airfoil theory is examined from this viewpoint in Figs. 5.26 and 5.3.

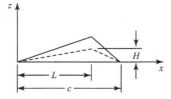

Figure 5.25. Simple cambered airfoil.

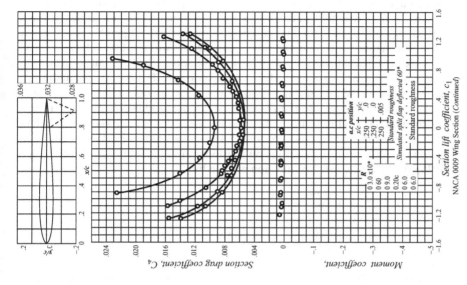

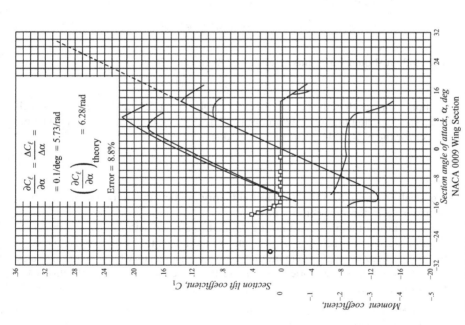

Figure 5.26. Behavior of symmetric thin airfoil (NACA 0009, from Abbott and Van Doenhoff (1959)).

Figure 5.26 shows the results of wind-tunnel tests on a NACA 0009 airfoil, which is a symmetrical airfoil with a 9 percent thickness ratio. The theory developed in Section 5.4 predicts that the lift-curve slope is 2π (per radian) and that the moment about the aerodynamic center is zero. The zero-moment prediction is exactly satisfied by the data because the moment coefficient lies exactly on the axis (for angles of attack in the range of -14 to $+14°$). The lift-curve slope is about 8.9 percent less than the theoretical value. Notice that two sets of data are shown. The lift coefficient versus the angle of attack with flap deflected shows the marked effect of camber on features such as the moment coefficient and α_{L_0}. This important effect is discussed in the next section. Flaps are the principal mechanism for generating control forces on airplanes.

Figure 5.3 shows similar test results for a cambered four-digit NACA airfoil of 15 percent thickness ratio. The predicted behavior of this NACA 4415 airfoil can be obtained by using the results in Section 5.4. Because thin-airfoil theory addresses only the camber function, this is the only airfoil information we need. From Glauert (1926), the mean-camber-line function is given by:

$$y_c = \frac{\tau(2xx_1 - x)x}{x_1^2} \qquad \text{for } x < x_1$$

$$y_c = \frac{\tau[(2xx_1 - x)x + (1 - 2x_1)]}{(1 - x_1)^2} \qquad \text{for } x \geq x_1$$

where, for the NACA 4415 airfoil, $\tau = 0.04$ (4 percent) and $x_1 = 0.4$ (40 percent). After differentiating and performing the integration in Eqs. 5.24 and 5.25:

$$A_0 = 0.06083 \text{ for, say, } \alpha = 4°$$
$$A_1 = 0.16299$$

The student should verify this result. Then, from Eq. 5.26:

$$C_\ell = \pi(2A_0 + A_1) = 0.894.$$

Compare this result with the experimental results in Fig. 5.3; the agreement is excellent. The experimental data also indicate that depending on Re, the aerodynamic center for the NACA 4415 airfoil is located at:

$$0.241 \leq \frac{x}{c} \leq 0.245.$$

We recall from Eq. 5.33 that thin-airfoil theory predicts the aerodynamic center to be at $x/c = 0.25$ for any arbitrary airfoil.

Thin-airfoil theory should not be applied to airfoils with too large a thickness ratio because it is based on a zero-thickness model. Nature provides assistance in this regard because increases in lift due to thickness effects—which are predicted by more accurate inviscid theories—are not fully realized in practical applications due to viscous effects.

EXAMPLE 5.1 *Given:* Consider a thin airfoil with a mean camber line given by the equation:

$$\overline{z} = 4h\left[\frac{x}{c} - \left(\frac{x}{c}\right)^2\right],$$

where h is the maximum camber.

Required: Find α_{L_0}, $C_{m_{ac}}$, and x_{CP}.

Approach: Use the appropriate equations from Section 5.4:

Solution:

(a) $\dfrac{d\bar{z}}{dx} = \dfrac{4h}{c}\left(1 - 2\dfrac{x}{c}\right)$. However $x = \dfrac{c}{2}(1 - \cos\phi) \;\rightarrow\; \dfrac{d\bar{z}}{dx} = 4\dfrac{h}{c}\cos\phi.$

Substituting and evaluating the integral in Eq. 5.29: $\alpha_{L_0} = -\dfrac{2h}{c}.$

(b) Evaluate A_1 and A_2 for substitution in Eq. 5.34. From Eq. 5.25, using the

results from (a), $A_1 = \dfrac{4h}{c}$ $A_n = 0$ for $A_n \geq 2 \Rightarrow C_{m_{ac}} = -\dfrac{\pi h}{c}.$

(c) Using Eq. 5.35 with the results from (b), and evaluating A_o from Eq. 5.24 for

later use in Eq. 5.26:

$$x_{CP} = \frac{h}{\left(2\alpha + \dfrac{4h}{c}\right)} + \frac{c}{4}.$$

Appraisal: The angle of zero lift is negative for positive camber, as it should be. The center of pressure is located downstream of the quarter-chord point.

EXAMPLE 5.2 *Given:* A thin airfoil has a mean camber line defined by:

$$\bar{z} = K\sin^2\phi,$$

where K is a small center.

Required: Find the moment about the aerodynamic center.

Approach: Evaluate Eq. 5.34 by first finding A_1 and A_2 using Eq. 5.25.

Solution:

$$\frac{d\bar{z}}{dx} = \frac{d\bar{z}}{d\phi}\frac{d\phi}{dx}$$

Now, $\dfrac{d\bar{z}}{d\phi} = 2K\sin\phi\cos\phi$ and

$$x = \frac{c}{2}(1 - \cos\phi) \;\rightarrow\; dx = \frac{c}{2}\sin\phi\,d\phi \;\rightarrow\; \frac{d\phi}{dx} = \frac{2}{c\sin\phi}.$$

Hence, $\dfrac{d\bar{z}}{dx} = \dfrac{4K}{c}\cos\phi$. Substitute in Eq. 5.25 to find A_1 and A_2; then, use Eq. 5.34 to find $C_{m_{ac}}$. The result is:

$$C_{m_{ac}} = -\frac{\pi K}{c}.$$

Appraisal: The pitching moment is a restoring moment. Also, because the camber, $\bar{z}$, has a length dimension, then K has a length dimension; when divided by the chord, the resulting moment is dimensionless, as it should be. For this airfoil with positive camber, the moment coefficient is negative, as it should be.

5.5 Thin Airfoil with a Flap

Movable surfaces on airfoils or wings are called *flaps* (i.e., high-lift devices) or *ailerons*. Ailerons provide roll or lateral control for the aircraft. Similar movable surfaces on the horizontal stabilizer and vertical fin supply pitch and yaw control. These are usually called the *elevator* and *rudder*, respectively.

We consider a two-dimensional airfoil at the angle of attack. Then:

$$L' = \frac{1}{2}\rho V_\infty^2 (c) C_\ell.$$

Now, think of this airfoil as being part of a wing. If the aircraft is slowing down in the landing process, then the velocity is decreasing. However, the lift must nearly equal the aircraft weight because the vertical acceleration is (hopefully) very small. This means that as the velocity decreases, the lift coefficient must increase (for constant wing area), which in turn means that the angle of attack must increase. This larger angle of attack may be dangerously close to the stalling angle, or at least result in an undesirable nose-up landing attitude. The alternative is a high-lift device—a flap—to generate a higher lift coefficient at the same angle of attack.

The effects of a 20 percent-chord simple split flap deflected 60° are shown in Fig. 5.26 for the NACA 0009 airfoil. Data points for this configuration are the inverted triangles. Tests were performed at a Re of $R = 6.0 \cdot 10^6$. One set of tests was arried out with "standard roughness," which indicates that the surface was not glass smooth as in most of the experiments. When the flap is deployed, the flow reacts as if the positive camber of the airfoil has increased. Because the flap is near the trailing edge, the effect of this increased camber is to make the angle of zero lift significantly larger in magnitude—in this case, $\alpha_{L_0} = -12°$—and thus to increase the lift coefficient at a particular geometric angle of attack. The value of the maximum lift coefficient ($C_{\ell \max}$) also is increased; notice the significant increase in $C_{\ell \max}$ (i.e., from about 1.3 to 2.1 for this airfoil). The results show that changing the effective camber of the airfoil has a minor effect on lift-curve slope; the primary effects are the change in α_{L_0} and the increased maximum lift coefficient. Figure 5.26 also indicates that the flap deflection causes the angle of stall to be reduced, but this reduction is not large enough to detract from the advantageous shift in the lift curve due to the larger negative-zero-lift angle.

In the case of differential deflection of the ailerons at the wing tips, the camber at one tip is increased while the camber at the other tip is decreased. Because the ailerons are placed in a high-sensitivity location near the trailing edge, and because they are situated near the tips of the wings with long lever arms about the fuselage axis, a small deflection of the ailerons is sufficient to cause a differential lift at the two wing tips that is large enough to roll the airplane.

There are many types of flaps (see Refs. 2–10). So-called slats located at the airfoil leading edge modify the airfoil camber and also force air *tangentially* along the upper surface of the airfoil, which delays airfoil stall. Devices located at the trailing edge are more effective than leading-edge devices (by about a factor of 3) because of the greater distance from the airfoil quarter-chord point. Flaps at the trailing edge may be single- or multi-element (e.g., a jet transport configured for landing). The geometry of the high-lift multi-element device provides an increase in camber and circulation, and the slots between the elements serve to duct high-pressure air from the lower surface to the upper surface of the flap, thereby delaying boundary-layer separation on the highly curved upper surface. These effects are discussed in detail in Chapter 8.

5.6 Distributed Singularity (Panel) Numerical Methods

Thin-airfoil theory is convenient and satisfactorily accurate; however, it has several limitations. It cannot be applied to arbitrarily thick airfoils with confidence because thickness effects were ignored in the derivation of the theory. If pressure-distribution information is required in addition to the airfoil lift and moment, then more of the coefficients in the Fourier-series representation must be found. Also, the result is Δp across the mean camber line and not the pressure distribution on the surface of the airfoil, as needed for a related boundary-layer solution.

The advent of digital computers offers the attractive alternative of a numerical rather than an analytical solution, and many of the assumptions required for the analytical solution can be dropped. In this section, we introduce one of these numerical approaches. It relies on superposition of singularities much like the analytical method; however, instead of a Fourier-series representation for the continuous variation of camber-surface vorticity, we seek a piecewise variation of the vorticity on discrete segments of the airfoil surface. Because the three-dimensional extension of one of these segments is a rectilinear "panel" on a wing surface, these methods are known as *panel methods*.

Panel methods are not restricted to the use of a vortex singularity to represent an airfoil or wing. Source and doublet singularities, discussed in Section 5.4, also may be used. At times, singularities are used in panel methods in combination. For example, it may be useful to represent an airfoil by a source distribution along the mean camber line to simulate the thickness effects plus a vortex distribution along the airfoil surface to account for the generated vorticity and lift.

There are a variety of panel methods. This introduction is intended to provide an idea of how the method is set up and how the aerodynamic performance of an aerodynamic surface is attained. The method outlined is *not* a so-called finite-difference or CFD scheme. Rather, panel methods use the power of a computer to solve large sets of simultaneous algebraic equations that are generated by repeated application of the tangency-boundary condition. Panel methods of solution are extended to three-dimensional wings and discussed in detail in Chapter 6. Eppler, 1990 is a benchmark presentation of panel methods, and several chapters of *Applied Computational Aerodynamics*, referenced in Chapter 6, present a good review of airfoil-panel methods.

The steps in any panel method of solution are the following:

1. *Choosing the number and type of singularities.* These singularities have an unknown strength associated with each one. For example, if there are N singularities, there are N unknown strengths. For this, we need N equations.
2. *Discretizing the aerodynamics surface into* N *segments or panels.* This simply means figuring out where on the surface we will place the singularities (Fig. 5.27). It is important that panels need not be all of the same size. They should be concentrated in regions where the variables are expected to undergo rapid changes. This most often happens in regions of rapid geometry change. Thus, for example, we should expect to see singularities clustered (i.e., small panels) near the leading edge of an airfoil where the curvature is greatest and near the trailing edge because that also is a region where there is a rapid change in flow properties.
3. *Placing the control points.* To devise N equations, recall that the singularities are solutions of the Laplace's Equation. In solving a differential equation, we must generate the generic solution and then impose the boundary conditions for the specific problem. Note that when we make use of the flow singularities, we already have the generic solution. The panel-solution method is based on satisfying the surface-tangency boundary condition. Because there are N unknowns, we must satisfy the surface-tangency condition at N points on the surface. These are called the *control points*, and placement follows logic similar to the placement of the singularities.
4. *Writing one equation for each control point.* Here, the influence of all of the singularities and the freestream at a fixed control point is summed. Then, it is required that the net local velocity be tangent to the surface at that control point or, equivalently, that the normal component of velocity at the control point be zero. To accomplish this, the relative position of the control point relative to the location of all of the singularities is required, as well as the geometrical slope of the surface. In addition, it must be decided how the singularities are to be distributed on their respective panels. In the simple concentrated vortex model illustrated herein, the continuous vortex sheet along the panel is combined into a single point.

   So-called higher-order vortex panel methods may use a curved panel. These methods also represent the vorticity as a linear or nonlinear variation along each panel. In general, higher-order panel methods achieve greater accuracy than using combined singularities on flat panels—but at the expense of a more complicated formulation and, often, with additional computation time required. The Program **AIRFOIL** (see subsequent description) provided for use with this textbook uses flat panels with a linear variation of vorticity from one end of the panel to the other.
5. *Imposing the Kutta condition (if necessary).* Lifting sharp-edge airfoils requires the Kutta condition to produce a realistic solution; other geometries, such as ellipses and cylinders, do not. Requiring that the velocity directions at the trailing edge are the same on the upper and lower surfaces at a cusped trailing edge, or that at a finite-angle trailing edge there must be a stagnation point, imposes the Kutta condition. This condition replaces one of the equations in Step 4, thereby reducing the required number of control points by one.

6. *Solving the system of equations generated by Steps 4 and 5.* Solving for N unknown singularity strengths requires the solution of a full $N \times N$ matrix. Clearly, the more panels that are used (i.e., the larger the value of N), the more accurately the method represents the continuous vorticity distribution along a continuous airfoil surface. Thus, the limitation on the panel method is how quickly we can solve such a system. As computational resources improve in processing speed, the number of panels used also can increase. For example, older PCs running at 33 MHz can solve a system with $N = 41$ in about 2 minutes. Far larger values of N can be handled on modern desktop computer systems.

We next demonstrate an application of the panel method, first by a simple example and then by using a computer code supplied with the text. Consider the NACA 0012 airfoil shown in Fig. 5.27, where the circles in the figure indicate the locations of the combined vortices ($N = 41$ unknowns) on the airfoil. The vortices define the end point of the panels and the control points are located midway between the ends of the 40 panels. Note the clustering of vortices (i.e., small panels) near the leading and trailing edges of the airfoil. Also note that there is one vortex at the leading edge and two coincident vortices at the trailing edge.

Because the airfoil shape is known, the coordinates of the control points can be generated approximately by a simple averaging of the coordinates of the vortices on either side. A slightly more accurate procedure is to average the x/c coordinate and then use the equation for the airfoil surface to generate the z/c coordinate of the control point. In either case, the result is the known positions of 40 control points, which are used to generate 40 equations. The remaining equation results from the Kutta condition.

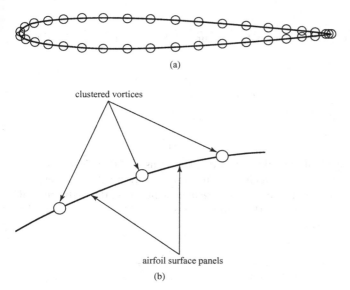

Figure 5.27. NACA 0012 airfoil with 41 vortices indicated.

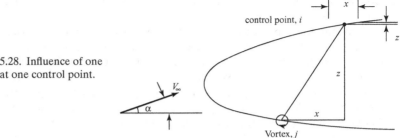

Figure 5.28. Influence of one
vortex at one control point.

Next, we impose the tangential-velocity boundary condition at the 40 control points.
To see how this is done, consider the velocity induced by one vortex at one control point,
as shown in Fig. 5.28. Also shown is the panel on which the control point is located. The
coordinates of the end points of the panels (i.e., the nearest vortices on either side of
the control points) are used to determine the unit tangent to the panel. The coordinates
of the control point are midway between the end points of the panel, called the *control
point* (x_i, z_i). The coordinates of the clustered vortex are (x_j, z_j). Recall from Chapter
4 that the velocity components induced by a vortex at a distance r from the vortex are:

$$u = \frac{\Gamma}{2\pi} \frac{z}{(x^2 + z^2)} \qquad w = -\frac{\Gamma}{2\pi} \frac{x}{(x^2 + z^2)},$$

where Γ is the strength of the vortex and x and z are the distances in the coordinate
directions from the center of the vortex to the control point. Now, the unit normal
to the surface is given by:

$$\hat{n} = \frac{-\Delta z_i \mathbf{i} + \Delta x_i \mathbf{k}}{\sqrt{\Delta x_i^2 + \Delta z_i^2}}.$$

Thus, the component of velocity normal to the surface at the control point, i, induced
by the vortex element, j, is:

$$(V_i)_{\text{Normal}} = -\frac{\Gamma_j}{2\pi} \frac{x\Delta x_i + z\Delta z_i}{(x^2 + z^2)\sqrt{\Delta x_i^2 + \Delta z_i^2}}, \tag{5.36}$$

where Γ_j is an as-yet-unknown vortex strength and:

$$x = x_i - x_j, \qquad z = z_i - z_j.$$

An equation including all of the unknown vortex strengths is obtained by summing
the contributions to the normal velocity at control point i from (a) all 41 vortices,
and (b) the normal component from the freestream velocity (with the body angle of
attack accounted for, if necessary). The sum is then set equal to zero to satisfy the
tangency boundary condition at control point i. Thus,

$$-\sum_{j=1}^{41} \frac{\Gamma_j}{2\pi} \frac{x\Delta x_i + z\Delta z_i}{(x^2 + z^2)\sqrt{\Delta x_i^2 + \Delta z_i^2}} - V_\infty \frac{\Delta z_i \cos\alpha - \Delta x_i \sin\alpha}{\sqrt{\Delta x_i^2 + \Delta z_i^2}} = 0.$$

There are 40 such control points at which this relationship must hold. Thus, simplifying and applying the expression at all of the control points requires that the system of 40 equations,

$$\sum_{i=1}^{40}\left\{\sum_{j=1}^{41}\Gamma_j\frac{x\Delta x_i+z\Delta z_i}{(x^2+z^2)\sqrt{\Delta x_i^2+\Delta z_i^2}}=2\pi V_\infty\frac{\Delta z_i\cos\alpha-\Delta x_i\sin\alpha}{\sqrt{\Delta x_i^2+\Delta z_i^2}}\right\}, \tag{5.37}$$

must be satisfied. Because there are only 40 control points and there are 41 unknown values of the vorticity, a closure equation is required. This is provided by the Kutta condition, which requires that the trailing edge be a stagnation point. Hence, the two vortices, which are coincident at the trailing edge, must have equal and opposite strengths and the equation for the Kutta condition is:

$$\Gamma_1+\Gamma_{41}=0. \tag{5.38}$$

The set consisting of Eqs. 5.37 and 5.38 then can be solved for the 41 unknowns. However, a difficulty is introduced by Eq. 5.38, which states that the strengths of the two vortices at the trailing edge self-cancel. Hence, their influence at each of the 40 control points will self-cancel. In effect, two vortices have been removed from the system, leaving a set of 40 independent equations with 41 unknowns. This difficulty can be resolved by leaving a small opening at the airfoil trailing edge (i.e., the upper and lower surfaces do not actually meet). As a result, the two vortices can be distinguished from one another. This sends the two vortices back to contribute to the summations at the control points in Eq. 5.37 and returns Eq. 5.38 as the 41st equation. The problem then becomes well posed, a system of 41 equations in 41 unknowns. As a result, the formulas for airfoil-thickness distributions (e.g., the NACA four-digit series) used in the calculation often are modified so as to leave the trailing edge slightly open.

Eqs. 5.37 and 5.38 constitute a linear system. The right side of Eq. 5.37 consists of known quantities once the freestream velocity and angle of attack are specified. For the example given here, the values of the 41 vortex strengths are found by solving this linear system. When the vortex strengths are known, the magnitude of the induced tangential velocity at each control point due to all 41 vortices may be calculated. The pressure distribution at each control point on the airfoil surface then follows from the Bernoulli Equation, where the local velocity at the control point is the sum of the induced tangential velocity and the tangential component of the freestream velocity at that control point. Knowing the airfoil-surface-pressure distribution, the lift and moment coefficients follow and/or a boundary-layer solution can be used if a more accurate calculation is required.

Note that although we used 41 vortices here, a much larger number must be used so that the predicted flow over the airfoil is smooth. To diminish this number, we may use the higher-order panel methods mentioned previously. Program **AIRFOIL** uses a linear variation in vortex strength over the panels, as shown in Fig. 5.29.

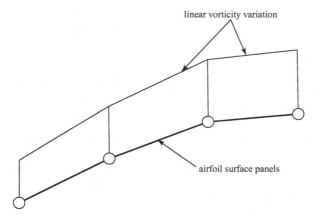

Figure 5.29. Linear variation of vortex strength in Program **AIRFOIL**.

Program AIRFOIL

This program computes the pressure coefficient, C_p, distribution on an arbitrary airfoil at an arbitrary angle of attack using the vorticity panel method discussed herein. It determines all of the values of the lift and moment coefficients, C_ℓ and C_m (for the moment taken about both the leading edge and the quarter-chord points). Additionally, the leading-edge stagnation point and the point-of-minimum-pressure locations are calculated. The location of the leading-edge stagnation point is important in boundary-layer calculations for the airfoil.

The user is presented with two options when executing the program. The first option has built-in information regarding the thickness and camber distribution for NACA four-digit-series airfoils; therefore, a separate data file need not be constructed. The second option allows the user to input the airfoil coordinates of any arbitrary airfoil found in the literature (e.g., Riegels, 1961) or on the Internet. The main ideas used in this code already were presented; an example application is provided as follows.

Example: NACA 0009 Airfoil

Using the first option in the program, a 9 percent-thick airfoil is chosen with zero camber and oriented at a 60-degree angle of attack. The program computes the vorticity distribution as described previously and presents the user with an output screen, which is reproduced here:

There are several things to notice in the figure. First, the graph of C_p versus x/c is usually presented with the z-axis reversed so that the C_p on the upper surface of the airfoil is on the upper part of the figure. The most striking feature of the graph is the large negative value of C_p at the leading edge. This corresponds to extremely low pressure and is the *leading edge suction peak*, which provides a major contribution to the overall lift.

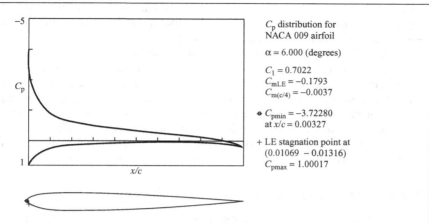

Figure 5.30. Output screen for Program **AIRFOIL**.

The theoretical value of $C_p = 1$ at the stagnation point is well captured. The coordinates of the stagnation point are given to the right side of the figure. Notice that the stagnation point is slightly below the geometrical leading edge of this symmetrical airfoil, as shown in the figure with the "+" sign, because the airfoil is at the angle of attack. The location of minimum pressure (close to the geometrical leading edge) is shown by a small circle. The coordinates and the value of C_p for this point are given to the right side of the graph as well. Finally, the values of C_l and C_m (both at the geometrical leading edge and at the quarter-chord) are given. The value of $C_l = 0.702$ is compared with the value of 0.65 shown in the experimental measurements in Fig. 5.26. The difference is due primarily to viscous effects, which tend to reduce the lift coefficient by lowering the suction peak.

It is interesting that thin-airfoil theory predicts a value of $C_l = 0.658$ for the NACA 0009 airfoil at a 6-degree angle of attack. This value is much closer to the experimental result than the prediction from the panel method Program **AIRFOIL**. The reason for this is that the approximations made in thin-airfoil theory lead to errors in the same direction as the viscous-flow correction required to compare the thin-airfoil prediction with the test result.

The panel method described here provides rapid and acceptably accurate pressure distributions for a given airfoil. The airfoil shape then can be modified, if necessary, to achieve a certain design or performance objective.

The main shortcoming of these methods is that the flows that are computed are necessarily inviscid and irrotational. Airfoils in a viscous flow but with little or no boundary-layer separation may be analyzed by patching together panel-method and boundary-layer solutions. Surface-pressure information from a panel method may be used in conjunction with viscous boundary-layer solutions to either estimate the skin friction on the airfoil or account for the flow-displacement effect due to the presence of the boundary layer interactively. This is illustrated in Chapter 8, where the frictional drag of a wing is evaluated by using an approach called *strip theory* (i.e., treat each section of the wing as two-dimensional). Many refinements are possible in panel methods to increase the accuracy and/or decrease the required computational time. The student is referred to Althaus and Wortmann, 1981, and to the literature for a discussion of these improvements.

Restricting the flow to an irrotational, inviscid model is unrealistic at high airfoil angles of attack or even at low angles of attack for thicker airfoils or for high-lift multi-element airfoils with flaps. In such cases, it becomes necessary to return to solving directly the governing partial-differential equations. Because these equations are nonlinear, superposition cannot be used. At the time of this writing, directly solving the nonlinear governing equations is still an evolving field (i.e., CFD). We introduce an important class of CFD methods—the finite-difference method—in Chapter 8 when flows with viscosity (which must be described by nonlinear equations) are considered.

5.7 Inverse Methods of Solution

The methods discussed here—superposition of distributed singularities (i.e., panel methods) and finite-difference solutions to the nonlinear differential equations,—are solutions to the *direct* problem: namely, given the airfoil shape, we find the pressure distribution. The *inverse*, or design, problem is the focus of much current analysis effort. Here, a streamwise velocity and/or pressure distribution along the airfoil surface is specified at the outset to meet certain performance criteria. Examples include the following:

1. A streamwise pressure distribution might be specified along the airfoil surface that would encourage the preservation of a laminar boundary layer over a considerable chordwise extent. This would result in a significant reduction in drag because the skin-friction drag due to a laminar boundary layer is much less than that due to a turbulent boundary layer.
2. A streamwise pressure distribution that would delay boundary-layer separation and hence reduce the form drag of the airfoil might be specified along the airfoil surface.
3. A streamwise pressure distribution with a relatively low suction peak on the upper surface of the airfoil might be chosen. This would result in a higher critical Mach number for the airfoil, meaning that it could operate efficiently at a higher subsonic cruise speed than an airfoil with a conventional pressure distribution. Increased speed and efficiency are important design objectives.

The required velocity or pressure distribution around the airfoil specifies an airfoil shape, which must be found. An iteration process involving variation of the airfoil shape with conformal-mapping techniques (see Section 5.2) is used in computer codes to find a configuration that is physically correct (i.e., meets the requirements that the shape is closed, with proper flow conditions at infinity). Such airfoils are said to be "tailored" because they were designed for a certain behavior specified in advance. For example, recent work at Delft University on low-drag sailplane airfoils specified a pressure distribution on the lower airfoil surface that was designed to promote laminar flow over the entire lower surface. Such airfoils yield performance close to the theoretical maximum in terms of the L/D ratio.

Motivation for many of the modern low-speed airfoil-design procedures comes from what might appear to be an unexpected source; namely, international competition soaring. The quest for improved performance from the standpoint of reduced L/D, (see Chapter 1), has led to improvements in every facet of low-speed airplane design. One of the most important ways to achieve increases in performance is through improvements in airfoil characteristics. Much of the early work to improve low-speed airfoils was conducted by German engineering students working at the *Akaflieg* groups in major universities. This work continues the tradition begun in the 1920s and 1930s, starting with the development of the Joukowski airfoils and their derivatives, the Göttingen airfoils. Two students, Richard Eppler and Franz Wortmann at the University of Stuttgart, were responsible for leading the revolution in laminar-flow airfoil design in the last several decades. Their airfoils are now in widespread use throughout the world, and their computer codes have been adapted by many institutions, including NASA. Figure 5.31 illustrates several of Eppler's airfoil designs for different aeronautical applications. We

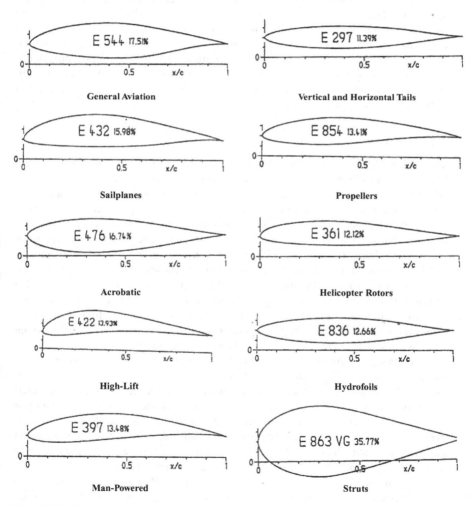

Figure 5.31. Computer-designed-airfoils for several applications (Eppler, 1990).

examine two of the airfoil shapes shown in Fig. 5.31. The discussion follows that given in Boermans (1997).

Consider the E.361 airfoil design for a helicopter rotor shown in Fig. 5.31. When a helicopter is in forward flight, the rotating blades experience a high resultant velocity (and low C_ℓ) as they sweep forward into the oncoming stream (i.e., "advancing" blades), and a low resultant velocity (and higher C_ℓ) as they rotate farther and then move in the streamwise direction (i.e., "retreating" blades). Compared to a similar conventional NACA airfoil, the pressure distribution on the E.361 at low C_ℓ ($\alpha = 1°$) is flattened. This reduction of the suction peak at about 10 percent chord allows for a delay in the onset of compressibility effects on the advancing helicopter blades, with a resulting increase in blade performance. The E.361 also has a more gradual onset of stall and a larger $C_{\ell max}$ compared to a conventional NACA section.

We now consider the Eppler E.476 airfoil tailored for an aerobatic aircraft application (Fig. 5.31). A desirable airfoil shape for an aerobatic aircraft is that the airfoil is symmetrical because the aircraft must have the same behavior in normal and in inverted flight. High-lift coefficients also are required for such an aircraft; the tailored airfoil has a higher $C_{\ell max}$ than a comparable NACA symmetrical airfoil. The E.476 has a gradual stall. However, an airfoil shape may be tailored to have a sharp (i.e., hard) stall because such stall behavior is desirable in an aerobatic aircraft when abrupt maneuvers are required.

5.8 Summary

The goal of this chapter is to provide the student with a solid grounding in the physical behavior of airfoils as well as an introduction to numerical methods, which should be studied in detail in other courses covering numerical analysis. Most modern airfoils are designed by using computer codes. The codes range in complexity from simple panel codes to CFD codes for solving nonlinear-flow problems. There is considerable emphasis on the inverse-airfoil problem. The future of aeronautical engineering clearly is *digital* regarding analysis, design, and production. It eventually will be possible to design reliably an airfoil or a complete wing by computer, with computer "experiments" taking the place (at least partially) of the typical expensive testing with physical hardware. Major benefits include the ability to vary the parameters over wider ranges than might be attainable in wind-tunnel testing. However, at present, there still is considerable dependence on wind-tunnel testing for verifying and "tweaking" or correcting results of the computational effort.

That this procedure works is demonstrated by the outstanding results achieved by Boermans at Delft University of Technology and by other investigators. New laminar-flow airfoils were developed by computational methods and then subjected to careful testing and tuning in a low turbulence wind tunnel at Delft University. For example, a new series of airfoils was developed that have laminar flow over 96 percent of the lower surface. These improvements led to sailplanes with demonstrated glide ratios greater than 60:1. This airfoil technology also is being applied currently in Europe to a new family of high-performance commercial aircraft.

Students reading this textbook who require detailed information regarding performance and, perhaps, the actual airfoil coordinates needed to reproduce an airfoil for applications are referred to Web sites that provide data for thousands of airfoil shapes. There also are Web sites that provide airfoil-design programs and a considerable selection of tutorial material. These items make it easy for students to supplement the material in this chapter and provide additional tools for the latter application of methods discussed herein. An interesting review of airfoil development is provided by Gregorek (1999).

PROBLEMS

5.1. Examine the data for the NACA 4415 airfoil shown in Fig. 5.3. Calculate the lift and pitching moment about the quarter-chord point for this airfoil when the angle of attack is 3.5° and the flow is at standard sea-level conditions with a velocity of 100 ft/s. The chord of the airfoil is 3 feet. Determine the required forces per unit span.

5.2 Consider the NACA 4415 airfoil with a 2m chord in an airstream with a velocity of 50 m/s at standard sea-level condtions. If the lift per unit span is 1,595 N, what is the angle of attack?

5.3 A thin symmetrical airfoil is held at an angle of attack of 2.5°. Use thin-airfoil theory to determine the lift coefficient and the moment coefficient about the *leading edge*.

5.4 The NACA 4412 airfoil has a mean camber line given by:

$$\frac{z}{c} = \begin{cases} 0.25\left[0.8\dfrac{x}{c} - \left(\dfrac{x}{c}\right)^2\right] & \text{for } 0 \le \dfrac{x}{c} \le 0.4 \\[2ex] 0.111\left[0.2 + 0.8\dfrac{x}{c} - \left(\dfrac{x}{c}\right)^2\right] & \text{for } 0.4 \le \dfrac{x}{c} \le 1.0. \end{cases}$$

Use thin-airfoil theory to calculate the lift coefficient as a function of the angle of attack. Plot these results for the range $-10° < \alpha < 10°$. Also calculate α_{L_0}. Compare the results to the plots in Fig. 5.3 for the thicker NACA 4415. Discuss the results.

5.5 Calculate the moment coefficient about the quarter-chord point and x_{cp}/c for the NACA 4412 airfoil for the angle-of-attack range $-10° < \alpha < 10°$. Plot these results and compare them to the plots in Fig. 5.3 for the thicker NACA 4415. Discuss the outcome.

5.6 Using the method described in Section 5.4, show that the expression for the moment about a three-quarter-chord point behind the leading edge of a symmetrical airfoil is:

$$\frac{1}{2}\rho V_\infty^2 \pi \alpha c^2.$$

Verify this result by using the fact that the center of pressure is at the quarter-chord point for any (small) angle of attack.

5.7 An airfoil has a mean camber line with the shape of a circular arc and a constant radius of curvature. The maximum mean camber is Kc, where K is a constant and c is the chord. The freestream velocity is V_∞ and the angle of attack is a. Using the assumption that $K \ll 1$, show that the circulation distribution is:

$$\gamma(\theta) = 2V_\infty \left[\left(\frac{1+\cos\theta}{\sin\theta} \right)\alpha - 4K\sin\theta \right].$$

5.8 For the circular-arc airfoil in Problem 5.7, show that the angle of attack of zero lift is $-2K$ radians and that the moment coefficient about the aerodynamic center is $-\pi K$.

5.9 A mean camber line with a reflexed trailing edge must have a point of inflection. Therefore, the simplest equation that can represent this geometry is a cubic. Four boundary condtions are required to determine the coefficients of such a cubic; two are the condition of zero camber at the leading and trailing edges. The camber-line equation contains two arbitrary constants, a and b, such that:

$$\frac{\bar{z}}{c} = a\left[(b-1)\left(\frac{x}{c}\right)^3 - b\left(\frac{x}{c}\right)^2 + \frac{x}{c} \right].$$

Show that $b = 7/3$ and the moment coefficient about the aerodynamic center is $c_{mac} = a\pi/24$ if the angle of zero lift of the airfoil is zero. Show that c_{mac} vanishes when $b = 15/7$. Plot the mean camber line for these two cases.

5.10 The mean camber line of an airfoil consists of two parabolas joined at their vertices. The maximum mean camber is 4 percent of the chord, and the position of the maximum mean camber behind the leading edge is 20 percent of the chord. Show that this airfoil has the characteristics of $c_{mac} = -0.133$ and $\alpha_{L_0} = -0.063$ radians.

When the geometric angle of attack is $3°$, show that the lift coefficient is 0.722 and the center of pressure is $0.352c$ behind the leading edge.

At the same angle of attack, a flap that is 15 percent of the airfoil chord is deflected $2°$ downward. Show that the lift coefficient becomes 0.828 and the center of pressure moves to a point that is $0.365c$ behind the leading edge.

5.11 Use Program **AIRFOIL** to determine the effect of thickness on the lift-curve slope, $\partial C_\ell / \partial \alpha$. First run the program for a small thickness (e.g., 0.1 percent) and an angle of attack of $1°$. Verify that the results give a lift slope close to the theoretical thin-airfoil result (2π). Then, run the program with a series of increasing thickness ratios up to 25 percent, for example. Is the increase in $\partial C_\ell / \partial \alpha$ linear? Verify the often-used approximation that:

$$\frac{\partial C_\ell}{\partial \alpha} = 2\pi + 0.052\tau$$

and find the range of applicability of this expression (i.e., how thick before nonlinearity is apparent). τ is the thickness in percent of chord; that is, when the thickness is 12 percent, $\tau = 12$.

5.12 Use Program **AIRFOIL** to determine the effect of camber on the lift coefficient C_ℓ. Run the program for several cambers and note that the lift curve is simply displaced upward depending on the maximum camber used. Hence, the expression for the variation of the lift with angle of attack often is given as:

$$C_\ell = \frac{\partial C_\ell}{\partial \alpha}\alpha + \kappa(\text{max camber}).$$

Determine the proportionality factor κ for the basic airfoil shape used.

REFERENCES AND SUGGESTED READING

Abbott, Ira H., and Van Doenhoff, Albert E., *Theory of Wing Sections*, Dover Publications, New York, 1959.

Althaus, D., and Wortmann, F. X., *Stuttgarter Profilkatalog I*, Friedr. Vieweg & Sohn, Branschweig/Wiesbanden, 1981.

Boermans, L. M. M., and van Garrel, A., *Design and Wind-tunnel Test Results of a Flapped Laminar Flow Airfoil for High-Performance Sailplane Applications*, Technical Soaring, Vol. 21, No. 1, January 1997.

Eppler, Richard, *Airfoil Design and Data*, Springer-Verlag, New York, 1990.

Glauert, H., *Elements of Aerofoil and Airscrew Theory*, Cambridge University Press, Cambridge, 1926/1947.

Gregorek, G., "A Century of Airfoil Development," AIAA Paper 99–0116, AIAA 37th Aerospace Sciences Meeting and Exhibit, Reno, Nevada, January 1999.

Hess, J. L., and Smith, A. M. O., Calculation of Potential Flow about Arbitrary Bodies, *Progress in Aeronautical Science*, Vol. 8, S. D. Kucheman (ed.), Pergamon Press, New York, 1967.

Karamcheti, Krishnamurty, *Principles of Ideal-Fluid Aerodynamics*, 2nd ed., Kreiger Publishing Co., Malabar, Florida, 1980.

Katz, Joseph and Plotkin, Allen, *Low-Speed Aerodynamics*, 2nd ed., McGraw-Hill Book Company, New York, 2001.

Lan, C.E. and Roskam, J., *Airplane Aerodynamics and Performance*, 4th ed., Revised Edition, DARcorporation, Lawrence, Kansas, 2000.

Raymer, D. P., *Aircraft Design: A Conceptual Approach*, AIAA Education Series, American Institute of Aeronautics and Astronautics , Washington, DC, 2001.

Riegels, Friedrich Wilhelm, *Airfoil Sections*, Butterworths, London, 1961.

Incompressible Flow about Wings
of Finite Span

In 1908, Lanchester visited Göttingen (University), Germany and fully discussed his
wing theory with Ludwig Prandtl and his student, Theodore von Kàrmàn. Prandtl
spoke no English, Lanchester spoke no German, and in light of Lanchester's unclear
ways of explaining his ideas, there appeared to be little chance of understanding between the two parties. However, in 1914, Prandtl set forth a simple, clear, and correct
theory for calculating the effect of tip vortices on the aerodynamic characteristics of
finite wings. It is virtually impossible to assess how much Prandtl was influenced by
Lanchester, but to Prandtl must go the credit . . .

John D. Anderson, Jr.
Introduction to Flight, 1978

6.1 Introduction

This chapter considers steady, inviscid, incompressible flow about a lifting wing of
arbitrary section and planform. Because the flow around a wing is not identical at all
stations between the two ends of the wing, the lifting finite wing constitutes a three-
dimensional flow problem. The two wing tips are located at distance $\pm b/2$, where b
is the *wing span*.

Certain terms must be defined before a study of finite wings can be begun
(Fig. 6.1). The coordinate axis system used is shown in Fig. 6.1a. A *wing section* is
defined as any cross section of a wing as viewed in any vertical plane parallel to the
x-z plane. It also is called an *airfoil section*. The wing may be of *constant section* or
variable section. If a wing is of constant section, wing sections at any spanwise station
have the same shape (e.g., NACA 2312). If a wing is of variable section, the wing-
section shape varies at different spanwise locations. For example, a wing of variable
section might have a NACA 0012 at the *root section* (i.e., the section in the plane
of symmetry at $y = 0$), then smoothly change in the spanwise direction until the wing
had, a NACA 2312 section at the tip.

The *planform area*, S, of a wing is the projected area of the wing at zero angle
of attack on a plane parallel to the x-y plane. If a wing has a *tapered planform*
(Fig. 6.1b), the section chord lengths vary along the span. The *taper ratio* is defined as
$\lambda = c_t/c_r$. The airfoil sections for a straight-tapered wing of constant section all have

the same descriptor (e.g., NACA 0012), but they do *not* have identical sizes from root to tip. If the tip chord is smaller than the root chord (the usual case), then the tip section has the same thickness *ratio* as the root chord but not the same thickness *dimension*.

The ratio of the square of the wing span divided by the wing-planform area appears so often in aerodynamic equations that this ratio has a name: the aspect ratio, $AR = b^2/S$ (Fig. 6.1c). For a rectangular planform, $AR = b/c$. The AR may be used to define a *mean wing chord* (or *geometric average chord*), $\bar{c}$; namely,

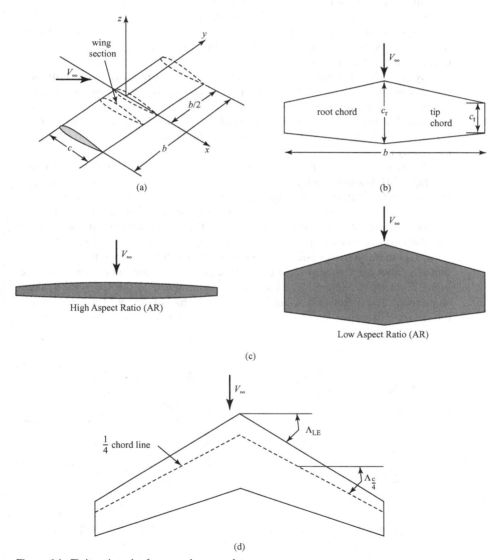

(a)

(b)

(c)

High Aspect Ratio (AR)

Low Aspect Ratio (AR)

(d)

Figure 6.1. Finite wing planforms and nomenclature.

$\bar{c} = b / \text{AR}$. This mean chord should not be confused with the mean aerodynamic chord used in performance calculations. The mean aerodynamic chord is defined by:

$$\text{mean aerodynamic chord} \ = \ \text{mac} = \frac{1}{S} \int_{-b/2}^{b/2} [c(y)]^2 \, dy$$

If the wing quarter-chord line is not parallel to the y-axis, then the wing is called a *swept wing* (Fig. 6.1d). The sweep angle to the quarter-chord line is used mostly in subsonic flow. Another sweep angle, the angle to the wing leading edge, is important for supersonic-flow considerations. If a swept leading edge and a straight trailing edge meet at a common point, the resulting planform is called a *delta wing*.

For convenience, a wing of rectangular planform is discussed initially; however, this chapter later addresses wings of arbitrary planform. Although the origin of coordinates may be taken at any chordwise station, it often is taken at the one-quarter chord point, as shown in Fig. 6.1a.

By definition, a finite wing has tips. If the wing is experiencing positive lift, then the average pressure on the lower surface of the wing is larger than on the upper surface. This pressure imbalance produces lift and also gives rise to a spanwise flow from the lower surface of the wing around the tips to the upper surface. Such a tip effect is not present when the wing is two-dimensional or of infinite span, with effectively no tips.

At the wing tip, there is a strong vortex set up that is due to the flow around the tip (Fig. 6.2a). Looking upstream, the sense of the tip vortex is clockwise at the left tip and counterclockwise at the right tip. These tip vortices trail downstream behind the wing and can be observed in wing-tunnel flow-visualization tests (Fig. 6.3a).

The spanwise flow field that is set up due to the flow around the wing tips means that the uniform flow that passes under the wing is given an outward velocity component so that the streamlines bend outboard (Fig. 6.2b). Similarly, the streamlines that pass above the wing surface experience an inward flow component and bend inboard. At the trailing edge, then, there is a mismatch in the spanwise velocity component. The spanwise flow just below the wing has an outboard component and the spanwise flow just above the wing has an inboard component. Such a velocity

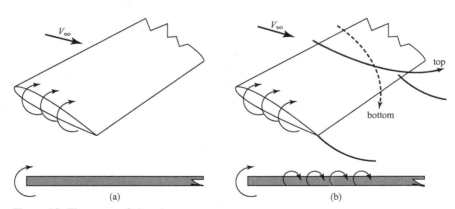

Figure 6.2. Flow over a finite wing.

discontinuity, occurring in essentially zero distance, was observed previously across a mean-camber line in two-dimensional thin-airfoil theory, where it was modeled by using a vortex. Likewise, the discontinuity in velocity that is present at the trailing edge of a finite wing is modeled by using vortices that begin at the wing trailing edge and trail downstream. The sense of the trailing vortices is clockwise for the left half-wing (looking upstream) and counterclockwise for the right half-wing. Each vortex filament trails downstream behind the wing like a thread . As might be expected, the magnitude of the spanwise-flow components and, hence, of the velocity discontinuity at the trailing edge and the related strength of the trailing-vortex filaments varies across the wing span. The spanwise-flow component, and the resulting velocity mismatch, is largest near the tip and zero at mid-span. Thus, the trailing vortices near the tip are much stronger than those farther inboard. It is shown later that the strength of the trailing-vortex system is closely related to the rate of change of circulation along the wing span.

All of the trailing-vortex filaments together are called a *vortex sheet*. Such a sheet is observed in practice as a thin viscous layer of high vorticity coming off the wing trailing edge. In practice, this vortex sheet is unstable and, at a distance downstream of the wing, the entire sheet rolls up into two large contra-rotating vortices that continue downstream and ultimately diffuse under the action of viscosity. These large vortices can be observed in vapor trails left behind aircraft flying at an altitude such that the low pressure in the core (i.e., center) of the vortex causes the water vapor in the atmosphere to condense and become visible, thus acting as a visualization medium (Fig. 6.3b). Figure 6.3(b) also shows some effects of compressibility made visible by the same atmospheric conditions that show the tip vortices.

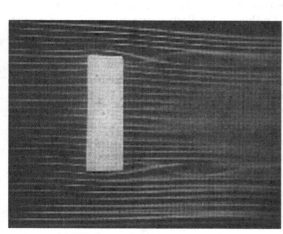

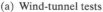
(a) Wind-tunnel tests   (b) Flight

Figure 6.3. Visualization of the tip vortex associated with flow over a finite wing.

For purposes of analysis, the vortex sheet can be approximated, with small error, as being planar. It is possible to ignore the downstream deformation of the vortex sheet because the velocity disturbance at the wing caused by a segment of the trailing-vortex filament decreases by the inverse-square of the distance from the segment to the wing. Thus, a few chord lengths downstream of the wing, the distortion of the vortex sheet has negligible effect at the wing. Numerical solutions allow the vortex sheet to deform naturally, if desired.

The existence of a trailing-vortex sheet, which is not present in the case of an infinite wing, has major significance. It is shown later that a rectangular wing of constant section does not exhibit the same properties as the individual airfoil sections that comprise it. Regarding lift, for example, the lift-curve slope of a finite wing of constant section is found to be less than the (two-dimensional) lift-curve slope of the airfoil sections. Thus, it is important to be able to predict wing lift and to optimize lift for given constraints. The primary task in this chapter, then, is to examine the behavior of a finite wing and to discover how forces on the wing depend on wing-shape parameters.

Recall that for a two-dimensional airfoil, the drag was predicted to be zero according to the inviscid-flow model. That is, when the streamwise component of pressure force acting on an element of airfoil surface was integrated around the airfoil, the net drag force was zero. However, the drag of a finite wing is found to be *nonzero*, even in an inviscid flow. The presence of the trailing-vortex sheet affected the pressure distribution on the wing such that there now is a streamwise-force component, or an *induced drag*. This drag is the price for generating lift with a finite wing, and it must be added to the drag due to viscous effects to arrive at the total drag of a wing. The functional dependence of this induced drag on wing geometry must be investigated because all forms of drag must be minimized.

In addition to finding how wing lift and drag depend on the geometry of the wing, it is important to determine the distribution of force (primarily the lift force) across the span for a given wing geometry. Imagine a wing to be a beam cantilevered from the fuselage. If a beam structure is to be designed successfully, the force distribution along the beam must be known. Similarly, a structural analysis of a wing requires that the wing-spanwise-lift distribution, or *spanwise loading*,* be specified by an aerodynamicist.

This chapter begins with a discussion of the Biot-Savart Law, which provides a mathematical relationship to account for the presence of the trailing-vortex sheet at the wing. Because the flow around a finite wing must satisfy the Laplace's Equation, it might be supposed that the finite-wing problem could be solved analytically by somehow distributing vortices over the wing surface, thereby paralleling the treatment of thin-airfoil theory in Chapter 5. This leads to a complex mathematical model. As a simplification, the vortices representing the wing are lumped together into a *lifting line*, which is a single finite-strength vortex filament extending spanwise from tip to tip and which represents the wing. The resulting theory is considered first in this chapter because it provides a fast and accurate way to uncover the variation of

---

* Be careful not to confuse this term with *wing loading*, which is an important parameter in calculating the performance of a flight vehicle. *Wing loading* is defined as the weight of the flight vehicle divided by the wing-planform area.

wing properties across the span and thus the basic dependence of wing behavior on wing geometry. Following this discussion, two numerical solutions—termed *vortex panel* and *vortex lattice* methods—are discussed. Both methods model the finite wing by distributing vortices, either on the wing surface or on the mean-camber surface (i.e., thin wing), over the entire wing planform. Strip theory, a method to account for viscous effects if the wing angle of attack is small, is introduced next. To conclude this finite-wing chapter, several related topics are outlined—namely: ground effect, winglets, vortex lift, and strakes and canards.

6.2 The Biot-Savart Law

Each filament of the trailing-vortex sheet "induces" velocity components at every point on and around a wing. As mentioned in Chapter 5, the vortex sheet does not actually *cause* a velocity to be present at a point. Rather, the flow field associated with the vorticity coming from the wing is configured such that it may be thought of as being induced by the vortices present in the flow model. A method of calculating these induced velocities is sought here as a generalization of the predicted behavior of the point vortex in two dimensions (see Chapter 4).

The general Biot-Savart Law mathematically expresses the velocity induced at any point in space by an element of a curved vortex filament. In this form, it is sometimes used to calculate the shape of the wakes of helicopter blades.

In the analytical and numerical models discussed herein, the vortex sheet is assumed to be planar, which allows the Biot-Savart Law to be considered in a simpler form. More complex analyses allow the trailing-vortex sheet to distort due to self-induced velocity disturbances. The Biot-Savart Law is stated here without proof. However, when applied to the familiar two-dimensional point vortex as a special case, the law provides the anticipated result.

We consider a vortex filament of length 1-2 with a constant strength, as shown in Fig. 6.4. Let ds be an increment of length of the filament.

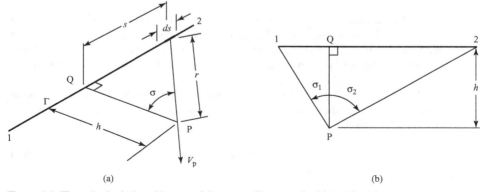

Figure 6.4. The velocity induced by a straight vortex filament of arbitrary length.

The Biot-Savart Law states (Fig. 6.4a) that if the filament and Point P are both in the same plane, then the velocity induced at the fixed but arbitrary Point P by an increment of filament, ds, is perpendicular to that plane and has a magnitude, dV_P, given by

$$dV_p = \frac{\Gamma}{4\pi} \frac{\cos\sigma}{r^2} ds, \qquad (6.1)$$

where r is the length of a line joining the increment of filament, ds, to Point P; PQ is the normal from Point P to the filament (of length h = constant); and the angle σ is the angle that r makes with PQ. We recognize from geometry (Fig. 6.4) that with s as the distance between Q and ds, then:

$$\frac{h}{r} = \cos\sigma$$

$$\text{and } \frac{s}{h} = \tan\sigma \rightarrow ds = h\sec^2\sigma \, d\sigma.$$

Substituting these relations as into Eq. 6.1 leads to an expression in terms of the variable σ, the length h being a constant. Summing the contributions of all of the increments ds comprises a filament 1-2 of arbitrary length and orientation, as illustrated in Fig. 6.4b.

$$V_P = \frac{\Gamma}{4\pi} \int_{\sigma}^{\sigma_2} \frac{\cos\sigma}{\left(\dfrac{h}{\cos\sigma}\right)^2}(h)\sec^2\sigma d\sigma = \frac{\Gamma}{4\pi h} \int_{\sigma_1}^{\sigma_2} \cos\sigma d\sigma$$

or

$$V_P = \frac{\Gamma}{4\pi h}[\sin\sigma_2 - \sin\sigma_1]. \qquad (6.2)$$

Note in Fig. 6.4b that if the end point(s) of the vortex filament lies to the left of the point of interest, P, then σ_2 is a negative angle.

If the vortex filament is infinite in extent, then $\sin\sigma_1 \rightarrow -1$ and $\sin\sigma_2 \rightarrow 1$. Equation 6.2 then states that:

$$V_P = \frac{\Gamma}{2\pi h}. \qquad (6.3)$$

Notice that this result agrees with the two-dimensional vortex behavior discussed in Chapter 4. The vortex filament in Eq. 6.3 is of constant strength, extends to infinity in both directions, and is perpendicular to the plane P–Q containing Point P. The Biot-Savart Law thus gives the expected result when applied to a problem corresponding to a two-dimensional point vortex.

If the vortex filament in Fig. 6.4 were only semi-infinite in length and extended from Q to infinity, then the limits of integration in Eq. 6.2 would be $\sigma_1 = 0$ and $\sigma_2 \rightarrow \pi/2$, leading to:

$$V_P = \frac{\Gamma}{4\pi h}. \qquad (6.4)$$

This result is useful later.

Example 6.1 *Given*: A rectangular vortex filament of strength = 200 m/s has the dimensions shown here. The rectangle is in a plane defined by A-B-C-D.

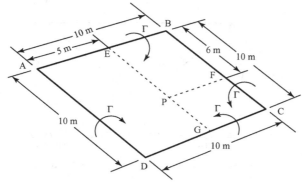

Required: Find the magnitude of the velocity induced by the vortex in a direction normal to the plane at Point P if Point P lies in the same plane.

Approach: The contribution of the four lengths of the vortex filament must be evaluated by applying the Biot-Savart Law, as expressed in Eq. 6.2. The limits of integration in this equation must be modified to reflect the different finite lengths of each segment of vortex filament.

Solution: Consider each segment in turn. The required angles are rounded off to the nearest degree, for convenience.

a. Segment A-B:

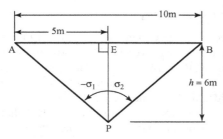

$$V_{P(a)} = \frac{\Gamma}{4\pi(6)}[\sin(40°) - \sin(-40°)] = \frac{\Gamma}{24\pi}(1.28)$$

b. Segments B-C and A-D have the same contribution by symmetry. Thus,

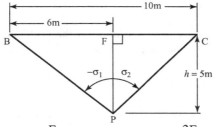

$$V_{P(b)} = (2)\frac{\Gamma}{4\pi(5)}[\sin(39°) - \sin(50°)] = \frac{2\Gamma}{20\pi}(1.40)$$

c. Contribution of segment C-D:

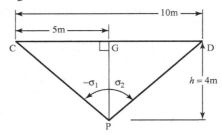

$$V_{P(c)} = (2)\frac{\Gamma}{4\pi(4)}[\sin(51°) - \sin(-51°)] = \frac{\Gamma}{16\pi}(1.56).$$

Adding the three results and substituting the given vortex strength:

$$V_P = 18.5 \text{ m/s}.$$

Appraisal: The given units may be verified by evaluating the units in Eq. 6.4 or in the vortex theory of lift, $L' = \rho V_\infty \Gamma$, where L' is the lift per unit span (meter) and ρ is the density mass per cubic meter.

Notice that P–E, P–F, and P–G in turn play the role of P–Q in Fig. 6.4, whereas A–P and B–P, for example, play the role of the line of variable length r in Fig. 6.4 at two extreme end points. Extreme care must be taken so as to use the correct signs for σ_1 and σ_2.

6.3 Prandtl Lifting-Line Theory

Ludwig Prandtl is considered by some to be the greatest aerodynamicist of all time. He was responsible for solving several of the most important theoretical problems in aeronautics, including the correct manner for treating viscous effects by means of boundary-layer theory and the fluid dynamics of finite wings by introduction of the elegant yet simple idea of the bound vortex.

Two important vortex theorems attributed to Helmholtz must be reviewed before the analysis may begin. These theorems state that:

(a) The strength of any vortex filament must be constant along the length of the filament.
(b) A vortex filament cannot end in the fluid; it must close on itself or end at a boundary in the fluid.

These ideas provide a useful visualization of the problem and motivate many of the mathematical steps in developing the theory for lift generation on a finite-wing surface.

Lifting-Line Concept

In both the thin-airfoil analysis and the numerical methods discussed in Chapter 5, the tangency-boundary condition was imposed along the mean-camber line or along the surface of the airfoil to generate a solution. The same is true in the numerical methods for three-dimensional wings, discussed later in this chapter. However, in the lifting-line theory that follows, a different approach is taken.

In the lifting-line analytical model, the vorticity representing a wing is collected into a single finite-strength vortex extending spanwise from wing tip to wing tip. Such a model leads to a simple solution of acceptable accuracy, providing that certain restrictions are placed on wing planform. If this analysis is to give satisfactorily accurate results, then the wing must be of relatively large aspect ratio (usually greater than 4 or 5) and the wing planform should not have a large taper or sweep. The reason for these restrictions becomes apparent as the theory develops. Because the wing is to be represented by a single spanwise vortex, the resulting theory does not supply any chordwise pressure-distribution information. However, the theory predicts forces on the wing satisfactorily through the vortex theory of lift.

The line vortex containing the combined chordwise circulation of the wing is termed a *lifting line*. It is also termed a *bound vortex* because it is fixed at the location of the wing in the stream and is not free to move. In accordance with the second Helmholtz theorem, this bound vortex cannot end at the wing tips. Thus, the lifting line is turned in a downstream direction at the tips and the two vortices trail downstream (Fig. 6.5). These two vortices have a physical counterpart (see Fig. 6.3): namely, trailing vortices generated at the wing tips. The two trailing vortices are termed *free vortices* because they are free to move and do not represent a fixed solid surface. The resulting U-shaped vortex system is termed a *horseshoe vortex* because of its shape.

The horseshoe vortex is designed to close on itself by closing the end of the horseshoe with another vortex (see the dashed line in Fig. 6.5). This has a physical reality, representing the starting vortex discussed in conjunction with two-dimensional airfoils and the Kutta condition in Chapter 5. In a steady-flow problem, this starting vortex is far downstream and does not affect results at the wing because of the $1/(\text{distance})^2$ influence. Thus, the horseshoe vortex alone may be used to model flow over a finite wing.

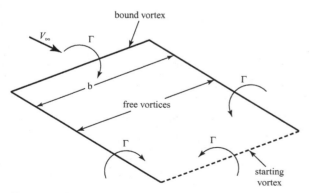

Figure 6.5. Simple horseshoe vortex representation of a wing.

Vortex Sheet

The vortex model in Fig. 6.5 is too simple and does not represent physical reality. If the circulation, Γ, is constant across the span—as it must be to satisfy the first Helmholtz theorem—then according to the Kutta–Joukouski Theorem (see Chapter 4), $L' = \rho V_\infty \Gamma$, and the lift per unit span must be constant across the span up to the wing tips. However, the lift is caused physically by a difference in surface pressure between the lower and upper wing surfaces; this pressure difference must be zero at the wing tips because there is no physical barrier between the two surfaces. Accordingly, the lift must go to zero at the wing tips. This contradiction is resolved by representing the wing by a lifting line consisting of a bundle of vortex filaments of different lengths (Fig. 6.6). The lifting line is now of variable strength spanwise. Each of the bundled filaments must be a horseshoe vortex, so that a number of vortex filaments trail downstream. In the final model, there is an infinite number of such filaments, each of infinitesimal strength, and the trailing filaments form a *trailing-vortex* sheet. In Fig. 6.6, only a few vortex filaments are shown, for clarity.

Before deriving the equation for calculating the induced velocity at the lifting line, the model given in Fig. 6.6 must be made more precise. Consider the lifting line as consisting of four vortex filaments (Fig. 6.7). Each bound-vortex filament has a different length and has a strength given by Γ_n. The lifting line then has a variable strength, ranging from Γ_1 at the wing tips to Γ_4 at the plane of symmetry, where:

$$\Gamma_1 = \Delta\Gamma_1$$
$$\Gamma_2 = \Delta\Gamma_1 + \Delta\Gamma_2$$
$$\Gamma_3 = \Delta\Gamma_1 + \Delta\Gamma_2 + \Delta\Gamma_3 \qquad\qquad (6.5)$$
$$\Gamma_4 = \Delta\Gamma_1 + \Delta\Gamma_2 + \Delta\Gamma_3 + \Delta\Gamma_4.$$

Because each bound filament is part of a horseshoe vortex, four vortex filament pairs trail downstream. Because all three legs of the horseshoe vortices must have the same strength, the trailing vortices each have the same strengths as the bound filaments from which they trail.

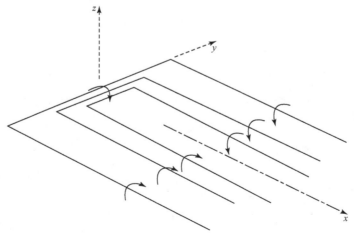

Figure 6.6. Lifting-line and trailing-vortex filaments.

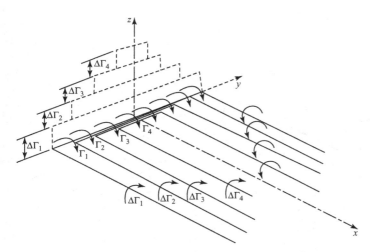

Figure 6.7. Details of lifting-line and trailing-vortex filaments.

Thus,

$$\Delta\Gamma_1 = \Gamma_1$$
$$\Delta\Gamma_2 = \Gamma_2 - \Gamma_1$$
$$\Delta\Gamma_3 = \Gamma_3 - \Gamma_2 \qquad\qquad (6.6)$$
$$\Delta\Gamma_4 = \Gamma_4 - \Gamma_3.$$

Now, we imagine that there are an infinite number of bound filaments, each of a different length and each of a different infinitesimal strength $d\Gamma$. In the limit, the stepwise distribution along the lifting line becomes continuous (Fig. 6.8) so that $\Gamma = \Gamma(y)$. We notice from Fig. 6.7 that the trailing-vortex filament strength is given by the *difference* in bound circulation across a spanwise interval at the point of origin of the trailing filament. With a continuous distribution of bound circulation, $\Delta\Gamma_n \to d\Gamma$. The difference in bound-circulation strength at any spanwise location for an incremental spanwise step dy is given by $\Delta\Gamma_n = \Delta\Gamma = (\Delta\Gamma/\Delta y)\Delta y$ or, in the limit, by:

$$d\Gamma = \frac{d\Gamma}{dy}dy. \qquad\qquad (6.7)$$

This equation states that the strength of the vortex filament depends on the spanwise gradient of the bound vorticity.

A continuous spanwise distribution of circulation (i.e., lift per unit span) and the resulting trailing-vortex sheet is shown in Fig. 6.8.

Although the distribution of $\Gamma(y)$ in Fig. 6.8 is yet to be found, if the wing is symmetrical about the root chord, then $\Gamma(y)$ must have the general shape shown in the figure because L' goes to zero at the wing tips, where the pressure difference across the wing bottom-to-top goes to zero. Equation 6.7 then makes physical sense: It states that the vortex filament trailing from the wing at $y = 0$ has zero strength (i.e., no spanwise-velocity-component mismatch, by symmetry), whereas the trailing

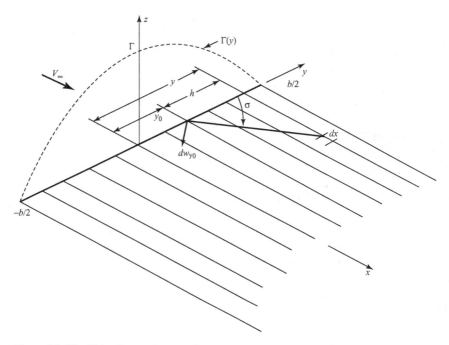

Figure 6.8. The lifting line and vortex sheet.

filaments near the wing tips have the greatest strength (i.e., largest spanwise velocity component mismatch at the trailing edge).

In Fig. 6.8, the spanwise distribution of circulation is shown as symmetrical about mid-span. This is the usual case because wing planforms normally are mirror images of one another about the centerline of a vehicle. An exception is the case in which the *ailerons* (i.e., small control surfaces located near the wing tips) are deflected asymmetrically to generate a rolling moment. Another exception is a study of the wing aerodynamics if one wing-flap actuator failed. These effects can be accounted for by a more elaborate lifting-line treatment. Only the symmetrical case is considered here.

Assumptions of the Lifting-Line Theory

In classic lifting-line theory, the wing is represented by a single finite-strength vortex or lifting line and the trailing-vortex sheet is assumed to be planar and parallel to the oncoming flow. Normally, the vortex sheet is assumed to be in the plane $z = 0$. Thus, the wake can induce velocity components in the x-y plane only in the z direction—that is, in a direction perpendicular to the lifting line. In particular, the wake cannot induce any velocity components in the spanwise (y) direction in the plane $z = 0$ containing the lifting line. Of course, the wake also can induce velocity components above, below, and beyond the span of the lifting line, but none of these are of interest in our flow model.

The basic lifting-line theory, then, neglects the spanwise velocity components induced at the lifting line by the trailing-vortex sheet. Thus, *each airfoil section behaves as if the flow were locally two-dimensional.* This assumption is at odds with the physical argument used previously that the spanwise velocity mismatch at the wing trailing edge gives rise to the trailing-vortex filaments. However, neglecting the spanwise-velocity components gives satisfactory results because they are small, typically a few percentage points of V_∞. The assumption of local two-dimensionality is worst at the wing tips, where the spanwise flow is large, but this area represents only a small fraction of the entire wing. One advantage of this assumption is that it allows a two-dimensional treatment of the downwash at the lifting line. Also, if each airfoil section behaves as if it were in a locally two-dimensional flow, then the chordwise pressure-distribution detail on the wing—which is lost in the lifting-line model—is available from two-dimensional airfoil theory or experimental data as in Chapter 5. The fact that the two-dimensional airfoil section is part of a finite wing is manifested by a modification in the angle of attack at which the section is operating. Finally, experimental data for two-dimensional airfoils may be used to predict the viscous drag of the wing. Thus, the advantages of using this assumption of negligible spanwise flow are many and the theory still exhibits satisfactory accuracy. If the wing is highly swept or of very low AR, then the spanwise-flow component is large, the assumption is invalid, and the simple lifting-line theory breaks down. The simple theory has been extended to treat swept wings, but this is not discussed here. The method becomes complicated and the problem is solved more easily numerically.

Because the lifting line represents the wing, it is positioned at the wing quarter-chord. This is because (1) each airfoil section comprising the wing acts as if the flow were two-dimensional (i.e., the trailing vortices present in the model induce no spanwise-velocity components); and (2) the lift force on an arbitrary two-dimensional airfoil in incompressible flow (see Chapter 5) acts at the aerodynamic center (i.e., at $c/4$), according to thin-airfoil theory. Recall that the assumed chordwise-vorticity distribution that led to this result (see Fig. 5.17) was chosen so as to satisfy the Kutta condition. Thus, placing the lifting line at the wing quarter-chord results in the Kutta condition being approximately satisfied all along the wing trailing edge.

Downwash

The induced velocity at the lifting line is given a special name: the *downwash*. It is assigned the velocity component symbol w and enters with a *negative* sign because the induced velocity is in the negative z-coordinate direction.

Recall that the Biot-Savart Law states that for a semi-infinite vortex filament extending from Point Q to infinity, as shown in Fig. 6.4 and Eq. 6.4, the induced velocity is given by $V_P = \Gamma/4\pi h$. This velocity is normal to the plane containing the filament and Point P. We apply this result to an arbitrary single filament in Fig. 6.8 and assume that the filament trails in the direction of V_∞. The induced velocity (i.e., downwash) at a fixed but arbitrary Point y_0 on the lifting line, due to a semi-infinite vortex filament of strength $d\Gamma$ that originates at the lifting line at a spanwise location y and trails downstream, is given by:

$$V_p = \frac{\Gamma}{4\pi h} \quad \rightarrow \quad dw_{y_0} = \frac{d\Gamma}{4\pi(y-y_0)}. \tag{6.8}$$

Points Q and 2 (at downstream infinity) in Fig. 6.8 aid in orientation with Fig. 6.4. This induced velocity is in the z direction. We note that because Eq. 6.8 is simply a repeat of Eq. 6.4, an integration over the streamwise length of the trailing-vortex filament already occurred. Eq. 6.8 is written as dw to emphasize the fact that this is the induced velocity due to only *one* trailing filament of infinitesimal strength.

Substituting the expression for $d\Gamma$ from Eq. 6.7,

$$dw_{y_0} = \frac{\left(\dfrac{d\Gamma}{dy}\right)}{4\pi(y - y_0)}\, dy. \tag{6.9}$$

Finally, we check the sign. Referring to Fig. 6.8, for the spanwise locations chosen, $y > y_0$ and the sense of the trailing-vortex filament is counterclockwise looking upstream. The velocity induced at y_0 then should be downward, or dw should be negative. Now, from Fig. 6.8, the distribution over the right half-wing is such that the rate of change with y is negative. Thus, Eq. 6.9 is physically correct as written.

Because Eq. 6.9 represents the downwash contribution at y_0 due to *one* semi-infinite filament, it remains to calculate the *total* downwash at y_0 due to *all* of the trailing-vortex filaments from the wing that comprise the vortex sheet. This amounts to a summing, or integration, over y; namely:

$$w_{y_0} = \frac{1}{4\pi} \int_{-b/2}^{b/2} \frac{(d\Gamma/dy)}{(y - y_0)}\, dy. \tag{6.10}$$

Notice that as we sum across the span, some filaments induce an upward component and others a downward component at Station y_0, depending on the relative magnitudes of y and y_0 and also on which half-wing is considered. Equation 6.10 is the sum total of all of these contributions and is directed downward for positive lift on the wing. The singular behavior of the integrand when the integration variable, y, has the specific value $y = y_0$ is addressed in due course.

Figure 6.9 shows a cross section of a wing according to the lifting-line model. The wing is in an oncoming freestream V_∞. Because a vortex filament does not induce any velocity (think of the vortex as exhibiting viscous-dominated, solid-body rotation at the center), the only other velocity component at the wing quarter-chord (i.e., at the lifting line) is the downwash due to the trailing-vortex wake, shown in Fig. 6.9. Because the vortex sheet is aligned with V_∞, the downwash component $w = V_S$ is perpendicular to V_∞. For positive lift, the downwash at the lifting line always is directed downward.

Figure 6.9. Velocity components at the lifting line.

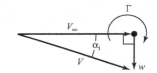

Induced Angle of Attack

Notice in Fig. 6.10 that the lifting line is now exposed to a *relative* wind, V, which is the vector sum of V_∞ and the downwash, w. Thus, the local airfoil section behaves as if the geometric angle of attack, α, has changed by an amount α_i, *the induced angle of attack*. This is the viewpoint that is carried forward—namely, that the angle of attack of the wing section was modified due to the presence of the vortex sheet.

Referring to Fig. 6.10, the concept of relative wind at the lifting line in Fig. 6.9 may be generalized to include the wing-section angle of attack and camber.

The angle of attack to which the wing section responds is no longer the geometric angle of attack, α, but rather the (smaller) effective angle of attack, α_{eff}. Camber may be included in the section shape by introducing the zero-lift angle, α_{L0}. Two-dimensional airfoil results may be used to specify α_{L0} for a particular wing section. When the line Z_{LL} (see Fig. 6.10) is aligned with V_∞ the lift of the airfoil section (behaving two-dimensionally) is zero. If a finite wing is set to a zero-lift condition, then a wing zero lift line (ZLLW) may be specified that is identical all across the span. A wing angle of attack, α_{aw}, then can be defined as the angle between the freestream and the ZLLW.

Now, we recall from two-dimensional thin-airfoil theory Eq. 5.74 that the lift coefficient of a thin airfoil is given by:

$$C_\ell = m_0(\alpha - \alpha_{L0}), \tag{6.11}$$

where the two-dimensional lift-curve slope, m_0 per radian, may be assigned the theoretical value of 2π or a value from experiment. Equation 6.11 also may be written with the lift-curve slope as a_0, where m_0 is per radian and a_0 is per degree.

Because a section of a three-dimensional wing is assumed to behave locally as a two-dimensional airfoil, it follows that for a finite-wing section:*

$$C_\ell = m_0(\alpha_{\text{eff}} - \alpha_{L0}). \tag{6.12}$$

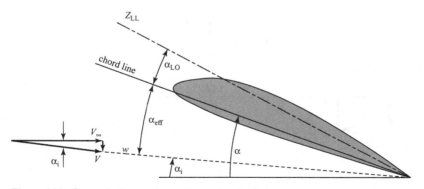

Figure 6.10. General wing section at the angle of attack.

---

* In some texts, the quantity $(\alpha_{\text{eff}} - \alpha_{L0})$ is written as α_0.

Finally, from the geometry shown in Fig. 6.12, at every spanwise station:

$$\alpha_{\text{eff}} = \alpha - \alpha_i, \tag{6.13}$$

where α_i is a positive angle. From Fig. 6.11:

$$\tan \alpha_i = -\frac{w}{V_\infty} \cong \alpha_i. \tag{6.14}$$

Recall that the downwash, w, is a downward-directed velocity component in Cartesian coordinates and is negative. Hence, a minus sign must be introduced in Eq. 6.14 so that the induced angle of attack, α_i, is a positive angle. In general, the induced angle of attack varies across the span.

The mathematical expression for the induced angle of attack follows from Eq. 6.14 and the equation for downwash, Eq. 6.10:

$$\alpha_i = -\frac{1}{4\pi V_\infty} \int_{-b/2}^{b/2} \frac{(d\Gamma/dy)}{(y-y_0)} dy \tag{6.15}$$

Lift and Induced Drag

The effect of the presence of downwash also must be examined from the viewpoint of the forces on a wing section. Consider Fig. 6.11. The Kutta–Joukouski Theorem states that the force generated by flow over a bound vortex per unit span is given by $F' = \rho V\Gamma$ and is normal to V—that is, normal to the *relative* wind. (In two dimensions, the relative wind is the freestream and the force perpendicular to this wind is then the lift force.) It follows from Fig. 6.11 that:

$$L' = F'\cos\alpha_i = \rho V\Gamma \cos\alpha_i = \rho V\Gamma \left[\frac{V_\infty}{V}\right], \tag{6.16}$$

or, finally

$$L' = \rho V_\infty \Gamma. \tag{6.17}$$

This is identical to the two-dimensional result.

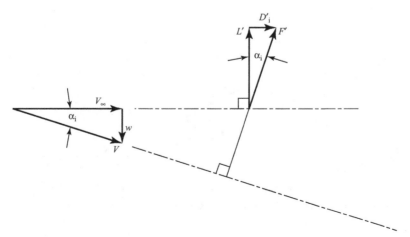

Figure 6.11. The forces on a wing section.

Notice, now, in Fig. 6.11 that the effect of the downwash is to tilt back the force vector, F', at the lifting line; it is no longer perpendicular to the freestream direction as it was in the two-dimensional case. There is now a component of F' in the freestream direction that, by definition, is a drag force. This drag force is called the *induced drag* because it arises due to the presence of the downwash, which is caused by the velocity induced by the trailing-vortex sheet. The induced drag per unit span is given the symbol D'_i. Then, because α_i is small (see Example 6.3):

$$D'_i = L' \tan \alpha_i \cong L' \alpha_i. \tag{6.18}$$

The induced drag is an important new concept. It was not present in two-dimensional airfoil theory because there was no trailing-vortex sheet present in that theory. Induced drag arises in *inviscid* finite-wing theory. It is the price for generating lift with a three-dimensional wing. Induced drag may be thought of physically as representing energy left behind an advancing wing. This energy is in the form of translational and rotational kinetic energy present in the trailing vortices. These vortices are deposited in the atmosphere and, ultimately, are dissipated by viscosity.

Induced drag is a significant drag contribution. Although α_i is small, in Eq. 6.18, it multiplies the lift per unit span, which is a large number proportional to the weight of the flight vehicle. Thus, induced drag must be studied and methods sought to minimize the magnitude of this drag. Recall from Chapter 1 that flight-vehicle performance and efficiency are directly dependent on making the drag from all sources as small as possible.

Wing Twist

Until now, the wing planform is assumed to be rectangular and α_{L0} is considered constant across the span. These restrictions are dropped before the fundamental wing equation is written.

If a wing planform is tapered, for example, then the wing chord is different at each spanwise location, y_0, of interest. This chord is denoted by $c(y_0)$. As noted previously, the induced angle of attack is generally different at each spanwise station of interest; hence, the notation, $\alpha_i(y_0)$ is used. If the wing has twist, then the geometric angle of attack, α, and the angle of zero lift, α_{L0} may be different at different spanwise stations, y_0. There are two types of wing twist as follows:

1. *Geometric twist.* Here, the wing has the same section shape from root to tip (e.g., NACA 0012 at every spanwise station), but the wing is physically deformed by twisting the tip relative to the root (Fig. 6.12a). This has the effect of changing the geometric angle of attack along the span so that $\alpha = \alpha(y_0)$. If the wing is twisted so that the geometric angle of attack of the tip is less than that of the root, this is termed *wash-out*. If the angle of attack at the tip is made greater than that at the tip, this is termed *wash-in*.
2. *Aerodynamic twist.* Here, the geometric angle of attack for each wing section is the same across the span, but the wing sections change. For example, the section could be a symmetrical NACA 0012 at the root, a NACA 1410 at mid-semi-span, and a cambered NACA 2408 at the tip (Fig. 6.12b). Of course, the spanwise

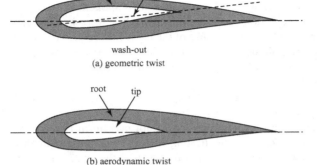

(a) geometric twist

Figure 6.12. Geometric and aerodynamic wing twist.

(b) aerodynamic twist

transition from one section to the next would be smooth. Aerodynamic twist has the effect of making α_{L0} take on a different value at a different spanwise station of interest; that is, $\alpha_{L0} = \alpha_{L0}(y_0)$.

A wing may have geometric or aerodynamic twist or both, in which case the effects are added. The amount and type of twist are specified for a given wing. Wing twist is used to modify the spanwise-loading distribution over a wing planform. For example, Fig. 6.13 shows the distribution of bound-vortex strength or, by definition, C_ℓ (see Eq. 6.19) along an untwisted wing of constant section with moderate taper. We notice in this figure that there are larger values of Γ at some spanwise stations than at others. Thus, at about 60 percent semi-span, the local airfoil section is operating at a comparatively large value of α_{eff} (see Eq. 6.14). The section at $2y/b = 0.60$ is "working harder" than, say, the section at mid-span. Thus, as the geometric angle of attack of the wing is increased, the wing stalls first at about 60 percent semi-span. This may render ineffective a portion of a control surface or flap located near that spanwise station because it would be immersed partially in a "dead air" separated flow. Incorporating wing twist into the wing design could modify the spanwise distribution of α_{eff} such that the wing would first stall farther inboard. An alternative method for tailoring the spanwise stall characteristics of a wing by modifying (i.e., "drooping") the shape of the wing leading edge is in Abbott and Van Doenhoff (1959).

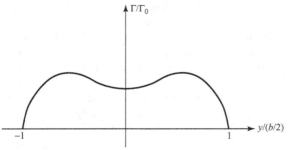

Figure 6.13. Spanwise loading on a tapered wing.

Fundamental Monoplane Wing Equation

This equation is written for a twisted wing of nonrectangular planform. It may be simplified as appropriate. First, by definition:

$$C_\ell(y_0) = \frac{L'(y_0)}{\frac{1}{2}\rho V_\infty^2 c(y_0)} = \frac{2\Gamma(y_0)}{V_\infty c(y_0)}. \tag{6.19}$$

Next, substituting Eq. 6.13 into the expression for C_ℓ in Eq. 6.12 and introducing the new notation:

$$C_\ell(y_0) = m_0[\alpha(y_0) - \alpha_i(y_0) - \alpha_{L0}(y_0)]. \tag{6.20}$$

Finally, we solve for α (for convenience):

$$\alpha(y_0) = \frac{C_\ell(y_0)}{m_0} + \alpha_{L0}(y_0) + \alpha_i(y_0)$$

and substitute for $C_\ell(y_0)$ from Eq. 6.19 and for $\alpha_i(y_0)$ from Eq. 6.15. The result is:

$$\alpha(y_0) = \frac{2\Gamma(y_0)}{m_0 V_\infty c(y_0)} + \alpha_{L_0}(y_0) - \frac{1}{4\pi V_\infty} \int_{-b/2}^{b/2} \frac{(d\Gamma/dy)|_{y_0}}{(y - y_0)} dy \tag{6.21}$$

where the speed and chord distribution, m_0 and α_{L_0}, are known for a given wing.

Eq. 6.21 is called the fundamental wing equation for a monoplane (i.e., for a single wing). This equation can be used to solve for $\Gamma(y_0)$—that is, for the distribution of bound vorticity—across the span. The solution process is not straightforward because Eq. 6.21 is an integro-differential equation for the unknown circulation strength.

The central problem of lifting-line theory is to determine the spanwise distribution of circulation—that is to find $\Gamma(y_0) = \Gamma(y)^*$—as a function of wing-shape parameters. If $\Gamma(y)$ can be found, $L'(y)$ follows directly. Then, the wing total lift is given by the integration of L' over the span, and the induced drag may be calculated from Eq. 6.18. Although circulation is not a physical variable, it is simplest to work with $\Gamma(y)$ as the unknown. Once $\Gamma(y)$ is found, the finite-wing problem is solved. A Fourier-series solution is discussed in the next section.

Spanwise-Lift Distribution on an Arbitrary Wing by Lifting-Line Theory

When faced with solving an integro-differential equation to find the chordwise-vorticity distribution in two-dimensional airfoil theory (see Chapter 5), a Fourier-series representation for the chordwise vortex-strength distribution, $\gamma(x)$, is used. Here, a Fourier series is used to represent the unknown bound-vortex-strength distribution, $\Gamma(y)$, in Eq. 6.21. As in Chapter 5, a transformation is made to relate a spanwise linear location to a location described by angular measure. For example, a circle of radius $b/2$ with the center at mid-span and measuring the angle θ from the positive y axis (Fig. 6.14). The required transformation is:

$$y = \frac{b}{2}\cos\theta. \tag{6.22}$$

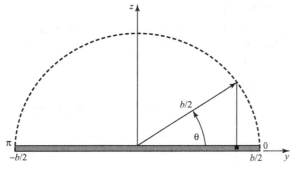

Figure 6.14. Transformation from linear to angular spanwise location.

Now, we assume a Fourier-series representation of the bound-vortex-strength distribution as given by:

$$\Gamma(\theta) = (2bV_\infty)\sum_{n=1}^{\infty} A_n \sin n\theta, \tag{6.23}$$

where A_n are unknown constants. Because L' (hence, Γ) must be zero at the wing tips by physical argument, there is no lead constant term in the series. For symmetrical spanwise loading, all of the cosine terms vanish because they never make a symmetrical contribution about the wing plane of symmetry, $y = 0$ or $\theta = \pi/2$ and because the cosine is positive for $0 < \theta < \pi/2$ and negative for $\pi/2 < \theta < \pi$. The coefficient $(2bV_\infty)$ is introduced for convenience from hindsight. Equation 6.23 is a solution only if a method for finding the A_n's can be established.

We examine the integral term in the defining equation, Eq. 6.21 and call it I. Writing:

$$\frac{d\Gamma}{dy} = \frac{d\Gamma}{d\theta}\frac{d\theta}{dy},$$

the integral term becomes, with the aid of Eq. 6.22:

$$I = \frac{1}{4\pi V_\infty}\int_{-b/2}^{b/2}\frac{(d\Gamma/dy)}{(y_0 - y)}\,dy = \int_{\pi}^{0}\frac{(d\Gamma/d\theta)}{\frac{b}{2}(\cos\theta_0 - \cos\theta)}\frac{d\theta}{dy}\,dy.$$

Using Eq. 6.23 to find $d\Gamma/d\theta$ in the integrand and inverting the limits of integration, this expression for I may be written as:

$$I = \frac{1}{\pi}\sum_{n=1}^{\infty} nA_n\int_{0}^{\pi}\frac{\cos n\theta}{(\cos\theta - \cos\theta_0)}\,d\theta. \tag{6.24}$$

As in two-dimensional thin-airfoil theory, this integrand has a singularity when the running variable of integration, θ, is equal to the fixed angle θ_0, which denotes the spanwise station of interest. The "principle value" of this integral is used here as in Section 5.14, Eq. 5.57; namely:

$$\int_{0}^{\pi}\frac{\cos n\theta\,d\theta}{(\cos\theta - \cos\theta_0)} = \frac{\pi\sin n\theta_0}{\sin\theta_0}. \tag{6.25}$$

Then, the integral in Eq. 6.24 becomes

$$I = -\sum_{n=1}^{\infty} nA_n \frac{\sin n\theta_0}{\sin \theta_0}. \qquad (6.26)$$

Finally, we substitute this expression for the integral in the defining equation, Eq. 6.21, and drop the subscript "0" because the equation must hold at any general spanwise station:

$$\alpha(\theta) = \frac{4b}{m_0 c(\theta)} \sum_{n=1}^{\infty} A_n \sin n\theta + \alpha_{L_0}(\theta) + \sum_{n=1}^{\infty} nA_n \frac{\sin n\theta}{\sin \theta}. \qquad (6.27)$$

This is the equation for determining the Fourier-series coefficients, A_n. The coefficients are found by applying Eq. 6.29 at several spanwise stations specified by different values of θ (see Example 6.2). At each station, c, α, and α_{L_0} are known. Thus, a set of simultaneous equations is generated, which can be solved for the unknown A_n's. The more terms that are taken in the series (i.e., the more spanwise stations at which Eq. 6.27 is applied), the more accurate are the coefficients and the theoretical results. With the A_n's known, the spanwise loading follows from Eq. 6.23.

If the spanwise loading is symmetrical, then $\Gamma(\theta) = \Gamma(-\theta)$. That is, symmetry demands that:

$$\sum_{n=1}^{\infty} A_n \sin n\theta = \sum_{n=1}^{\infty} A_n \sin n(\pi - \theta).$$

Expanding the right side of this equality:

$$\sum_{n=1}^{N} A_n \sin n\theta = \sum_{n=1}^{N} A_n [\sin n\pi \cos n\theta - \cos n\pi \sin n\theta).$$

Now, $\sin n\pi$ is zero for all integers n, but $\cos n\pi = -1$ only for n odd. So, if the identity is to be satisfied, there can be no terms appearing with n even. This means that *for symmetrical loading the A_n's must be zero for all even values of n*. Likewise, for symmetrical loading, we need only calculate the spanwise loading for one semi-span. The loading on the other semi-span is simply a mirror image in the plane of symmetry.

Example 6.2 *Given:* A wing planform is shown here. This wing has a constant section with a section lift-curve slope of $m_0 = 6.7$ per radian and an angle of zero lift of -1.5°. The wing is symmetrical about $y = 0$. The wing has geometric twist, with a geometric angle of attack of 4° at the wing root, decreasing linearly with semi-span to 2° at the tips.

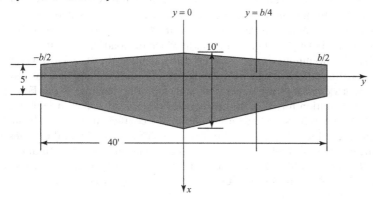

Required: Calculate the spanwise loading on this wing. For convenience, calculate only two terms in the Fourier-series expression for $\Gamma(y)$. Evaluate the coefficients of these terms at mid-span and at half-semi-span.

Approach: Write two simultaneous equations for the two unknown Fourier-series coefficients. Because the wing is symmetrical, all of the even A_n's are zero. The two simultaneous equations for A_1 and A_3 are obtained by writing Eq. 6.23 twice, once at the spanwise location at $y = b/4$ ($\theta = \pi/3$) and once at mid-span ($\theta = \pi/2$).

Solution: Here, the angle of zero lift is a constant across the span. Because the geometric angle of attack varies linearly, we write the following from the given:

$$\alpha = 4° - 2° \left[\frac{y}{b/2} \right] \Rightarrow y = 0, \ \alpha = 4°; \ y = \frac{b}{2}, \alpha = 2°.$$

Regarding the chord, from the given geometry of the planform, it is 10 ft. at the root and 7.5 ft. at mid-semi-span. Then, writing Eq. 6.23 at mid-span, $y = 0$, $\theta = \pi/2$, $c = 10$:

$$4 \left(\frac{\pi}{180} \right) = \frac{(4)(40)}{5.7(10)} [A_1 \sin(\pi/2) + A_3 \sin 3(\pi/2)] +$$

$$+ (-1.5) \left(\frac{\pi}{180} \right) + \left[A_1 + 3A_3 \frac{\sin 3(\pi/2)}{\sin(\pi/2)} \right].$$

Again, at mid-semi-span, $y = b/4$, $\theta = \pi/3$, $c = 7.5$:

$$3 \left(\frac{\pi}{180} \right) = \frac{(4)(40)}{(5.7)(7.5)} [A_1 \sin(\pi/3) + A_3 \sin 3(\pi/3)] +$$

$$+ (-1.5) \left(\frac{\pi}{180} \right) + \left[A_1 + 3A_3 \frac{\sin 3(\pi/3)}{\sin(\pi/3)} \right],$$

solving for the two unknowns, $A_1 = 0.018$ and $A_3 = -0.0047$.

Thus, to two terms:

$$\Gamma = 2(40) V_\infty [(0.018) \sin \theta - (0.0047) \sin 3\theta]$$

from Eq. (6.25).

Appraisal: The angles that appear in the defining equation must be expressed in radians. The resulting expression for $\Gamma(\theta)$ can be expressed as $\Gamma(y)$ through the transformation, Eq. 6.24, and evaluated at several values of y, giving the spanwise loading L' across the span using Eq. 6.18. The loading is symmetrical about $y = 0$. The number of stations at which the loading is evaluated is independent of the number of stations used to calculate the A_n's. However, the more spanwise stations at which the fundamental equation is applied, the more simultaneous equations are generated, and the more accurate is the Fourier series representation for $\Gamma(y)$. In the previous example, if a third station was used to generate a third simultaneous equation, the coefficient A_5 would be introduced and the values of A_1 and A_3 would change. The sensitivity of the values of A_n to the number of simultaneous equations used may be studied by running Program **PRANDTL**, which is introduced shortly. In this program, the user selects the number of Fourier-series coefficients desired and the computer program solves the simultaneous equation set.

A method for determining the spanwise-lift distribution on a wing of arbitrary planform is now established. Fig. 6.13 shows a calculated spanwise loading, $\Gamma(y)$, for an untwisted tapered wing of constant section at a fixed geometric angle of attack. The loading is nondimensionalized by the value of the circulation at mid-span $(y = 0), \Gamma = \Gamma_0$.

Recalling the results in two-dimensional airfoil theory, we might suspect that an integration to find the total forces on the wing would result in a dependence on only a few of the Fourier-series coefficients. This is investigated next.

Forces on an Arbitrary Finite Wing from Lifting-Line Theory

We now apply the lifting-line theory to estimate important aerodynamic properties of a finite wing of arbitrary shape and airfoil.

1. *Lift Force*

   From the Kutta–Joukouski Theorem, the lift per unit span is related to the circulation by $L' = \rho V_\infty \Gamma$. Thus, the total lift on the wing is obtained by integrating the lift per unit span across the span; namely:

$$L = \int_{-b/2}^{b/2} L' dy \Rightarrow L' = \rho V_\infty \int_{-b/2}^{b/2} \Gamma(y) dy. \tag{6.28}$$

The wing-lift coefficient is given by:

$$C_L \equiv \frac{L}{1/2 \rho V_\infty^2 S} = \frac{2}{V_\infty S} \int_{-b/2}^{b/2} \Gamma(y) dy. \tag{6.29}$$

Standard notation is that the forces on a three-dimensional wing are unprimed, and the wing-force coefficients are written with an upper-case subscript. The nondimensionalization is with the wings planform area, S.

Transforming to the angular-measure variable, θ, and integrating:

$$C_L = \frac{2}{V_\infty S} \int_\pi^0 \left[(2b V_\infty) \sum_{n-1}^\infty A_n \sin n\theta \right] \left[\left(\frac{b}{2} \right) (-\sin\theta d\theta) \right].$$

The product $[(\sin(n\theta))(\sin)]$, when integrated between the limits π and 0, is zero for $n \neq 1$. Thus, only the A_1 term remains and:

$$C_L = \pi A_1 \text{AR}. \tag{6.30}$$

Recall that the aspect ratio, AR, is a wing-planform property and $\text{AR} = b^2/S$.

Eq. 6.30 shows that the wing-lift coefficient depends on only *one* Fourier-series coefficient. This does *not* mean that the fundamental monoplane equation must be applied at only one spanwise station. The more simultaneous equations that are generated for the A_n's, the more accurate is the value for A_1.

The variation of wing-lift coefficient with AR for a tapered wing is shown in Fig. 6.15. This is not simply a linear plot, as Eq. 6.30 might suggest, because the value of A_1 changes with AR. Wing lift and the wing lift-curve slope are discussed later.

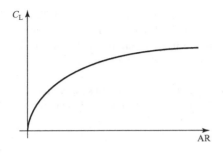

Figure 6.15. Variation of the wing-lift coefficient with the AR for a tapered wing.

2. *Induced Drag Force*

From Eq. (6.18), the induced drag is given by $D'_i = L'\alpha_i$. Then, L' may be written as a Fourier-series expression by using Eqs. 6.17 and 6.23. The expression for α_i is given by Eq. 6.15 and the integral in Eq. 6.15 already was evaluated as Eq. 6.26. Dropping the subscripts in Eq. 6.26:

$$D_i = \int_{-b/2}^{b/2} [L'][\alpha_i](dy) =$$

$$= \int_{\pi}^{0} \left[(\rho V_\infty)(2bV_\infty) \sum_{n=1}^{\infty} A_n \sin n\theta \right] \left[\sum_{n=1}^{\infty} nA_n \frac{\sin n\theta}{\sin\theta} \right] \left(-\frac{b}{2}\sin\theta d\theta \right)$$

and

$$C_{D_i} = \frac{D}{1/2\rho V_\infty^2 S} = 2\frac{b^2}{S} \int_0^\pi \left[\sum_{n=1}^{\infty} A_n \sin n\theta \right] \left[\sum_{m=1}^{\infty} A_m \sin m\theta \right] d\theta.$$

We note that the integrand is the product of two summations. The running subscript in the second summation is changed to m to emphasize this point. Now:

$$\int_0^\pi \sin n\theta \sin m\theta d\theta = \begin{cases} 0 & n \neq m \\ \dfrac{\pi}{2} & n = m. \end{cases}$$

Expanding the two Fourier-series, multiplying term, by term, and then integrating yields:

$$C_{D_i} = 2AR\left(\frac{\pi}{2}\right)\left[A_1^2 + \sum_{n=2}^{\infty} nA_n^2 \right] = \pi AR(A_1^2)\left[1 + \sum_{n=2}^{\infty} n\left(\frac{A_n^2}{A_1^2}\right) \right].$$

Notice that the summation is always a positive quantity, because all of the A_n's appear as the square. This summation has a value that depends on the values of the Fourier-series coefficients and, hence, on the wing planform. We give this summation the symbol δ, which is known for a specified wing planform. Finally, we use Eq. 6.30 to substitute for A_1. The result is:

$$C_{D_i} = \frac{C_L^2}{\pi AR}(1+\delta), \qquad (6.31)$$

where

$$\delta \equiv \sum_{n=2}^{\infty} n \left(\frac{A_n}{A_1} \right)^2.$$

This is an important result. It states that the induced-drag coefficient depends on the lift coefficient squared, so that a portion of the drag of an airplane is the result of the production of lift. Notice the importance of the AR. Recall the discussion in Chapter 1 about the Voyager airplane used for the first nonstop flight around the earth and the illustrations showing high-performance sailplanes. ARs that sometimes exceed 50 are used as one of several drag-reducing design features for such specialized aircraft, which require the lowest possible drag in the lower speed range. (Boermans, 2006 and Dillinger and Boermans, 2006).

Because $\delta \geq 0$, then for a given AR, the induced drag is a minimum when $\delta = 0$. The conditions necessary for this to be so are studied shortly. The quantity $(1 + \delta)$ often is written as $1/e$, where e is called the *span efficiency factor*. This parameter is used frequently in the aeronautical industry as a measure of the efficiency of the wing design in reducing induced drag. For minimum induced drag, $e = 1.0$. As induced drag increases, the value of e decreases. Then, Eq. 6.31 may be written alternately as:

$$C_{D_i} = \frac{C_L^2}{\pi e \mathrm{AR}}, \tag{6.32}$$

where

$$e = \frac{1}{[1+\delta]}.$$

For a typical wing, δ is a small quantity on the order of 0.05 (corresponding to $e = 0.952$). Thus, the AR has a much larger role in setting the magnitude of the induced drag than the span-efficiency factor. Induced drag is discussed further in a subsequent section.

Now, we run Program **PRANDTL** to explore the dependence of lift and induced drag on all of the parameters that describe a tapered wing.

Program PRANDTL. This program solves the monoplane-wing equation for straight tapered-wings. The user selects the number of Fourier-series coefficients to be used in the solution as well as the wing AR, taper ratio, twist, and angle of attack. The program calculates the values of the Fourier-series coefficients, the wing lift and induced drag, and the span-efficiency factor. Examine the effects of these parameters by running several cases of your choice, compare the results, and note the trends.

Elliptic-Lift Distribution

The results for the calculation of the induced drag of an arbitrary finite wing, Eq. 6.32, indicated that the drag was least when all of the A_n's in the Fourier series were zero for $n > 1$. This section examines the performance of a wing when the induced drag is minimum and then investigates the wing geometry that is necessary to achieve this minimum induced-drag condition.

1. *Elliptic Loading*

The general Fourier-series expression for $\Gamma(\theta)$, the spanwise circulation distribution, is given by Eq. (6.23). With all of the A_n's set equal to zero except A_1, this equation reduces to:

$$\Gamma = 2bV_\infty A_1 \sin\theta \tag{6.33}$$

At the root section $y = 0$ ($\theta = \pi/2$), let $\Gamma = \Gamma_0$, where:

$$\Gamma_0 = 2bV_\infty A_1. \tag{6.34}$$

Then, Eq. 6.34 may be written as:

$$\Gamma = \Gamma_0 \sin\theta. \tag{6.35}$$

Now, we recall from a trigonometric identity that $\sin^2 + \cos^2 = 1$. We also recall from the spanwise location transformation, Eq. 6.22, that $y = (b/2)\cos\theta$. It follows that:

$$\sin^2\theta + \left(\frac{y}{b/2}\right)^2 = 1 \quad \Rightarrow \quad \sin\theta = \left[1 - \left(\frac{y}{b/2}\right)^2\right]^{1/2}.$$

Then, substituting this relationship into Eq. 6.35 and rearranging:

$$\frac{\Gamma^2}{\Gamma_0^2} + \frac{y^2}{(b/2)^2} = 1, \tag{6.36}$$

which is the equation for an ellipse in y-Γ coordinates, as illustrated in Fig. 6.16. This result states that if the bound vorticity, or circulation Γ, and therefore the lift per unit span, L', is distributed eliptically across the wing span (i.e., *elliptical loading*) then the induced drag is a minimum.

The elliptical-loading results that follow are useful because they represent a standard for comparison. They also are used as a basis to express results for arbitrary spanwise loading by means of a correction factor (compare Eqs. 6.32 and 6.44).

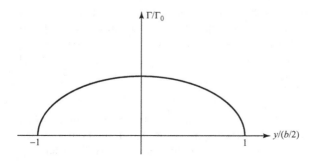

Figure 6.16. Elliptical spanwise loading for minimum induced drag.

In the remainder of this section, notice the repeated appearance of the ratio b^2/S, which is the reason that the special name *aspect ratio* was given to this quantity. Also notice the primary role of the AR in both the elliptical-spanwise-loading results and those for arbitrary spanwise loading.

2. *Wing Lift*
Recall that $L' = \rho V_\infty \Gamma$ and use the spanwise-location transformation, Eq. 6.22. Then, for elliptical loading, Eq. 6.35, the wing lift is given by:

$$L = \int_{-b/2}^{b/2} L' \, dy = \int_\pi^0 \rho V_\infty [\Gamma_0 \sin\theta] \left[-\frac{b}{2}\cos\theta \, d\theta \right],$$

or

$$L = \frac{\Gamma_0 \pi b \rho V_\infty}{4} \tag{6.37}$$

and

$$C_L = \frac{L}{1/2 \rho V_\infty^2 S} \Rightarrow C_L = \frac{\Gamma_0 \pi b}{2 V_\infty S}. \tag{6.38}$$

Finally, the circulation at the wing root may be expressed as:

$$\Gamma_0 = \frac{2 C_L V_\infty S}{\pi b}. \tag{6.39}$$

3 *Downwash*
The general expression for the downwash is given by Eq. 6.10, where $d\Gamma/dy$ may be written as $(d\Gamma/d\theta)(d\theta/dy)$. For elliptical loading, $d\Gamma/d\theta = \Gamma_0 \cos\theta$ from Eq. 6.35. Thus, Eq. 6.10 becomes:

$$W_{y_0} = \frac{1}{4\pi} \frac{\Gamma_0}{b/2} \int_\pi^0 \frac{\cos\theta}{[\cos\theta - \cos\theta_0]} \, d\theta.$$

Inverting the limits and placing a minus sign in front of the integral to compensate, the integral has a principle value of π, given by Eq. 6.25 with $n = 1$. The downwash at any point along the span then is given by:

$$w = -\frac{\Gamma_0}{2b}. \tag{6.40}$$

This equation indicates that in the special case of elliptical loading, the downwash is a *constant* all along the span. Thus, for a wing with elliptic loading, stall occurs all along the span at the same time. Now, using Eq. 6.39:

$$w = -\frac{C_L V_\infty S}{\pi b^2} = -\frac{C_L V_\infty}{\pi AR}. \tag{6.41}$$

The magnitude of the downwash thus depends on the magnitude of the wing-lift coefficient and varies inversely as the wing AR.

4. *Induced Angle of Attack*
The expression for the induced angle of attack follows from Eq. 6.43 and the definition of α_i, Eq. 6.14. Thus,

$$\alpha_i = \frac{-w}{V_\infty} = \frac{\Gamma_0}{2bV_\infty} = \frac{C_L}{\pi AR} \tag{6.42}$$

and the induced angle of attack also is *constant* across the span.

5. *Induced Drag*
From Eq. 6.18, $D'_i = L'_i \alpha_i$ so that:

$$C_{D_i} = \frac{1}{q_\infty S} \int_{-b/2}^{b/2} D_i \, dy = \frac{\alpha_i}{q_\infty S} \int_{-b/2}^{b/2} L' \, dy = C_L \alpha_i \tag{6.43}$$

and using Eq. (6.42):

$$C_{D_i} = C_{D_{i_{MIN}}} = \frac{C_L^2}{\pi AR}. \tag{6.44}$$

The induced drag increases as the lift coefficient is squared.

6. *Relations ships Between Section and Wing Coefficients*
From Eqs. 6.12 and 6.13, the section-lift coefficient for a finite wing may be written as:

$$C_\ell = m_0(\alpha - \alpha_i - \alpha_{L_0}). \tag{6.45}$$

Now, we assume an untwisted wing of constant section with elliptical loading. Under this condition:

α = constant across the span, untwisted (no geometric twist) wing
α_{L0} = constant across the span, wing of constant section
α_i = constant across the span, elliptical loading

The two-dimensional lift-curve slope, m_0, has the theoretical value 2π, and experimental values are not greatly different from that value. Then, *if m_0 is assumed constant*, for an untwisted wing of constant section with elliptical spanwise-lift distribution, Eq. 6.45 states that the section-lift coefficient, C_ℓ, must be constant across the span and further more that:

$$C_L = \frac{1}{q_\infty S} \int_{-b/2}^{b/2} L' \, dy = \frac{1}{q_\infty S} \int_{-b/2}^{b/2} [q_\infty C_1 c] \, dy = \frac{C_1}{S} \int_{-b/2}^{b/2} c \, dy = C_1 \tag{6.46}$$

because the integral of the chord over the span is simply the planform area. Thus, under these assumptions, the numerical value of the section and wing-lift coefficients is the same. Similarly,

$$C_{D_i} = \frac{1}{q_\infty S} \int_{-b/2}^{b/2} D'_i \, dy = \frac{1}{q_\infty S} \int_{-b/2}^{b/2} [L' \alpha_i] \, dy = C_1 \alpha_i = C_{d_i} \tag{6.47}$$

so that under these same assumptions, the numerical value of the section and wing induced-drag coefficients are the same.

Wing Geometry Required for Elliptical Loading, Lifting-Line Theory

It is shown in the previous section that elliptical-spanwise loading corresponds to minimum induced drag. There are three ways to design a wing to achieve elliptical-spanwise loading: (1) using a suitable wing planform, or (2) using wing twist, and it can be approximated (3) using a simple tapered planform.

1. *Planform*

   Consider an untwisted wing of constant section with all sections having the same two-dimensional lift-curve slope, m_0. It follows from Eq. 6.45 that the lift coefficient C_ℓ is constant across the span. Thus, from Eq. 6.36:

$$C_\ell = \text{constant} = \frac{L'}{q_\infty c} = \frac{\Gamma_0\left[1-\left(\frac{y}{b/2}\right)^2\right]^{1/2}}{q_\infty c} \quad \Rightarrow \quad c = K\left[1-\left(\frac{y}{b/2}\right)^2\right]^{1/2}$$

where $K = (\Gamma_0/q_\infty C_\ell)$, which is a constant. Thus, the variation of the wing chord, c, with span, y, is given by:

$$\frac{c^2}{K^2} + \frac{y^2}{(b/2)^2} = 1, \tag{6.48}$$

which is the equation for an ellipse. The constant K is independent of lift because from Eq. 6.39:

$$K = \frac{\Gamma_0}{q_\infty C_1} = \left(\frac{1}{q_\infty C_1}\right)\left(\frac{2C_L V_\infty S}{\pi b}\right) = \left(\frac{4S}{\rho V_\infty \pi b}\right)\left(\frac{C_L}{C_L}\right) = \frac{4S}{\rho V_\infty \pi b}$$

Eq. 6.48 states that if a wing is untwisted and of constant section with constant lift-curve slope, then an elliptical-lift distribution is generated if the wing has an elliptical planform. Furthermore, the elliptical-lift distribution occurs at *any* value of wing-lift coefficient. Thus, elliptical-lift distribution can be achieved in practice but at the cost of complexity of manufacture.

2. *Twist*

   Assume (for simplicity) only that m_0 = constant for any wing section. A geometric twist (varies with span) or an aerodynamic twist (α_{L_0} varies with span) or both are permitted. The induced angle of attack, α_i, is a constant because of the elliptical-loading requirement. Then, from Eq. 6.45:

$$C_\ell = m_0(\alpha - \alpha_i - \alpha_{L_0}),$$

and using Eqs. 6.36 and 6.42:

$$C_\ell = \frac{L'}{q_\infty c} = \frac{\rho V_\infty \Gamma_0\left[1-\left(\frac{y}{b/2}\right)^2\right]^{1/2}}{q_\infty c(y)} = m_0\left[\alpha(y) - \frac{C_L}{\pi AR} - \alpha_{L_0}(y)\right]$$

or $\quad c(y)\left[\alpha(y) - \alpha_{L_0}(y) - \left(\frac{C_L}{\pi AR}\right)\right] = \frac{(4bC_L)}{(\pi m_0 AR)}\left[1-\left(\frac{y}{b/2}\right)^2\right]^{1/2}. \tag{6.49}$

The local variation of wing chord as expressed by $c(y)$ and wing twist as expressed by the absolute angle of attack $\alpha_a(y) = [\alpha(y) - \alpha_{L_0}(y)]$ must be such as to satisfy Eq. 6.49 for all values of y. In particular, for a rectangular wing of constant section, Eq. 6.49 shows that an elliptical spanwise loading can be achieved provided that the geometric twist of the wing also varies elliptically with span. However, because C_L does not cancel out in Eq. 6.49, this twist provides elliptical spanwise loading at only one (design) value of wing-lift coefficient. Again, as in the case of an elliptical planform, spanwise elliptical loading can be achieved at the cost of fabrication complexity.

3. *Tapered Wing*
   Recall from Eqs. 6.31 and 6.44 that a wing of arbitrary planform with arbitrary spanwise loading has an induced drag that is a factor $(1 + \delta)$ larger than that for elliptical loading. A straight-tapered wing with a taper ratio of about 0.4 has a value of δ or about 0.01. Thus, this wing—which is relatively straightforward to manufacture—has an induced-drag coefficient that is only about 1 percent above the minimum. We verify this conclusion by running Program **PRANDTL** for various taper ratios and examining the values of induced drag in the neighborhood of a taper ratio of 0.4.

Remarks on Lift and Drag of a Finite Wing from Lifting-Line Theory

It is now useful to summarize and focus on important features of the lifting-line theory.

1. *Lift-Curve Slope*
   Recall from Eq. 5.74 that the two-dimensional lift coefficient is given by:

$$C_\ell = m_0(\alpha - \alpha_{L_0}) = m_0\alpha_a,$$

where α_a is the absolute angle of attack. Because the angle of zero lift of the section is a function of the airfoil shape and, hence, is independent of α, the two-dimensional lift-curve slope is given by:

$$\frac{dC_\ell}{d\alpha} = m_0 \quad \text{per radian} \quad \text{(two-dimensional)} \qquad (6.50)$$

Recall that $m_0 = 2\pi$ from thin airfoil theory.

Recalling Eqs. 6.42, 6.45, and 6.46, which are valid for an untwisted finite wing of constant section with an elliptical-lift distribution and constant α_i, we write:

$$C_L = C_\ell = m_0[(\alpha - \alpha_{L_0}) - \alpha_i] = m_0\left[\alpha_a - \frac{C_L}{\pi AR}\right]. \qquad (6.51)$$

Solving:

$$C_L = \frac{m_0\alpha_a}{1 + \dfrac{m_0}{\pi AR}}.$$

Again, the zero-lift angle is independent of the geometric angle of attack so that:

$$\frac{dC_L}{d\alpha} \equiv m = \frac{m_0}{1 + \frac{m_0}{\pi AR}} \qquad \text{(three-dimensional, elliptical loading).} \qquad (6.52)$$

This is the lift-curve slope for a finite wing. Using the thin-airfoil theory result for m_0,

$$\frac{dC_L}{d\alpha} = m = \frac{2\pi}{1 + \frac{2}{AR}}. \qquad (6.53)$$

These results, which specify the lift-curve slope for a finite wing with elliptical loading, may be modified to describe a wing with arbitrary spanwise loading by inserting a correction factor, τ. Thus,

$$\frac{dC_L}{d\alpha} = \frac{m_0}{1 + \frac{m_0}{\pi AR}(1 + \tau)} \qquad \text{(arbitrary spanwise loading)} \qquad (6.54)$$

or

$$\frac{dC_L}{d\alpha} = \frac{2\pi}{1 + \frac{2}{AR}(1 + \tau)}. \qquad (6.55)$$

The correction factor, τ, depends on the values of the A_n's in the Fourier series and has a typical value of $0.05 < \tau < 0.20$.

The lift-curve slopes for a two-dimensional airfoil and a three-dimensional untwisted wing of constant section with elliptical loading and the same airfoil section are compared in Fig. 6.17. Notice that the finite-wing effect is to reduce the lift-curve slope of the wing compared to that for the airfoil sections that comprise the wing.

When the airfoil lift coefficient is zero for a two-dimensional airfoil, Eq. 5.74, then $\alpha = \alpha_{L0}$. For an untwisted wing of constant section with elliptical loading, Eq. 6.51 states that when $C_L = 0$, then $\alpha = \alpha_{L0}$ because $\alpha = 0$. Because the airfoil and the wing in Fig. 6.17 have the same section, the intercept at $C_\ell = C_L = 0$ is the same for both curves.

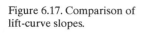
Figure 6.17. Comparison of lift-curve slopes.

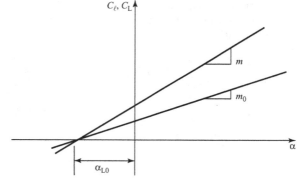

2. *Wing Drag*

The finite wing has an induced drag even in an inviscid flow because the drag is caused by the trailing-vortex sheet. In a viscous flow, there also is a drag on the wing due to the presence of viscosity. This drag is termed *profile drag* and represents the sum of the drag due to friction on the wing surface and the form drag due to separation of the boundary layer on the airfoil. (An extreme example of the drag due to separation is found in the case of a two-dimensional cylinder.) Even at a small angle of attack, there is some separation over the surface of a wing near the trailing edge. The drag coefficient for a wing then is written properly as:

$$C_{Dt} \equiv C_{D_{total}} = C_{D_{visc}} + C_{D_i}. \tag{6.56}$$

The value of the viscous-drag coefficient for the finite wing can be found from experimental two-dimensional airfoil data (see Chapter 5) or numerically by using *strip theory* in (see Section 6.5). This two-dimensional viscous-drag information may be used directly in Eq. 6.56 if the two-dimensional drag coefficient is perceived as based on a mean chord defined by the wing geometry as:

$$\bar{c} \equiv \frac{AR}{b}. \tag{6.57}$$

Then, if the profile drag per unit span, D'_{visc}, is assumed to be essentially constant across the span:

$$C_{d_{visc}} = \frac{D'_{visc}}{q_\infty \bar{c}}\left[\frac{b}{b}\right] = \frac{D_{visc}}{q_\infty S} = C_{D_{visc}}.$$

An important device for representing the aerodynamic performance of a wing is the *drag polar*. This is simply a plot of the drag coefficient versus the lift coefficient, as shown schematically in Fig. 6.18. The dashed curve represents the part of the drag due to the pressure distribution and viscous effects (usually called the *profile drag*). This part of the drag changes slowly as lift coefficient (and, hence, the angle of attack) increases. Notice that the induced drag increases rapidly with lift coefficient. At low flight speeds (requiring a high lift coefficient), the induced drag is the dominant part. Thus, it is important in an efficient design to keep the induced drag as small as possible. Recall that a parameter that strongly influences the magnitude of the induced drag is the AR of the wing planform. These effects are evaluated in detail in Chapter 9 for actual airplane configurations.

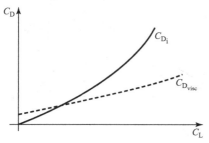

Figure 6.18. Drag polar for profile and induced drag.

Example 6.3 *Given*: Consider a wing with a rectangular planform. The wing is untwisted and of constant section, with a lift-curve slope correction factor of $\tau = 0.05$. The wing span is 16 ft. and the chord is 2 ft. The measured profile-drag coefficient of the wing section is 0.005, the angle of zero lift of the section is $-1.3°$, and the section lift-curve slope is assumed to be 2π/radian.

The wing is to be tested in a wind tunnel at a geometric angle of attack of $7°$ at a test-section dynamic pressure of 1.5 psia.

Required: Find
(a) lift-curve slope of the wing, per degree
(b) lift coefficient of the wing
(c) lift force on the wing during test (lbs)
(d) drag coefficient of the wing
(e) approximate magnitude of the downwash

Approach: Use the various equations in this chapter, as appropriate.
Solution: Either the value of m must be converted to "per degree" at the end or m_0 must be used as "per degree" at the outset. The latter choice is not as convenient but is used here to emphasize a point. Thus, $m_0 = 2\pi$/radian $= (2\pi)/(180/\pi) = 0.1096$/degree. As a further preliminary, $AR = b^2/S = b2/(b)(c) = b/c$ for a rectangular wing; hence, $AR = 16/2 = 8$ for this wing.

(a) Using Eq. 6.45:

$$ m = \frac{m_0}{1 + \dfrac{m_0(180/\pi)(1+\tau)}{\pi AR}} = \frac{0.1096}{1 + \dfrac{(0.1096)(57.3)(1.05)}{\pi(8)}} = 0.087 \,/\, \text{degree}. $$

(b) $C_L = m(\alpha - \alpha_{L0}) = (0.087)[7 - (-1.3)] = 0.72$.
(c) $L = C_L \, q_\infty \, S = (0.72)(1.5)(144)[(16)(2)] = 4{,}978$ lbf.
(d) $C_{Dt} = C_{Dvisc} + C_L^2/e\pi AR = 0.005 + (0.72)^2/(0.95)(8)\pi = 0.0267$.
(e) The downwash may be estimated by using Eq. 6.43, valid for elliptical loading. Substituting $w/V_\infty = \tan(1.3°) \cong 1.3° = 0.0227$ radian, so that the downwash is 2.3 percent of the freestream velocity.

Appraisal: (a) Notice that with the choice of m_0 in units of "per degree," the numerator of Eq. 6.45 is expressed in "per degree," which makes the answer for m as "per degree." The ratio in the denominator of Eq. 6.45 must be dimensionless, so the value of m_0 /radian must be used in the denominator.
(b) The magnitude of C_L is reasonable and it is positive.
(c) Dynamic pressure must be converted to lbf/ft^2.
(d) This is a typical value for using only drag.
(e) The downwash and the spanwise-flow components associated with the trailing vortex sheet are comparable in magnitude. Thus, assuming the flow over the wing to be locally two-dimensional is a good assumption.

Figure 6.19 is a comparison of experimental data and prediction using lifting-line theory. The theory was validated by many experiments and gives satisfactory results if the wing AR is not too small and the wing sweep is not excessive. When properly

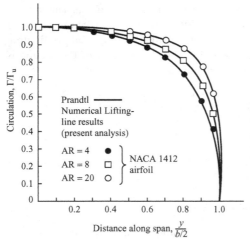

Figure 6.19. Comparison of lifting-line theory with experimental results (see Anderson, 1984).

applied, lifting-line theory provides a fast and accurate method for obtaining the continuous spanwise distribution of several quantities, such as lift and induced angle of attack.

Finally, we note that moments about finite wings and the location of the wing aerodynamic center and center of pressure can be calculated as weighted averages of the wing-section properties. These topics are not discussed here.

6.4 Wing-Panel Methods

The advent of the digital computer enabled the development of solution methods for finite wings that are superior to methods that represent the wing by a lifting line. These solution methods are an extension of the airfoil-panel methods discussed in Chapter 5, with singularities here being distributed over the wing planform. The singularities may be sources, doublets, vortices, or certain combinations of these. A source distribution splits the streamlines and thus represents wing thickness, whereas a doublet or vortex distribution gives rise to lift. Vortex elements are discussed here for a wing with an arbitrary planform. Appeal is made to the tangency (i.e., "no-flow") boundary condition, as in Chapter 5, and a system of simultaneous linear-algebraic equations is developed by applying the tangency-boundary condition at numerous points on the wing surface. The wing is assumed to be at a small angle of attack relative to the freestream flow because the theory is inviscid and irrotational and large areas of separation cannot be accommodated. Unlike lifting-line theory, panel methods require no restrictions on wing sweep or AR, but they do rely on the principles of superposition.

The two panel methods described herein, the vortex panel method (VPM) and the vortex lattice method (VLM), can be used to model inviscid flow over wings with airfoil sections of large or small thickness ratios. The VPM is illustrated for a wing of arbitrary-thickness ratio, and the surface-pressure distribution on the wing is found. The VLM is discussed for a thin wing, that is, the wing thickness is assumed to be small and its effect is neglected. When the wing thickness is neglected, such panel methods often are called *lifting-surface methods*. Lifting-surface methods apply the tangency-boundary condition on the camber surface of the wing rather than on the surface of the wing. The advantage of a lifting-surface analysis is that it is easier to program because far fewer panels are needed and the influence equations are simpler while satisfactory accuracy is maintained, as it was in thin-airfoil theory. The disadvantage of a lifting-surface analysis is that it provides the pressure difference (i.e., pressure loading) across the camber surface rather than the pressure distribution on the wing surface. The pressure distribution on the wing surface is needed as input for any related boundary-layer solution. However, lifting-surface methods provide force and moment data.

In both of the panel methods described herein, the wing surface must be subdivided into a suitable number of small quadrilateral panels that reflect the wing shape and planform. These panels need not be of the same size, usually being smallest in regions of rapidly varying flow properties. Thus, given wing surface must be discretized into panels as a preliminary step in the calculation. This grid generation is itself an important area of study. Each panel has vorticity distributed over the surface. In the following discussion, the vorticity is combined with vortex elements for convenience. Higher-order solutions with curved panels and distributed vorticity are found in the literature.

Vortex Panel Method

The wing top and bottom surfaces are divided into N flat quadrilateral panels. Each panel is assigned a closed vortex ring consisting of four vortex elements, each of strength Γ_n, as shown in Fig. 6.20a. The magnitude of the unknown Γ_n's is evaluated during the numerical solution. The Helmholtz theorem demands that each vortex ring be of constant strength along the length. The four sides of each vortex ring are placed just inside the four sides of the associated panels. The control point, or collocation point (i.e., the point on the panel where the tangency-boundary condition is to be applied), is located along the three-quarter chord line at the mid-span of the panel.

Regarding the Kutta condition, in general, the strengths of the transverse vortices at the wing trailing edge (e.g., the downstream elements in Fig. 6.20) are unequal and nonzero. Thus, $\left|\Gamma_{3,U}\right| \neq \left|\Gamma_{3,L}\right| \neq 0$. The simplest approach is to apply the two-dimensional Kutta condition all along the trailing edge—namely, that the transverse-filament-vortex strength must be zero all along the trailing edge. To satisfy this condition, free wake panels are added to the model. Each wake panel must have a vortex-ring strength that is equal and opposite to the net strength—say, $\Delta\Gamma_3$—of the two transverse vortices at the trailing-edge location from which the wake trails. Thus, for example, the Kutta condition requires that $\Delta\Gamma_{3,\text{ wake}} = -\Delta\Gamma_3$. Something also must be stated about the wake geometry. Fig. 6.20 shows a typical wake panel extending to downstream infinity from a pair of wing trailing-edge panels. Because the wake panels cannot support any pressure difference, they must be aligned with the local flow streamlines. This streamline shape may be obtained from experiment. However, as an approximation, the wake panels may be aligned with the freestream direction, with the wing chord, or with an other convenient reference line. The wing angle

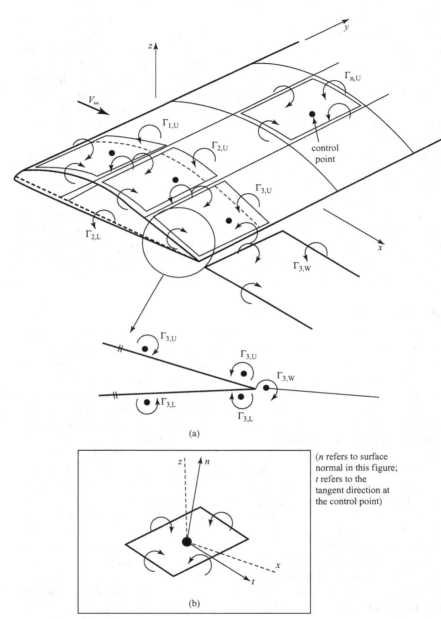

Figure 6.20. Vortex-panel geometry.

of attack is small, so that this type of approximation is usually satisfactory. If we desire to accommodate wake curvature, it is useful to divide each semi-infinite wake panel into several panels of arbitrary length but all with the same strength. The self-induced distortion of the wake (i.e., wake rollup) can be included in more complicated numerical models, if desired.

The numerical solution requires that a geometric description of the problem be established before calculations can begin. Regarding a reference system, the location of each control point must be described, as well as the corner points of each vortex ring and the direction of the normal to each panel. Finally, the straight-line distance from each control point to the end points of each of the four vortex filaments, that comprise each vortex ring must be calculated and stored.

One linear-algebraic equation is written for each control point on the surface panels by applying the surface-boundary condition. This requires the calculation of the velocity induced at each control point by all of the vortex segments on all of the surface panels (including the contribution of the vortex rectangle located on the surface panel containing the control point itself) and by all of the vortex segments on all of the wake panels. Recall from Section 6.2 that each induced velocity is perpendicular to a plane containing the particular element at the vortex ring and the control point in question. Hence, the direction, as well as the magnitude, of each induced-velocity contribution is known so that a resultant velocity normal to the panel containing the control point may be established. Now, because the orientation of each surface panel relative to the freestream direction is known for any given wing and angle of attack, the normal component of freestream velocity can be found at each control point as well. The defining equation at the control point reflects the application of the no-flow through boundary condition: The normal velocity induced at the control point by all of the vortex elements on all of the surface and wake panels must be equal and opposite to the normal component of freestream velocity at that same control point.

As an example, we consider the control point on Panel 1 in Fig. 6.20a. The velocity, V_i, induced at that control point by the four vortex elements of strength Γ_1 on Panel 1 is calculated, as in Example 6.1. The resultant is perpendicular to the Panel. The velocity induced by each of the four vortex elements on each of the remaining surface panels can be calculated using the Biot-Savart Law, as given in Eq. 6.2. Each induced velocity, V_i, is perpendicular to a plane containing the control point and the particular vortex element in question. Because the wing geometry is known, the component of V_i acting perpendicular to Panel 1—namely, $V_i|_{s=1}$ may be calculated. The process is repeated for each vortex element of each wake panel. When all of the calculations are made for the normal component of velocity induced at the control point on panel $s = 1$, the result may be represented as follows:

$$V_P|_{S=1} \sum_{n=1}^{N} \sum_{m=1}^{4} (V_i|_{s=1})_{n,m} + \sum (V_i|_{s=1})_{wake}, \qquad (6.58)$$

where V_P is the resultant normal component of the velocity induced at Point P on Panel 1 by all of the N surface panels and all of the wake panels. The resultant normal component, V_P, then is set equal to the freestream-velocity normal component at the control point on panel $s = 1$ and one algebraic equation is written. Thus,

$$V_P \big|_1 = V_{\infty \text{normal}} \big|_1. \tag{6.59}$$

The entire process is repeated for Panel 2, $s = 2$, and so on, until ultimately a total of N linear-algebraic equations in the form of Eq. 6.58 have been written that contain N unknowns, Γ_1 through Γ_N. It is apparent that there is much geometry bookkeeping to be done, which is why the method requires a computer. Usually, the Biot-Savart Law is expressed and used in a more general vector form. Also, the induced velocity and the freestream velocity may be calculated in coordinate-component form at each control point and then nulled rather than applying the boundary condition in terms of two velocity components perpendicular to the surface panel containing the control point, as illustrated herein.

Notice that a vortex element located on one side of the wing influences the flow at a control point on the opposite side of the wing. Furthermore, the appropriate distances between the two ends of the vortex element and the control point, to be used in the Biot-Savart Law, is the straight-line distance and not the distance as measured along the wing surface. The "$1/r$" influence of the vortex filament is still valid here and is confirmed by experiment.

The array of simultaneous algebraic equations is solved for the unknown element strengths, Γ_n. The pressure distribution on the wing surface may be found in the following way, referring to Fig. 6.20b. Define an orthogonal and tangential coordinate direction at each control point. Then, use the velocity information generated when applying the no-flow-boundary condition at each control point to find the resultant velocity in the newly defined tangential-coordinate direction. Do this at every control point, accounting for the contributions due to all of the induced velocities as well as to the freestream velocity. This is the magnitude of the tangential velocity, $V_{n,t}$, at a particular control point. The local static pressure at a control point, p_n, then may be found by using the Bernoulli Equation. Thus,

$$p_0 = p + \left[\frac{V^2}{2} \right] \Rightarrow p_n = p_0 - \left[\frac{V_{n,t}^2}{2} \right], \tag{6.60}$$

where p_0 is the (known) stagnation pressure of the oncoming flow.

If we prefer, the magnitude of the local pressure coefficient at a control point may be found instead because for incompressible irrotational flow, it was shown in Chapter 4 that:

$$C_{p_n} = 1 - \left(\frac{V_{n,t}}{V_\infty} \right)^2. \tag{6.61}$$

The predicted static pressure or pressure coefficient is assumed to be constant over each of the panels, n, corresponding to the n control points.

The pressure-distribution predictions from the VPM agree satisfactorily with experiments. However, unless numerous panels are used to represent the wing, the chordwise pressure distribution on the wing surface is not sufficiently detailed (or is too bumpy) to be used as input to a boundary-layer program. This is the motivation for the strip-theory approach, which is described later.

The pressure acting at the control point of each surface panel acts normal to the panel and inward on the panel surface. As noted previously, this pressure is assumed to be constant over the panel. Thus, there is a force acting normal to and on the surface of the nth panel, which is given by:

$$\Delta F_n = (p_n)(\Delta A_n),\tag{6.62}$$

where ΔA_n is the surface area of the nth panel and there are $n = N$ panels.

Finally, the total lift force on the wing can be determined by first finding the component of ΔF_n, which is perpendicular to the freestream direction (i.e., the panel contribution to lift, ΔL_n). Then, summing over all of the panels on both the top and bottom surfaces of the wing, taking care that the contribution of ΔL_n is signed positive upward:

$$L = \sum_{n=1}^{N} [\Delta L_n].\tag{6.63}$$

Alternately, the spanwise lift distribution may be found by summing the contribution to lift of corresponding panels on the top and bottom surfaces of a wing at a particular spanwise location. The accuracy of the VPM is enhanced as the number of panels is increased. Results from a typical panel-code solution for a finite wing are compared with experiments in Fig. 6.21. Agreement with experimental data is excellent until viscous effects begin to dominate at large angles of attack.

In the preceding discussion of the VPM and in the following discussion of the VLM, the intent is to physically describe the method in words. There are many variations of the methodology, and the prediction codes are much more efficient than the outline described here suggests. Variations in the prediction-code flow chart and in the panel-generation code, which defines the location and shape of the surface panels, can significantly affect solution accuracy and efficiency. The student is referred to the current literature and to Katz and Plotkin (1991). Panel methods circa 1985 are compared in Margason et al. (1985).

Vortex-Lattice Method

In the VLM, the finite wing again is represented by N flat panels as defined by a preliminary grid generation. In the following discussion, it is assumed that thickness effects are negligible. What follows, then, is a so-called lifting-surface theory, although the VLM may be applied to wings with nonzero thickness. The wing is represented by a camber surface and the tangency-boundary condition is applied on the camber surface rather than on the wing surface. The wing may have arbitrary camber and planform. The angle of attack is assumed to be small; therefore, this inviscid theory describes a thin wing with negligibly small regions of separation.

Again, vortex singularities are distributed over a surface. However, in contrast to the VPM described herein, in the VLM discussion, each panel is assigned a horseshoe vortex rather than a vortex ring. The placement of the transverse element of the horseshoe vortex and the control point is suggested by the following: Consider a wing panel as a flat plate in an effectively two-dimensional flow. Such a flat plate of chord c is shown at the angle of attack in Fig. 6.22.

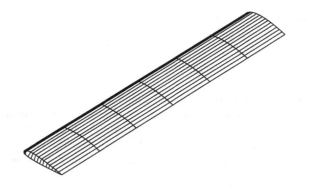

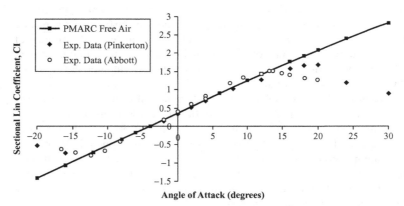

Fig. 6.21. Panel-code prediction for section of finite wing at mid-span compared with experimental results for a two-dimensional NACA 4412 airfoil (Bangasser, 1993).

Represent the lift on this plate by a combined vortex of strength Γ located at $c/4$, which is the center of pressure for an airfoil according to thin-airfoil theory. From the Biot-Savart Law, at some distance, h, along the plate, the vortex induces a velocity $V = \Gamma/2\pi h$. Now, we recall from Eq. 5.14 that the circulation around a two-dimensional flat-plate airfoil is given by $\Gamma = \pi c\alpha V_\infty$. Finally, the tangency-boundary condition (i.e., no flow through the surface) requires that V must be equal and opposite to a component of freestream velocity given by $V_\infty(\sin\alpha) \cong V_\infty\alpha$. Appealing to tangency:

$$\frac{\Gamma}{2\pi h} = V_\infty\alpha \text{ or } \frac{(\pi c\alpha V_\infty)}{2\pi h} V_\infty\alpha \tag{6.64}$$

or, solving for h:

$$h = c/2. \tag{6.65}$$

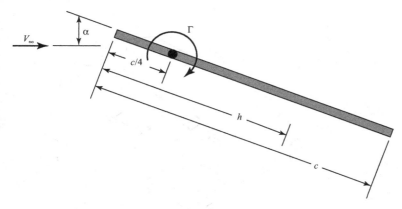

Figure. 6.22. Two-dimensional flat plate at the angle of attack.

Thus, with the transverse-bound vortex on a panel at $c/4$, the tangency-boundary condition is satisfied at one point, $c/4 + h = 3c/4$. We choose to locate the control point on each panel at this location and we apply the tangency-boundary condition at this control point. If the wing leading edge is swept, the bound vortex is skewed at the sweep angle relative to the y-axis. The bound vortices have different (i.e., unknown) strengths that vary in both a spanwise and chordwise direction.

Two free-vortex filaments always trail downstream from the ends of the transverse element on the panel (Fig. 6.23). Each pair of (free) trailing-vortex elements must have the same strength as the bound vortex from which it originates (i.e., Helmholtz). The free-vortex elements cannot support any pressure difference, so they must trail off of the bound vortex and away from the wing in a direction parallel to the local streamlines; that is, they must follow a curved path. However, because the wing angle of attack is small the free vortices may be assumed to follow a straight line at the freestream or another convenient direction. If desired, the trailing vortices may be divided into straight-line segments so as to better model the physical reality and, if desired, may follow the wing surface until the trailing edge is reached. As in the lifting-line theory, the influence of the starting vortex portion of each horseshoe at the wing is ignored as being negligible.

Carefully compare Figs. 6.20 and 6.23 to understand better the difference between the vortex-panel and the vortex-lattice formulation of the finite-wing problem. In particular, contrast the modeling of the trailing vortices. In the VPM, the trailing-vortex pairs behind the wing have the same strength as the bound vortex on the trailing-edge panel from which they originated. In the VLM, the trailing-vortex pairs have the same strength as the bound vortex on the individual chordwise panel from which they originated.

A control point is located at the three-quarter chord of each panel and the Biot-Savart Law is used to calculate the velocity induced at each control point by all of the other horseshoe vortices. The bound vortex located on the panel that contains the particular control point in question contributes to the induced velocity at that control point because the point of interest and the bound vortex do not coincide, as in the lifting-line theory. Thus, each horseshoe vortex contributes three velocity

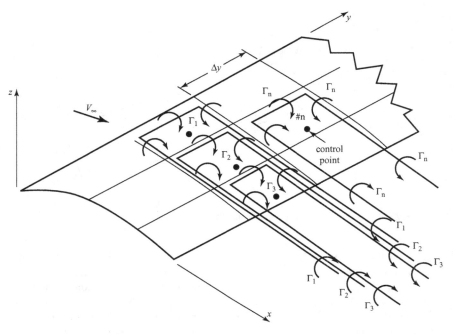

Figure 6.23. VLM.

components at each control point, and the induced velocities may be computed by using Eq. 6.2. The induced-velocity components are normal to a plane containing the control point and the vortex element, so that all of the induced velocities finally may be assembled into the resultant velocity, V_p, which is perpendicular to the panel in question. Thus, the resultant normal velocity, induced at any panel, s, due to all of the horseshoe vortices located on that panel and on every other panel is given by:

$$V_P\big|_s = \sum_{n=1}^{N} \sum_{m=1}^{3} \left(V_i\big|\right)_{sn.m} \qquad (6.66)$$

where each horseshoe vortex makes three contributions; one due to the bound vortex and two due to the free vortices.

The resultant induced velocity at each panel control point must be equal and opposite to the normal component of freestream velocity at each control point if the tangency-boundary condition (i.e., no flowthrough) is to be satisfied. The result is a set of N simultaneous linear-algebraic equations, where N is the number of panels. There is only one unknown vortex strength associated with each panel because the free vortices trailing from each panel have the same strength as the bound vortex associated with that panel. The array of N simultaneous equations then can be solved for the N unknown values of Γ.

As in the VPM there is significant geometry for the computer code to handle; however a general code may be written to describe the location of the control points and the horseshoe vortices, and the calculations are repetitive. With the bound-vortex strength at each panel now known, and the effect of the wake accounted for through the influence of the trailing-vortex filaments, the chordwise and spanwise lift distribution on the wing may be calculated, as well as the total wing lift.

In particular, as shown in Fig. 6.23, the lift contribution of bound Vortex 1, is given by:

$$L_1 = L_i'(\Delta y) = \rho V_\infty \Gamma_1(\Delta y). \tag{6.67}$$

Similarly, the lift on Panel 2 is simply:

$$L_2 = \rho V_\infty \Gamma_2(\Delta y) \tag{6.68}$$

and so on. The chordwise and spanwise distributions of lift follow. Finally, the total lift on the wing is:

$$L = \sum_{n=1}^{N} L_n. \tag{6.69}$$

Likewise, the pressure distribution on the wing can be found because:

$$\Delta p \big|_n = \frac{L_n}{A_n}, \quad n = 1, 2, ..., N, \tag{6.70}$$

where A_n is the panel area. Alternately, the pressure distribution on the camber surface may be found by calculating the magnitude of the local tangential velocity at the control point on each panel and then appealing to the Bernoulli Equation, as outlined in the prior VPM discussion.

Notice that the pressure calculated on each panel by the VLM illustrated herein is actually the pressure difference across the zero-thickness representation of the wing and not the pressure on the wing surface. Hence, although the lifting-surface solution is easier to program than a thick-wing solution, it does not yield pressure-distribution information that is compatible with a coupled boundary-layer solution.

The induced drag of the wing may be found from the VLM by finding the downwash and the induced angle of attack for each horseshoe vortex, as in the lifting-line theory. Once the strengths, Γ, of all of the trailing-vortex filament pairs are found, the velocity component in the z direction at each control point follows by appealing to the Biot-Savart Law. The velocity component in the z direction at a control point is, effectively, the downwash at the bound-vortex filament on that panel because the two are close together. Thus, the induced-drag contribution of each panel, n, is given by:

$$D_i \big|_n = -\left[L_n \right]\left[(\alpha_i)_n \right] = -\left[L_n \right]\left[\frac{w_n}{V_\infty} \right] = -\rho V_\infty \Gamma_n (\Delta y)_n \left[\frac{w_n}{V_\infty} \right], \tag{6.71}$$

and the induced drag of the finite wing is:

$$D_i = -\rho \sum_{n=1}^{N} \Gamma_n (\Delta y)_n (w)_n. \tag{6.72}$$

Another method for evaluating the induced drag of the wing may be used if the trailing-vortex filaments are assumed to leave the wing in the freestream direction.

Under this assumption, the wake-vortex sheet has a small pressure difference across it (not physically correct because the sheet does not have the correct physical shape). However, under this assumption, the wake sheet contributes zero drag because the wake vorticity is parallel to the freestream. In this case, the induced drag of the wing from which the wake originates can be evaluated by integration of the wake properties in the two-dimensional flow occurring in a cross plane far downstream of the wing and perpendicular to the wake sheet. This plane is called the *Trefftz plane*.

A detailed numerical example of the VLM as applied to a swept wing with zero thickness and camber (i.e., a flat plate) is in Thomas, 1976. Close study of this example is of great assistance if the student must write a VLM program. Results of the VLM as applied to a rectangular wing and a comparison with experiment, are shown in Fig. 6.24. In this figure, the wing is not taken to be of zero thickness, as discussed, but rather is modeled by a combination of sources and vortex lattices. The predicted surface-pressure distribution on the wing at two spanwise stations and the predicted induced drag of the wing shows excellent agreement when compared with experimental data.

Program VLM

Program VLM applies the VLM for the case of straight-or swept-wing planforms with or without twist. We choose the incompressible case. The program accepts the following from the user: AR, taper ratio, and root-chord and tip-chord angles of attack. Linear twist is assumed if the two latter values are different. The program calculates the required horseshoe-vortex strengths to satisfy the tangency-boundary condition and then returns to the user the wing-lift coefficient, the induced-drag coefficient, and the moment coefficient. The induced drag is calculated by evaluating the downwash and the induced angle of attack. In the case of highly swept wings, the Polhamus leading-edge-suction analogy is used to simulate the vortex lift (see Section 6.8). When this occurs, the output screen makes note of this lift contribution. The following exercises are suggested to the student:

1. Determine the coefficients in the formula $C_L = (A\alpha + B)$ for a wing of the user's choice, where is the root absolute angle of attack: Note that B is zero for an uncambered wing but nonzero if the wing is twisted.
2. Examine the effect of the AR on L/D_i. First, consider the case of a wing with a taper ratio (TR = (tip chord)/(root chord)) of 0.7, untwisted, at an angle of attack of 5°. Then, vary the wing AR from 1 to 20.
3. Examine the effect of sweep on L/D_i. Consider the case of a wing with an AR of 6, TR = 0.3, and zero twist. Run the program for several sweep angles up to 45°.
4. Examine the effect of wing twist on L/D_i. Consider a wing with an AR of 6, TR = 0.3, and a leading-edge sweep angle of 35°. The following twist examples are suggested:
   (a) root $\alpha = 4°$; tip $\alpha = 4°$;
   (b) root $\alpha = 4.5°$; tip $\alpha = 3°$;
   (c) root $\alpha = 4.75°$; tip $\alpha = 3°$;
   (d) root $\alpha = 5°$; tip $\alpha = 3°$.

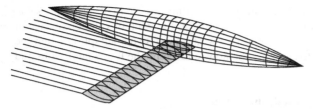

(a) Representative paneling for three-dimensional wing vortex lattice

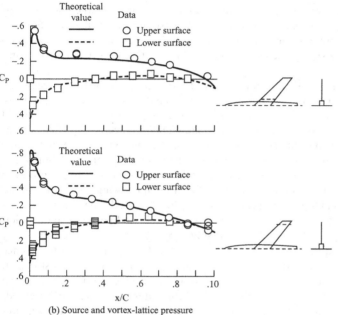

(b) Source and vortex-lattice pressure
coefficients on a wing

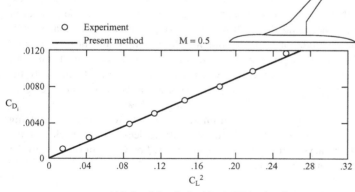

(c) Induced drag for Lockheed ATT-95 aircraft

Figure 6.24. VLM compared with experimental data (Thomas, 1976).

The panel methods described herein are valid only for inviscid, incompressible flow. Furthermore, the solutions are found only on the wing or camber surface. If flow details away from the wing are desired within these two assumptions, then the methods can be extended to compute the flow induced by the flow singularities and the freestream.

6.5 Comments on Wing-Analysis Methods

Lifting-line theory provides a simple and accurate method for determining the spanwise variation of properties (e.g., spanwise loading) for wings of large AR with zero or small sweep angle. A numerical scheme may be set up to calculate the Fourier-series coefficients required for the solution if we desire. Such schemes also can keep track of the relative size of the coefficients so that the Fourier series may be truncated as desired.

In industry today, the VPM and the VLM are used extensively for predicting behavior at cruise, as long as there are no large boundary-layer separation effects. Either method may be used with a fuselage panel code (see Chapter 7). An inviscid solution for cruise provides spanwise wing-loading and induced drag. Then, a skin-friction estimate may be added or a viscous boundary-layer solution may be patched onto the inviscid-flow solution to account for the viscous effects. This boundary-layer solution can be made interactive (i.e., displacement effect due to the presence of the boundary layer accounted for as a modification of the wing section) or noninteractive (i.e., inviscid flow and viscous boundary layer treated separately).

To model a viscous flow around a wing completely and correctly, the Navier-Stokes Equations (see Chapter 8) must be solved by CFD methods. The complexity of the equations and the uncertainty of how to treat the appearance of turbulence in the boundary-layer equations render such an approach of marginal cost benefit to industry at this writing. However, CFD methods are used extensively in research and, as the methods become more advanced and computers get faster, CFD methods will soon become routine. Currently, the panel methods discussed in this chapter constitute important tools for predicting the behavior of wings.

The availability of all of the various computer-solution methods makes new demands on the physical insight of designers. They must understand the physics of a particular flow problem and the accuracy of the required results so that a computer code can be selected (or written) that will be as simple and fast as possible while incorporating all of the phenomena of importance. Unfortunately, it is not yet possible to write a simple code to include everything. Tradeoffs must be made. If this task is performed properly, a solution is forthcoming that provides the necessary information with suitable accuracy. For example, simple, low-order panel codes are inexpensive to run and are valuable for parametric studies. Flow problems involving wings at high angle of attack or wings with high-lift devices—both of which have large separated boundary-layer regions—must be solved by CFD methods with the assistance of suitable wind-tunnel tests.

6.6 Aerodynamic Strip Theory

Thus far, we examine rapid-solution methods using panels and we see that we can quickly obtain a good estimate of the overall inviscid performance (i.e., lift coefficient, induced-drag coefficient, and moment coefficient) of any candidate wing. We can even obtain rather good spanwise information regarding the variation of several quantities, such as induced angle of attack and downwash. However, chordwise information is limited. In the wing-panel methods, we obtain information only at a few chordwise stations corresponding to locations of the singularities; in lifting-line theory, there is no chordwise information at all. As discussed in Chapter 8, such information is necessary to determine the effects of the boundary layer to compute the viscous drag. Chapter 8 also points out that the local surface-pressure gradient has a key role in the behavior of the boundary layer and that it should be known accurately. Thus, we discuss here a relatively fast method of supplementing the information from a panel method, or even from lifting-line theory, to obtain such detailed information. Using the effective angle-of-attack information from either a three-dimensional panel code or lifting-line theory, aerodynamic strip theory describes the chordwise pressure distribution on a three-dimensional wing by computing the chordwise pressure distribution at separate spanwise stations. Because this approach necessarily neglects spanwise flow, the method works best on high AR wings or wings with little sweep. Notice that the three-dimensional effect is not completely eliminated because we use an effective angle of attack that was determined by accounting for three-dimensional effects.

Virtually any three-dimensional panel method combined with any airfoil pressure-distribution method can be used in this strip-theory approach. Here, we demonstrate the idea by using the VLM and the AIRFOIL codes introduced previously. The basic steps are as follows:

1. Use a three-dimensional panel code (e.g., VLM) to compute the spanwise distribution of effective angle of attack, α_{eff}, for a given wing. Recall from Eq. 6.13 that $\alpha_{eff} = \alpha - \alpha_i$.
2. At selected spanwise stations, use the local wing-section information and the effective angle of attack to determine the section-pressure distribution (i.e., **AIRFOIL**).

Consider a straight-tapered wing with an AR of 6.4 and a taper ratio of 0.4; use the NACA 0012 airfoil as the wing section. Let the wing have a geometric angle of attack of 5° at the plane of symmetry and a linear twist (i.e., root to tip) of 3°. Using the VLM program illustrated previously, with 11 spanwise stations and 7 chordwise stations, the following results are obtained:

Spanwise Station, $2y/b$	α_{eff}
0.0455	3.477
0.1364	3.420
0.2273	3.352
0.3182	3.255
0.4091	3.136
0.5000	2.996
0.5909	2.835
0.6818	2.651
0.7727	2.439
0.8636	2.247
0.9545	1.177

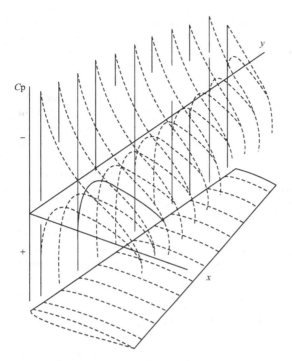

Figure 6.25. C_p distribution over semi-span of a finite wing using strip theory.

For each spanwise station, the effective angle of attack is used in Program **AIRFOIL**, along with the section information, to obtain a detailed surface-pressure distribution at the station. The result is shown in Figure 6.25. Shown in the figure are the detailed pressure distributions from the Program **AIRFOIL** for each of the 11 spanwise stations used by the VLM program. The figure was generated by reading the 11 C_p data files generated by the Program **AIRFOIL** and making a plot of the wing and the C_p distributions. The C_p data files also would be used by a boundary-layer program to compute the viscous drag over the wing (see Chapter 8).

6.7 Ground Effect

When a finite wing (and airplane) operates near the ground within a distance of approximately 20 percent of the wing span, as in landing or takeoff, the wing behavior is modified from that observed in an unrestricted freestream. This is called *ground effect*, and it may be explained by referring to the wing-lifting-line model.

Let a lifting wing be represented by a horseshoe vortex, as shown in Fig. 6.26a. For simplicity, we consider a vertical plane that encloses one trailing-tip vortex and let the plane be sufficiently far downstream so that the flow in the plane may be assumed as two-dimensional. If a solid surface is inserted below the trailing vortex as in Fig. 6.26b, the boundary condition to be satisfied is that the flow must be tangent to that surface. However, the trailing vortex induces a vertical-velocity component at the surface and the boundary condition is not satisfied. This means that the flow physically must adjust to be tangent to the surface; however, a single vortex is too simple a model to satisfy this requirement. This problem must be resolved by introducing an *image* vortex at the same distance below the surface as the trailing vortex is above. The image vortex has the same strength as the trailing vortex but the oppo-site sense, as shown in Fig. 6.26c.

At any Point A on the surface, the superposing of the induced velocity V_1 due to the trailing vortex Γ_1 with the velocity V_2 due to the image vortex results in a cancel-lation of the vertical-velocity component at the surface, and the tangency-boundary condition thereby is satisfied. Notice, however, that the velocity at any Point B within the flow field is modified by the presence of the image vortex so that the velocity at Point B is not identical to what it would be in an unconfined flow.

Now, visualize this picture extended into three dimensions, with the lifting wing close to a surface being represented by a lifting line and a trailing-vortex sheet. A mechanism still must be present to satisfy the tangency-boundary condition at the ground plane. The required image-vortex sheet modifies the unconfined flow; in particular, it significantly modifies the downwash at the lifting line.

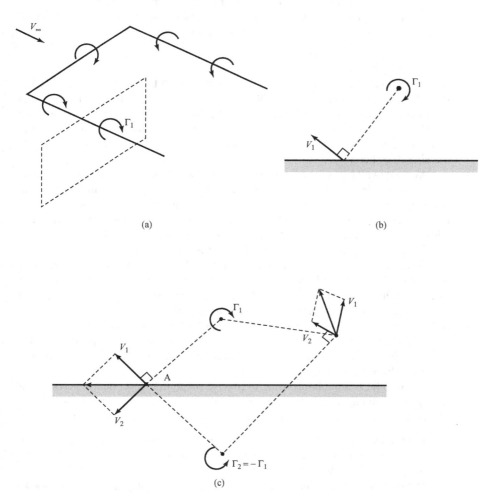

Figure 6.26. Modeling of ground effect by image vortices.

The effect of the image-vortex sheet is that the downwash at the wing is considerably reduced and the wing lift-curve slope is increased. Because the downwash (and induced angle of attack) is decreased, the induced drag also is reduced. The net effect on the aircraft of the increase in lift (at the same angle of attack) and decrease in drag is that just before touchdown, the aircraft briefly seems to be buoyed up and floating. During the interaction, the aircraft is said to "flare," or manifest ground effect. Sometimes the flare maneuver is supplemented by an increase in the angle of attack that can exceed the stalling angle at altitude. This allows touchdown at the lowest possible speed so that landing loads are minimized and rollout thereby is shortened.

The concept of using image vortices to represent the effect of a solid wall may be used to correct for the effect of the solid wall in a wind-tunnel test section. When an overly large model is tested, the requirement that the flow be tangent to the solid wall induces significant changes in the flow field compared to that of flight in the atmosphere in which the flow is not so constrained. By representing the effect of the wind-tunnel walls with image vortices, the effect of the solid wall on the measured data can be estimated and the data corrected accordingly.

6.8 Winglets

After examining results from the Program **PRANDTL**, it is desirable to make the AR of a wing as large as possible. However, there are practical limitations to increasing the wing span and, hence, the AR. The larger the span, the greater the length of the cantilever beam represented by the half-wing and the greater the structural problems. Also, a large wing span makes ground-maneuvering and parking of the aircraft difficult.

Aircraft of all types have been designed with winglets instead of increased span. For example, Fig. 6.27 shows a modern sailpane with winglets. In this application, the benefit comes without increased span. (Span is limited by international competition rules; the glider shown is a 15-meter-span "standard class" racing machine.) The glide ratio is increased by at least one L/D point, from 42 to about 43, by this modification. The equivalence of glide and L/D ratios is explained in Chapter 1.

Winglets increasingly are used in commercial aviation because the improvement in effective AR leads directly to reduced operational costs. The benefits result from the influence that the winglets have on the behavior of the tip vortices and the interaction of these vortices with the wing flow field. Thus, they originally were called *vortex diffusers* by the inventor, Richard Whitcomb of NASA (Whitcomb, 1976, Flechner, etal, 1976). There is evidence that the basic concept already was known before manned flight actually was accomplished, but Whitcomb receives the credit for the winglet in its modern form.

The winglet is a miniature wing of precise shape that is set nearly vertically at the tip of the main wing, with the winglet leading edge inboard of the trailing edge. The corner is carefully faired. The winglet reduces drag in two ways as follows:

1. The relative wind at the winglet, which is the vector sum of the oncoming stream and the flow around the wing tip, acts on the winglet so as to produce a force component in the negative drag (i.e., thrust) direction.
2. The presence of the winglet distorts the shape of the wing-tip vortex and modifies the strength of the trailing-vortex sheet.

Figure 6.27. Schleicher ASW-24B with winglets.

Winglet design currently is accomplished by means of numerical programs such as the panel methods discussed in the previous sections. As always, we must consider the tradeoff between reduced induced drag and additional form drag resulting from the increased surface area. The numerical results then are verified by careful wind-tunnel testing, and an iterative process leads to the best compromise between performance needs across the operational speed range of the aircraft.

6.9 Vortex Lift

Previous discussions regarding the effects of viscosity on the behavior of two-dimensional airfoils and finite wings of conventional planform emphasized that operating these devices above moderate angles of attack results in flow separation at or near the leading edge, with a resulting catastrophic loss of lift. A delta wing with sharp leading edges, when operated at relatively small angles of attack, exhibits separated flow over the upper surface without a loss of lift until a very large angle of attack has been reached. This behavior is explained by the generation of vortex lift. Steady-flow vortex lift is discussed in this section.

A delta-wing planform is attractive for supersonic flight because it has a small wave drag. However, it still must operate at subsonic speeds, particularly on takeoff and landing. If the delta wing has a sharp leading edge (desirable for supersonic flight), then at subsonic speeds and at small angles of attack, the flow separates at the leading edge and forms two large and dominant spiral vortices, shown in Fig. 6.28a with strong axial convection along the vortex core. These two spiral vortices alter the pressure distribution on the upper surface of the delta wing, causing a large suction pressure to be established on the wing surface almost directly beneath the centers of the spiral vortices, as shown in Fig. 6.28b.

These large suction pressures on the wing upper surface furnish a so-called vortex lift, which at large angles of attack provides a significant increment to the potential flow (i.e., no separation) lift on the delta wing (Fig. 6.29). The spiral vortex

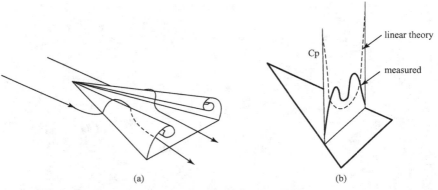

Figure 6.28. Delta wing at large angle of attack (Polhamus, 1966).

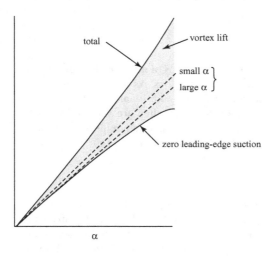

Figure 6.29. Behavior of delta wing at angle of attack. (Polhamus, 1966).

is sensitive to the geometry of the wing leading edge. For example, if the leading edge is rounded, the vortex-lift effect is reduced.

It is a challenging problem to model the details of the spiral vortices. However, a method was developed by Polhamus, 1966, which treats the vortices by an analogy and provides a surprisingly accurate prediction of the vortex lift and the associated drag due to lift. The analogy lies in the treatment of the leading-edge suction on a delta wing. Recall that leading-edge suction was introduced in the discussion of flat-plate airfoils in Chapter 5. When a near-zero-thickness airfoil is at an angle of attack, as shown in Fig. 6.30a, the flow upstream of the stagnation streamline reverses and flows around the sharp leading edge, causing a large suction as the streamline traverses a near-zero radius in inviscid flow. In a viscous fluid, the flow around the leading edge separates, as shown in Fig. 6.30b, and the streamlines upstream of the stagnation point behave as if they were flowing around a very blunt leading edge in an inviscid flow, as shown in Fig. 6.30c.

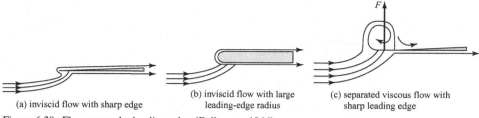

(a) inviscid flow with sharp edge

(b) inviscid flow with large leading-edge radius

(c) separated viscous flow with sharp leading edge

Figure 6.30. Flow around a leading edge (Polhamus, 1966).

Polhamus Method

The Polhamus method assumes that the lift on a delta wing consists of two parts: potential flow lift and vortex lift. The potential flow lift is calculated by assuming that there is no leading-edge suction developed at the leading edge of the delta wing because the streamlines pass smoothly around the leading-edge separation bubble, as shown in Fig. 6.30c. It is further assumed that the flow reattaches on the upper surface of the wing after passing around the separation vortex. Thus, the flow model assumes that the potential flow lift is decreased only by the loss of the leading-edge suction force. Lifting-surface theory may be used to calculate the potential flow lift.

As noted previously, the vortex-lift method of Polhamus does not attempt to model flow details. Rather, it recognizes that there must be a force on the wing arising from the pressure required to keep the centrifugal force in equilibrium as the flow passes around the separation vortex. It is assumed that this force is of the same magnitude as the leading-edge-suction force required to sustain attached flow around a large leading-edge radius (see Fig. 6.30b). The difference is that the force due to separation acts primarily on the upper surface (see Fig. 6.30c) rather than on the wing leading edge (see Fig 6.30b). In effect, the leading-edge-suction force is assumed to rotate 90° and become perpendicular to the wing-chord plane. The result is that there is a normal force exerted on the wing (force F in Fig. 6.30c) that adds to wing lift.

According to the Polhamus model, the lift on a delta wing becomes:

$$C_L = (\text{potential flow lift}) + (\text{vortex lift})$$

or

$$C_L = K_P \sin\alpha \cos^2\alpha + K_v \cos\alpha \sin^2\alpha \qquad (6.73)$$

where K_P and K_v are constants that can be found from lifting-surface theory and α is the angle of attack of the delta wing. The constants for a delta wing in incompressible flow are shown in Fig. 6.31 as a function of the AR of a delta wing at a very low speed.

The loss of leading-edge-suction force due to the presence of the separation vortex results in a drag penalty. The same simple approach also can be used to predict the inviscid drag due to lift of a sharp-edged delta wing at low speeds with vortex lift and zero leading-edge suction (Polhamus, 1968). Namely:

$$D_{\text{Lift}} = C_L \tan\alpha = K_P \sin^2\alpha \cos\alpha + K_v \sin^3\alpha, \qquad (6.74)$$

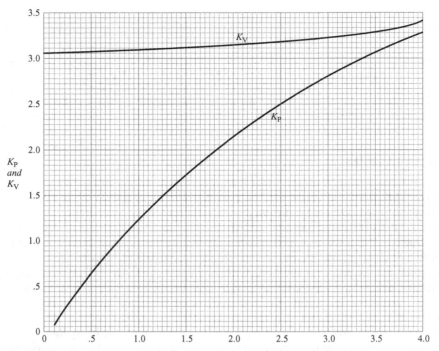

Figure 6.31. Values of K_P and K_v for a delta wing at $M = 0$ (Polhamus, 1968).

where D_{Lift} is the drag due to lift (not written as D_i because it is not associated with a trailing-vortex sheet).

This theory (i.e., Polhamus) predicts results with excellent agreement with experimental data up to large angles of attack (Fig. 6.32). The theory describes the behavior of a delta wing to very large angles of attack. Ultimately, at an angle of attack that depends on the AR of the delta wing; the theory begins to over-predict the lift on the wing as the measured lift begins to drop off. At this angle of attack, the spiral vortex has started to break down (i.e., the vortex begins to experience what is called *vortex burst*). The vortex breakdown proceeds from the trailing edge of the wing to the front. At some point, the vortex completely breaks down, the upper surface of the wing is a turbulent separated region, and the delta wing experiences stall.

The leading-edge-suction analogy described previously was extended to predict the behavior of arrow- and diamond-wing planforms. Charts for evaluating K_P and K_v needed to calculate the potential flow and vortex-lift terms for these planforms are in work by Polhamus, 1971. A subsonic-compressibility procedure is included as well.

The previous discussion applied only to steady flight. Vortex lift also is important in fighter operations at high subsonic speed when a high value of lift is desired for maneuvering and control, which corresponds to a time-dependent problem. During maneuvers, the spiral vortex from the leading edge changes strength with time and the location of the vortex-bursting is time-dependent. The unsteady vortex-lift problem is discussed in Nelson, 1991.

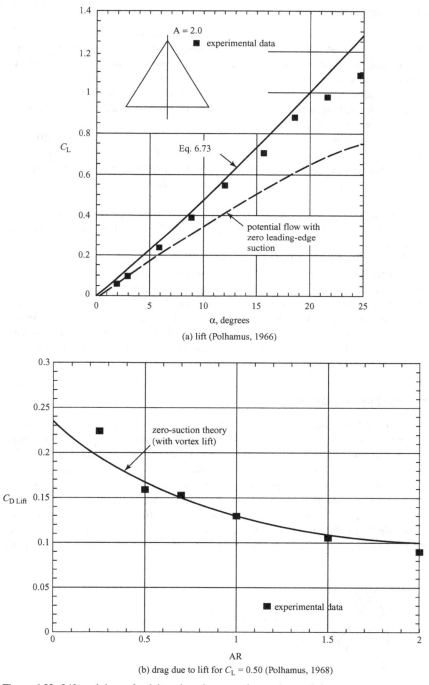

(a) lift (Polhamus, 1966)

(b) drag due to lift for $C_L = 0.50$ (Polhamus, 1968)

Figure 6.32. Lift and drag of a delta wing; theory and experimental data.

Leading-Edge Vortex Flap

As noted previously, the delta wing is a planform, that was developed for supersonic flight, but it also must operate at subsonic speeds. At high angle of attack, the wing lift is increased by virtue of vortex lift (see Fig. 6.28), but the wing drag also increases with increasing vortex lift. This is because the presence of a separation vortex at the wing leading edge effectively rotates the leading-edge suction through 90° (see Fig. 6.30c). This means that there is no longer a leading-edge suction to supply thrust and alleviate some of the drag.

The L/D ratio is a measure of the aerodynamic efficiency of a wing. Although vortex lift increases C_L, the simultaneous *increase* in C_D can result in the effective wing L/D ratio being lower. This reduced L/D may have a major impact on perform- ance parameters such as takeoff and climb, where the L/D ratio has an important role.

The leading-edge vortex flap (LEVF) is a device that improves L/D for a delta wing at a high angle of attack at some penalty in lift. The LEVF is a small deflecting surface mounted at the leading edge of a delta wing, as shown in Fig. 6.33.

When the LEVF is deflected downward, the leading-edge separation acts on the flap surface and a thrust component is generated (compare Figs. 6.30c and 6.33). Thus, wing drag is decreased (i.e., thrust is increased) at the cost of a decrease in vortex lift. At very high values of C_L, the vortex on the flap moves downstream and the full suction force F in Fig. 6.33 is not recovered. The experimental results in the paper by Rinoie and Stollery, 1994, indicate that at low speed, a 60° delta wing with sharp leading and trailing edges has a dramatic (i.e., 40%) improvement in L/D at $C_L = 0.45$ with a vortex flap deflection of 30°.

6.10 Strakes and Canards

Strakes, also called leading-edge extensions are highly swept surfaces that are added to the wing at the wing root, shown in Fig. 6.34a. Their purpose is to cause leading- edge separation, resulting in a strong spiral vortex that sweeps back and over the wing, thereby providing vortex lift on the strake and added lift on the wing.

A *canard*, Fig. 6.34b is a separate miniature lifting surface placed forward of the main wing on an aircraft. Delta-wing aircraft must operate at a high angle of attack at low speeds (i.e., low dynamic pressure) to generate a sufficiently large lift coefficient to maintain flight. (The Concorde has a separate nose section that is angled down- ward for landing to improve pilot visibility.) At operational values of lift coefficient,

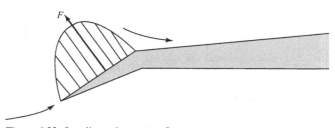

Figure 6.33. Leading-edge vortex flap.

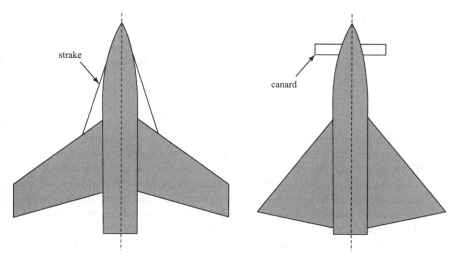

Figure 6.34. Strake and canard planforms.

the delta-wing configuration exhibits a large nose-down pitching moment. The lifting canard supplies a nose-up pitching moment to counteract this behavior. A lifting canard also is useful for trim when flap deflection leads to the nose-down pitching moment of an aircraft. Careful canard design and placement can lead to beneficial mutual-interference effects between the canard and the main wing.

PROBLEMS

6.1. Consider the drag force induced by the starting vortex filament alone. The lifting line has span b and constant circulation Γ. If a is the distance between the starting vortex and the lifting line, show that the drag due to the starting vortex filament alone is:

$$D = \frac{\rho \Gamma^2}{2\pi}\left(\sqrt{1+(b/a)^2}-1\right).$$

Should this drag increment be added to the others required in estimating the total lift on the wing? If not, why not?

6.2. Why is the downwash on the tail of an airplane resulting from the wing wake almost twice as large as the downwash on the wing resulting from the wing wake?

6.3. The lift distribution on a certain wing varies linearly from root to tip, as follows:

$$L' = L'_C\left(1-\frac{|y|}{b}\right),$$

where L'_C is the value of lift per unit span at the centerline and b is the span. Consider a unit area, dA, of the vortex sheet that forms the wake located at distance b downstream of the lifting line and at distance $b/4$ from the x-axis.

Determine the downwash velocity induced at the right wing-tip ($x = 0$, $y = b/2$, $z = 0$) by this increment of the vortex sheet.

6.4. For the linear-lift distribution in Problem 6.3, show that the induced angle of attack at spanwise location y_0, located somewhere between the tip and the root, is given by:

$$\alpha_i(y_0) = \frac{L_C'}{4\pi\rho V_\infty^2 b} \ln\left(\frac{y_0^2}{b^2/4 - y_0^2}\right).$$

6.5. An elliptical wing is moving through sea-level air at a speed of 100 mph. The wing loading is 8 lbf/ft². The wing is untwisted and has the same airfoil section from root to tip. The lift-curve slope m_0 of the section is 5.7 per degree. The wing-span is 35 ft. and the AR is 6. Calculate the section-lift coeffient and the induced-drag coefficient. Show that the effective, induced, and absolute angles of attack are constant along the span. Find the values of these angles of attack.

6.6. Two identical wings arranged in tandem were used by John Montgomery in his successful 1905 balloon-launched glider, the *Santa Clara*. Which of the two wings produces the largest drag? Explain.

6.7. A finite wing of span b = 60 ft. is represented by a lifting line with constant circulation, Γ, as shown in Fig. 6.1. Assume the freestream velocity to be 150 ft/s and the freestream density to be $\rho = 0.002$ slugs/ft³. The wing-lift coefficient is $C_L = 0.3$. Find:

(a) L

(b) L′

(c) Γ

Also, at a spanwise location $y = b/4$ and at the lifting line, find the magnitudes of:

(d) downwash, w, in ft/s

(e) induced angle of attack, in degrees

6.8. A rectangular wing of AR 7 has a uniform section whose lift-curve slope m_0 = 5.6/radian and is in an incompressible flow. The wing is symmetrical about $y = 0$. The angle of attack, measured from the flight path to the ZLL of the local section, is 5° at the root and 2° at both wing tips and varies linearly with y over each half-span.

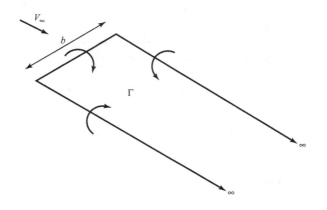

Diagram for Problem 6.7

(a) Calculate the induced-drag coefficient of this wing using the fundamental monoplane equation. Use only two terms of the series, setting up the equation at $\theta = \pi/3$ and $\theta = \pi/2$.

(b) Compare the result in (a) to the case of an elliptical-lift distribution at the same wing-lift coefficient.

6.9. Data for a tapered wing are given here. The wing is symmetrical about mid-span, and all variations in parameters are linear from the mid-span to the wing tips. The span of the wing is 12m and the flight velocity is 50m/s.

Parameter	Mid-Span Value	Wing-Tip Value
chord m	3	1.5
section lift-curve slope (per radian)	5.8	5.8
absolute angle of attack (degrees)	5.5	3.5

(a) Find the spanwise distribution of circulation about the wing, using two terms of the Fourier series and satisfying the finite-wing equation at $\theta = \pi/3$ and $\theta = \pi/6$.

(b) Find the trailing-vortex-sheet strength at a point one chord length downstream of the wing semi-span and give units with the answer.

6.10. A finite wing is represented by a lifting line with a parabolic-bound-vortex strength distribution—namely:

$$\Gamma(y) = \Gamma_s \left(1 - \frac{y^2}{(b/2)^2} \right).$$

(a) Find the value of C_L if the wing is rectangular with span 20 ft, chord 3 ft, freestream velocity 50 ft/s, and $\Gamma_s = 40$ ft^2/s.

(b) What is the trailing-vortex-filament strength, γ, at $y = b/8$ for $\Gamma_s = 40$ ft^2/s and span $b = 20$ ft?

6.11. An untwisted wing has a uniform section and an elliptic planform (i.e., the wing has elliptic loading). The section lift-curve slope is $m_0 = 2$/radian and the wing is operating at $C_L = 0.9$. Find the magnitude of the angle between the freestream (i.e., flight path) and the ZLL of a section. This angle is sometimes called the absolute angle of attack, a, by analogy with two-dimensional airfoils, but it is not as useful in three dimensions because the section responds to the relative wind. Notice that with elliptic loading, the required angle is constant across the span (i.e., the ZLL of the section is the same as the ZLL of the wing (ZLL = ZLLW).

6.12. Calculate the downwash, w, at Point P on the lifting line (see Fig. 6.6) at $y = b/4$ due to all of the trailing vortices between $3b/8$ and the wing tip if the circulation distribution across the span is given by:

$$\Gamma(y) = \Gamma_s \left(1 - \frac{|y|}{(b/2)} \right).$$

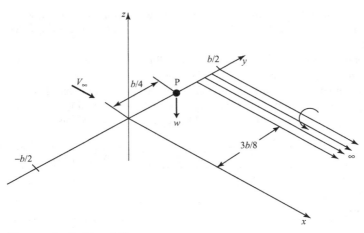

Diagram for Problem 6.12

6.13. A rectangular wing of span 24 in. and chord 6 in. is mounted in a low-speed wind tunnel where it is tested at standard sea-level conditions at a test-section velocity of 150 ft/s. At a certain angle of attack, the lift is measured and the induced drag is deduced with the help of prior data. The results are $L = 16.0$ lb$_f$, $D_i = 0.5$ lb$_f$. The lift measurement is known to be accurate but the strain gauges on the drag balance are suspect. Assuming that the lift value is correct, comment on your confidence in the drag measurement. Justify your answer.

6.14. Recall that the total drag of a finite wing is the sum of the drag due to viscosity and induced drag. A wing with a span of 4 ft. and a planform area of 2 ft.^{2} has a span efficiency factor, e, of 0.91. It is tested in a wind tunnel ($\rho = 0.002$ slugs/ ft^3) at a freestream velocity of 100 ft/s (i.e., incompressible). The measured lift for a certain angle of attack is 16 lbs and the corresponding total drag is 0.80 lb. Calculate the ratio of the induced drag to the friction drag for the wing at this test condition, that is, What is the ratio D_i/D_f?

6.15. Suppose that a twisted rectangular wing of span 50 ft has, at a certain angle of attack, a spanwise loading given by:

$$\frac{\Gamma}{V_\infty} = \sin\theta + 0.3\sin(2\theta) + 0.2\sin(3\theta) + 0.1\sin(5\theta) \text{ ft.}$$

(a) Is this wing elliptically loaded? Why or why not?
(b) Is this wing symmetrically loaded? Why or why not?
(c) Is $C_{Di} = C_L^2/AR$? If not, evaluate the ratio C_L^2/AR.

6.16. An untwisted wing of constant section has an elliptical planform. The wing loading is 900 N/m^2. The wing has a span of 10 m and an AR of 6. It is in steady flight at a height of 2,000 m on a standard day at a speed of 50 m/s. Assume the section-lift-curve slope to be $m_0 = 6.0$/radian.
(a) Find the section-lift coefficient.
(b) Find the induced-drag coefficient.
(c) Find the power expended to overcome the induced drag of the wing.

6.17. An untwisted wing of constant section has a span of 12 ft., an area of 36 ft.^{2}, and an elliptic planform. It develops 139 lbf of lift when tested at a velocity of 60 mph in a wind tunnel (take = 0.002 slugs/ft^3). The two-dimensional lift-curve slope of the wing section is 6.10/radian and the angle of zero lift of the section is −3°.
(a) Find the magnitude of the circulation at the wing root (plane of symmetry).
(b) Find the magnitude of the geometric angle of attack of this wing under test.

6.18. A wing is in steady flight and supporting a weight of 10,000 lbs at a speed of 200 ft/s and at an altitude where = 0.002 slugs/ft^3. The wing is of constant section, is untwisted, and has an elliptic planform with a span of 45 ft. and a planform area of 289 ft^2. The wing section lift-curve slope is 6.0/radian and the zero lift angle of the section is -3°. For this wing:
(a) Find the downwash, w (ft/s).
(b) Find the induced-drag coefficient.
(c) Find the effective angle of attack, α_{eff}.
(d) Find the geometric angle of attack, α.

6.19. A wing with elliptic loading is at a geometric angle of attack of 7° and is developing lift such that the wing-lift coefficient is $C_L = 0.9$. The wind-induced drag coefficient is 0.05. Find the angle of zero lift of the wing section if the two-dimensional lift-curve slope of the section is 6.2/radian.

6.20. A certain wing is tested in a flow with a freestream velocity of 50 ft/s and develops 92 lbs of lift. The wing is untwisted and has the same symmetrical section from tip to tip. The wing planform is tapered, with a planform area of 24 ft^2 and a span of 12 ft. The viscous drag of the wing is 70 percent of the total wing drag. The span efficiency factor is 0.909, the wing-lift-curve slope is 4.5/radian, and the lift-curve slope of the section is 6.0/radian.
(a) Find the value of the L/D ratio at which this wing is operating.
(b) Find the angle of attack at which the wing is operating.

6.21. A model rectangular wing with span 18 in. and chord 6 in. is tested at 200 ft/s in a wind tunnel with $\rho = 0.002$ slugs/ft^3. The lift is measured at 60 lbs and the induced drag is deduced to be 10 lbs. What is the span efficiency factor, e, for this wing?

REFERENCES AND SUGGESTED READING

Anderson, J. D., *Fundamentals of Aerodynamics,* McGraw-Hill, New York, 1984.

Applied Computational Aerodynamics, Henne, P. A. (Editor), AIAA Progress in Astronautics and Aeronautics Series, Vol. 125, 1990.

Bangasser, C. T., "An Investigation of Ground Effect on Airfoils Using a Panel Method," Masters Thesis, University of Tennessee Space Institute, August 1993.

Boermans, L. M. M., "Research on Sailplane Aerodynamics at Delft University of Technology. Recent and Present Developments," Report NV vL 2006, Netherlands Association of Aeronautical Engineers, June 2006.

Dillinger, J. and Boermans, L. M. M., "Aerodynamic Design of the Open Class Sailplane Concordia," XXVIII OSTIV Congress, Eskilstuna, Sweden, June 2006.

Flechner, S. G., Jacobs, P. F., and Whitcomb, R. T., "A High Subsonic Speed Wind-Tunnel Investigation of a Representative Second-Generation Jet Transport Wing," NASA TN D-8264 (N76-26264), July 1976.

Katz, J., and Plotkin, A., *Low Speed Aerodynamics—From Wing Theory to Panel Methods,* McGraw Hill, New York, 1991.

Margason, R. J., Kjelgaard, S. O., Sellers, W. L., Morris, C. E., Walkey, K. B., and Shields, E. W., "Subsonic Panel Methods—A Comparison of Several Production Codes,". AIAA Paper 85-0280, 23rd Aerospace Sciences Meeting, 1985.

Nelson, R. C., "Unsteady Aerodynamics of Slender Wings," in AGARD Special Course on Aircraft Aerodynamics at High Angle of Attack, N91-22105/1/XAB, March 1991.

Polhamus, E. C., "A Concept for the Vortex Lift of Sharp-Edge Delta Wings Based on a Leading Edge Suction Analogy," NASA TN D-3767, (N67-13171) December 1966.

Polhamus, E. C., "Application of the Leading-Edge Suction Analogy of Vortex Lift to the Drag Due to Lift of Sharp-Edged Delta Wings," NASA TN D-4739, (N68-21990) August 1968.

Polhamus, E. C., "Charts for Predicting the Subsonic Vortex-Lift Characteristics of Arrow, Delta, and Diamond Wings," NASA TN D-6243, (N71-21973) April 1971.

Ross, H. M., and Perkins, J. N., "Tailoring Stall Characteristics Using Leading Edge Droop Modifications." AIAA Journal of Aircraft, Vol. 31, No. 4, July/August 1994, p. 767.

Rinoie, K., and Stollery, J. L., "Experimental Studies of Vortex Flaps and Vortex Plates," AIAA Journal of Aircraft, Vol. 31, No. 2, (pp. 322–329) March/April 1994.

Serrin J., "Mathematical Principles of Classic Fluid Dynamics," Flugge, S. (ed.) *Handbuch der Physik* VIII/1, Springer-Verlag, Berlin, Heidelberg, New York, (pp. 125–263), 1959.

Thomas, J. L., "Subsonic Finite Elements for Wing-Body Combinations," NASA SP-405 (pp. 11–26), 1976.

Whitcomb, R. T., "A Design Approach and Selected Wind-Tunnel Results at High Subsonic Speeds for Wing-Tip Mounted Winglets," NASA TN D-8260, (N76-26163) July 1976.

7 Axisymmetric, Incompressible Flow around a Body of Revolution

7.1 Introduction

The flow considered in this chapter is assumed to be steady, incompressible, inviscid, and irrotational. The body immersed in the flow is assumed to be a body of revolution at zero angle of attack. An understanding of incompressible flow around bodies of revolution at zero or small angle of attack is important in several practical applications, including airships, aircraft and cruise-missile fuselages, submarine hulls, and torpedoes, as well as flows around aircraft engine nacelles and inlets. This type of flow problem is best handled in cylindrical coordinates (x, r), as shown in Fig. 7.1. Recall that r and θ lie in the y-z plane.

Because the flow fields discussed in this chapter are axisymmetric, the flow properties depend on only the axial distance x from the nose of the body (assumed to be at the origin in most cases) and the radial distance, r, away from this axis of symmetry. The flow properties are independent of the angle θ. As a result, we may examine the flow in any $(x$-$r)$ plane because the flow in all such planes is identical due to the axial symmetry. It is convenient to develop the defining equations initially in cylindrical coordinates (i.e., dependence on x, r, and θ) and then to simplify them for axisymmetric flow (i.e., dependence on x, r only).

Although there are only two independent variables (x, r) in axisymmetric flow, there are significant differences between such flows and two-dimensional, planar flows with two independent variables (see Chapter 4). As discussed later, an important three-dimensional relief effect is present in axisymmetric flow that is absent in two-dimensional, planar flows.

This chapter begins with a derivation of the continuity equation in cylindrical coordinates, starting with the general continuity equation in three-dimensional vector form and using vector identities. Then, the derivation of the continuity equation is repeated from a physical approach using Conservation of Mass principles. An extra term appears in the continuity equation in cylindrical coordinates that is not present in the continuity equation for two-dimensional, planar flow. This extra term is significant in the results of this chapter. It is important that the student fully understand why this additional term is present; hence, the repetition.

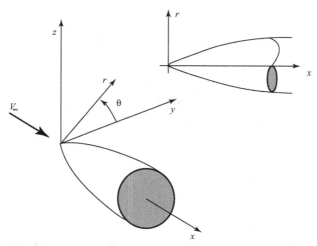

Figure 7.1. Cylindrical coordinates.

The momentum equation for axisymmetric flow reduces to the Bernoulli Equation as in the case of Cartesian coordinates. For incompressible flow, the energy equation is not needed.

Following the derivations of the continuity and momentum equations, the defining equations for the velocity potential and the stream function are developed. The problem of axisymmetric flow around a body of revolution may be treated either analytically or numerically. An appeal to the superposition of elementary solutions for the stream function (as carried out for planar flow in Chapter 6) provides analytical solutions for the flow around axisymmetric bodies of varying geometries. Such analytical results lead to physical insights about the flow field, as well as serving as benchmarks for numerical solutions. After a discussion of analytical methods, the chapter addresses the subject of numerical solutions of the axisymmetric, incompressible flow problem. Panel methods similar to those described in Chapter 6 are considered, as well as numerical methods, which use a distribution of singularities on the body axis. These methods usually are accompanied by a coupled boundary-layer analysis of some type to be able to predict the frictional drag (due to viscous-shear stresses at the body surface) and the form drag (due to boundary-layer separation) of a body of revolution. Normally, there is no well-defined trailing edge so that no Kutta condition is imposed and there is no trailing-vortex wake. If the aft portion of a fuselage is swept up (e.g., Lockheed C-130), then trailing vortices are formed that must be accounted for in a mathematical model.

In addition to the numerical solution of the direct problem (i.e., For a given body shape, what is the pressure distribution and boundary-layer behavior?), an important practical problem in numerical analysis is centered on finding a minimum-drag body shape. This is important because body shapes represented by the body of revolution have a major role in the total drag produced by several types of flight vehicles. For example, in the case of an airship, Lutz, 1998, points out that the drag of the airship hull accounts for about 66 percent of the total airship drag. Dodbele et al. state that for a transport aircraft, the fuselage drag contributes about 48 percent of the total aircraft drag when a turbulent boundary layer is present on all surfaces of the aircraft.

This percentage changes dramatically when a laminar boundary layer is assumed to exist on the lifting surfaces of a vehicle. (Recall the discussion of laminar flow airfoils in Chapter 5.) If the boundary layer on the wing and tail surfaces can be kept laminar (i.e., low frictional drag), then the fuselage drag becomes responsible for up to 70 percent of the total vehicle drag. It follows that a major payoff is possible if a fuselage can be shaped so as to maintain a laminar boundary layer, with a resulting lower skin friction, over as large a region as possible while also avoiding boundary-layer separation farther aft. The discussion of numerical solutions for axisymmetric flows concludes with remarks on the numerical analysis of a complete aircraft shape.

7.2 Axisymmetric Continuity and Momentum Equations

The study of the defining equations begins with the equations in vector form, as derived in Chapter 3. Thus, assuming zero body force and no viscous effects, then:

continuity: $\nabla \cdot V = 0$ (3.52)

momentum: $\rho \frac{\partial V}{\partial t} + \rho(V \cdot \nabla)V + \nabla p = 0$ (3.66)

The notation to be used is illustrated in Fig. 7.2. The subscripts on the velocity components emphasize *direction*—they do *not* denote partial differentiation of the quantity, as used in some texts. The vector operations in Eqs. 3.52 and 3.66 now are expanded in cylindrical coordinates and then reduced by the assumption of axial symmetry. Because the flow is incompressible, the density, ρ, is assumed to be constant. We let the local velocity vector be given by:

$$V = u_x e_x + u_r e_r + u_\theta e_\theta,$$ (7.1)

where u_x, u_r, and u_θ are the velocity components in the cylindrical coordinate directions and e_x, e_r, and e_θ are the unit vectors in those directions (see Fig. 7.2).

The Continuity Equation

Writing the velocity vector and the divergence operator in cylindrical coordinate notation, Eq. 3.52 becomes:

$$\left[e_x \frac{\partial}{\partial x} + e_r \frac{\partial}{\partial x} + e_\theta \frac{1}{r} \frac{\partial}{\partial \theta} \right] \cdot \left[u_x e_x + u_r e_r + u_\theta e_\theta \right] = 0$$ (7.2)

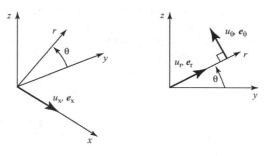

Figure 7.2. Cylindrical coordinate notation.

Notice that the derivatives in the first set of brackets are taken relative to three two-term products in the second set. Thus, applying the chain rule to the $\partial/\partial x$ derivative, for example:

$$\left(e_x \frac{\partial}{\partial x}\right) \cdot (u_x e_x) = e_x \cdot \left(e_x \frac{\partial u_x}{\partial x}\right) + e_x \cdot \left(u_x \frac{\partial e_x}{\partial x}\right). \tag{7.3}$$

A complete expansion of Eq. 7.2 thus contains 18 terms (the student should verify this). These terms fall into the following five different categories:

(1) Three terms contain a vector-dot product of unity, which then multiplies the derivative of a velocity component relative to its own coordinate direction. For example:

$$e_x \cdot e_x \frac{\partial u_x}{\partial x} = \frac{\partial u_x}{\partial x}.$$

These three terms are nonzero and are written later.

(2) Six terms contain the derivative of a velocity component relative to another coordinate direction; for example, $\partial u_\theta / \partial r$. These six terms are nonzero, but they all appear as the coefficient of a unit vector, which then is dotted into a unit vector at right angles to itself, yielding zero. For example:

$$e_r \cdot \left(\frac{\partial u_\theta}{\partial r} e_\theta\right) = 0.$$

Thus, all of these terms drop out in the expansion of Eq. 7.2.

(3) Three terms contain the derivative of a unit vector relative to its own coordinate direction; namely, $\partial e_x / \partial x$, $\partial e_\theta / \partial \theta$, and $\partial e_r / \partial r$. Because the unit vectors are all of constant length, $\partial e_x / \partial x$ and $\partial e_r / \partial r$ are zero. The exception is $\partial e_\theta / \partial \theta$. Derivatives relative to θ denote changes in direction. Analogous to (5), it may be shown that:

$$\frac{\partial e_\theta}{\partial \theta} = -e_r.$$

However, this term then is dotted with e_θ in Eq. 7.2 so that the contribution of this term is zero as well.

(4) Five terms contain the derivative of a unit vector relative to another coordinate direction. These terms are:

$$\frac{\partial e_x}{\partial r}, \frac{\partial e_r}{\partial x}, \frac{\partial e_x}{\partial \theta}, \frac{\partial e_\theta}{\partial x}, \frac{\partial e_\theta}{\partial r}$$

All of these terms are zero because for a differential change in the magnitude of the denominator, there is no change in the magnitude or direction of the numerator unit vector.

(5) One term is an exception to Category (4) namely $\partial e_r / \partial \theta$, This term makes a non zero contribution to the equation. To see why, examine Fig. 7.3.

$$\frac{\partial e_r}{\partial \theta} = \lim_{\Delta\theta \to 0} \frac{(e_r)_2 - (e_r)_1}{\Delta\theta} = \frac{[(e_r)_1 + B] - (e_r)}{\Delta\theta}$$

$$= \frac{[(e_r)_1 + (1)(\Delta\theta)(e_\theta)] - (e_r)_1}{\Delta\theta} = e_\theta \tag{7.4}$$

Figure 7.3. Rate of change of unit vector.

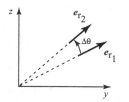

When introduced into Eq. 7.2 the term containing this derivative then becomes:

$$\frac{1}{r}e_\theta \cdot \left[u_r \frac{\partial e_r}{\partial \theta} \right] = \frac{1}{r}e_\theta \cdot [u_r e_\theta] = \frac{u_r}{r} . \tag{7.5}$$

Thus, Eq. 7.2 contains four nonzero scalar terms and the continuity equation in cylindrical coordinates becomes:

$$\frac{\partial u_x}{\partial x} + \frac{\partial u_r}{\partial r} + \frac{1}{r}\frac{\partial u_\theta}{\partial \theta} + \frac{u_r}{r} = 0. \tag{7.6}$$

Continuity Equation—Incompressible Flow—Cylindrical Coordinates

Setting $\partial / \partial \theta = 0$ to reduce Eq. (7.6) to axisymmetric flow:

$$\frac{\partial u_x}{\partial x} + \frac{\partial u_r}{\partial r} + \frac{u_r}{r} = 0 . \tag{7.7}$$

Continuity Equation—Axisymmetric Incompressible Flow

The continuity equation, Eq. 7.7, is sometimes written in a more convenient form (see Eq. 7.18). The student should expand Eq. 7.2, examine each term, and then verify Eq. 7.7.

Now, we contrast the continuity equation in cylindrical coordinates, Eq. 7.7, with the continuity equation in two-dimensional Cartesian coordinates obtained by expanding Eq. 3.52; namely:

$$\frac{\partial u}{\partial x} + \frac{\partial w}{\partial z} = 0, \tag{7.8}$$

where u and w are the velocity components in the x and z directions, respectively. The presence of the additional term in the axisymmetric case leads to important differences between two-dimensional planar and axisymmetric problems even though each problem is a function of only two independent variables. This is equally true for a compressible flow.

To verify the presence of this extra term, we rederive the continuity equation for the axisymmetric case by applying the Conservation of Mass principle to a fixed control volume of differential size. The flow is assumed to be steady but compressible for later use. The resulting equation then is simplified so as to describe incompressible flow. (Fig. 7.4).

The mass flux in through faces 1, 2, and 3 is:

(1) $\rho u_x(rd\theta dr)$

(2) $\rho u_\theta(dx dr)$

(3) $\rho u_r(rd\theta dx)$

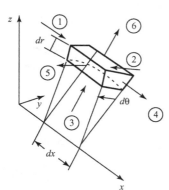

Figure 7.4. Differential control volume in cylindrical coordinates.

and the mass flux out through faces 4, 5, and 6 is:

(4) $-\left[\rho u_x + \dfrac{\partial}{\partial x}(\rho u_x)rd\theta dr\right]$

(5) $-\left[\rho u_\theta + \dfrac{\partial}{\partial \theta}(\rho u_\theta)\right]dx dr$

(6) $-\left[\rho u_r + \dfrac{\partial}{\partial r}(\rho u_r)\right](r+dr)d\theta dx$

Terms (2) and (5) are equal in axisymmetric flow because there is no change in flow properties relative to the angle θ when axial symmetry is assumed (i.e., $\dfrac{\partial}{\partial \theta}(\rho u_\theta) = 0$). Setting the sum of all of these six terms equal to zero by virtue of the conservation of mass for steady flow and then canceling Terms (2) and (5), as well as all like terms of opposite sign, the final result is:

$$-\frac{\partial}{\partial r}(\rho u_x)rdxdrd\theta - \rho u_r drd\theta dx - r\frac{\partial}{\partial r}(\rho u_r) - \rho\frac{\partial u_r}{\partial r}dr^2 d\theta dx = 0. \qquad (7.9)$$

The last term in this equation is of higher order than the other terms because it contains a product of differential magnitudes to the fourth power, whereas the three other terms in the equation contain only products of these quantities to the third power. Thus, the last term may be considered negligible (i.e., one order smaller) compared to the other terms in the equation in the limit as dx, dr, and $d\theta$ become very small. Recognizing this and then dividing by the common coefficient (i.e., $-dxdrd\theta$), Eq. 7.9 becomes:

$$r\frac{\partial}{\partial x}(\rho u_x) + \rho u_r + r\frac{\partial}{\partial \theta}(\rho u_r) = 0, \qquad (7.10)$$

which is valid for steady, compressible, axisymmetric flow. Finally, assuming $\rho = $ constant, dividing by r, and rearranging:

$$\frac{\partial u_x}{\partial x} + \frac{\partial u_r}{\partial r} + \frac{u_r}{r} = 0. \qquad (7.11)$$

Notice that Eq. 7.11 obtained by the application of the Conservation of Mass physical concept for an axially symmetric flow, is identical to Eq. 7.7 obtained by expanding the vector formulation of the general Conservation of Mass equation (Eq. 3.52) for incompressible flow in cylindrical coordinates and then assuming axial symmetry.

The Momentum Equation

As in the two-dimensional, planar case, the vector-momentum equation (Eq. 3.66) written for axisymmetric flow may be reduced to the Bernoulli Equation for the special case of incompressible, inviscid flow. To understand this, we expand Eq. 3.66 in cylindrical coordinates. Then, we simplify this vector equation by assuming axial symmetry. Next, we apply this vector equation along a streamline by taking the dot product of each term with an incremental streamline length $ds = dx e_x + dr e_r$. Then, we appeal to the fact that a streamline is defined by:

$$\frac{u_r}{u_x} = \left[\frac{dr}{dx}\right]_{streamline}$$

and substitute this relationship as appropriate. The final result is:

$$\rho d\left[\frac{u_x^2}{2} + \frac{u_r^2}{2}\right] + dp = 0.$$

Assuming incompressible flow, this may be integrated to give:

$$\frac{1}{2}\rho V^2 + p = \text{constant}, \tag{7.12}$$

which simply is Bernoulli's Equation, as seen in Chapter 4 in Cartesian coordinates. The student should follow the procedure outlined to verify Eq. 7.12.

This result, Eq. 7.12, could have been deduced directly by realizing that the Bernoulli Equation represents the momentum equation for any steady, inviscid, incompressible flow. Because the Bernoulli Equation is an *algebraic* equation containing two scalar quantities (i.e., velocity magnitude and pressure magnitude at a point) and because scalar quantities are independent of any coordinate system, then it is valid for any coordinate system.

Eq. 7.12 indicates that if the magnitude of the velocity at any point in an axisymmetric flow field can be found, then the static pressure at that point may be found directly. As in two-dimensional, planar flow, it is most convenient to introduce scalar functions to find the velocity components and, hence, the required velocity magnitude.

7.3 Defining Equation for the Velocity Potential

This development parallels that for two-dimensional, planar flows in Chapter 4. If the flow is assumed to be irrotational, then the curl of the velocity vector is zero, $\nabla \times V = 0$. From this, it follows that a velocity potential, φ, exists such that $V = \nabla\varphi$. Expanding the gradient operator in cylindrical coordinates, the velocity vector is given by:

$$V = u_x e_x + u_r e_r + u_\theta e_\theta = e_x \frac{\partial\varphi}{\partial x} + e_r \frac{\partial\varphi}{\partial r} + e_\theta \frac{1}{r}\frac{\partial\varphi}{\partial\theta}, \tag{7.13}$$

from which it follows that in axisymmetric flow:

$$u_x \equiv \frac{\partial \varphi}{\partial x}, \quad u_r \equiv \frac{\partial \varphi}{\partial r} \qquad (7.14)$$

For incompressible flow, the continuity equation is given by $\nabla \cdot V = 0$ so that for irrotational flow $\nabla \cdot (\nabla \varphi) \equiv \nabla^2 \varphi = 0$. Writing the Laplacian operator in cylindrical coordinates, it follows that:

$$\frac{\partial^2 \varphi}{\partial x^2} + \frac{1}{r}\frac{\partial}{\partial r}\left[r\frac{\partial \varphi}{\partial r}\right] + \frac{1}{r^2}\frac{\partial^2 \varphi}{\partial \theta^2} = 0, \qquad (7.15)$$

which for an axisymmetric flow reduces to:

$$\frac{\partial^2 \varphi}{\partial x^2} + \frac{\partial^2 \varphi}{\partial r^2} + \frac{1}{r}\frac{\partial \varphi}{\partial r} = 0, \qquad (7.16)$$

where the velocity potential $\varphi = \varphi(x, r)$. Thus, as in the two-dimensional planar flow case, the velocity potential for axisymmetric flow satisfies the Laplace's Equation. Because the Laplace's Equation is linear, superposition techniques may be used to construct solutions.

7.4 Defining Equation for the Stream Function

Recall from Eq. 7.7 that the continuity equation for axisymmetric flow is:

$$\frac{\partial u_x}{\partial x} + \frac{\partial u_r}{\partial r} + \frac{u_r}{r} = 0, \qquad (7.17)$$

which may be written as follows:

$$\frac{\partial u_x}{\partial x} + \frac{1}{r}\frac{\partial}{\partial r}(ru_r) = 0. \qquad (7.18)$$

Now, we multiply Eq. 7.18 through by r. Because x and r are independent variables, the coefficient r in the first term may be included within the x-derivative, and Eq. 7.18 becomes:

$$\frac{\partial(ru_x)}{\partial x} + \frac{\partial(ru_r)}{\partial r} = 0. \qquad (7.19)$$

Eq. 7.19 is simply an alternate way of writing the continuity equation Eq. 7.18. Examination of Eq. 7.19 shows that it is satisfied by a scalar function (i.e., the stream function) such that:

$$\frac{\partial \psi}{\partial r} \equiv ru_x, \quad \frac{\partial \psi}{\partial x} \equiv -ru_r, \qquad (7.20)$$

which defines the stream function, $\psi(x, r)$ for axisymmetric flow. As in the two-dimensional, planar case, the derivative of the stream function yields a velocity component orthogonal to the derivative direction. Because ψ is related to the volume flow rate, a negative sign goes with the $\partial \psi / \partial x$ expression in Eq. 7.20 so that the radial-velocity u_r component has the correct sign to properly reflect continuity.

For an irrotational flow, the curl of the velocity vector (i.e., the vorticity) is zero. Applying the curl operator for axisymmetric flow, it follows that:

$$\frac{\partial u_x}{\partial r} - \frac{\partial u_r}{\partial x} = 0. \tag{7.21}$$

Substituting the definition of the stream function, Eq. 7.20, into the irrotationality condition, Eq. 7.21, yields:

$$\frac{\partial}{\partial r}\left(\frac{1}{r}\frac{\partial \psi}{\partial r}\right) - \frac{\partial}{\partial x}\left(\frac{1}{r}\frac{\partial \psi}{\partial x}\right) = 0. \tag{7.22}$$

Finally, expanding Eq. 7.22 and rearranging:

$$\frac{\partial^2 \psi}{\partial x^2} + \frac{\partial^2 \psi}{\partial r^2} - \frac{1}{r}\frac{\partial \psi}{\partial r} = 0. \tag{7.23}$$

Eq. 7.23 is the defining equation for the stream function. If solutions for this equation can be found, then the velocity components can be determined by using Eq. 7.23, and the Bernoulli Equation provides the corresponding pressure distribution.

As in the two-dimensional, planar case (see Chapter 4), Eq. 7.23 is a linear equation so that elementary solutions may be superposed to generate solutions for more complex flows. As before, it is not necessary to solve Eq. 7.23 directly because useful elementary solutions for the stream function may be constructed, as demonstrated later.

Notice that the defining equation for the stream-function equation, Eq. 7.23, is *not* the Laplace's Equation in the case of axisymmetric, incompressible, irrotational flow. Compare Eq. 7.23 with Eq. 7.16; the sign before the third term is not the same. Contrast this with two-dimensional, planar, incompressible, irrotational flow, where the velocity potential and the stream function *both* satisfy the Laplace's Equation.

Following the same procedure used in the planar case, elementary solutions for the stream function are constructed next and then superposed to generate more complex flow fields. The process begins with consideration of a three-dimensional point-source flow. A brief digression follows to examine the flow about a sphere in uniform flow. Then, the source flow and a uniform flow are superposed so as to construct the solution for a stream function that describes the flow around an axisymmetric body.

7.5 Three-Dimensional Point Source at the Origin of Coordinates

The point source referred to herein is a true point source in that flow exits from a point radially in all directions in three-space. The point source previously discussed in two-dimensional flow (see Chapter 5) is actually the cross section of a line source of infinite length that is perpendicular to the two-dimensional plane under study. Recall that source flow also is radial in two-dimensional, planar flow.

Consider a point source at the origin of a three-dimensional, Cartesian-coordinate system, as shown in Fig. 7.5.

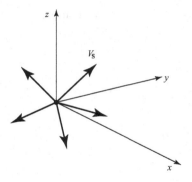

Figure 7.5. Source flow at the origin.

By definition, the flow is purely radial, with a velocity given by V_s. We consider a fixed spherical control volume surrounding the origin of coordinates, with the center of the sphere at the origin (Fig. 7.6). Then, we make a calculation for the mass flux through the surface of the control volume. Recall that the surface area of a sphere of radius a is given by $4\pi a^2$.

We let Λ represent the volume flow out of the point source per unit time. Because the source flow is purely radial, the outflow velocity vector is always perpendicular to the spherical control surface and has a constant magnitude over the surface. Thus, the mass flow rate out of the source is:

$$\dot{m} = \rho\Lambda = \rho \iint_{\hat{S}} V \cdot n d\hat{S} = \rho V_s \iint_{\hat{S}} d\hat{S} = \rho V_s \,(4\pi a^2). \tag{7.24}$$

Solving for V_s gives:

$$V_s = \frac{\Lambda}{4\pi a^2}. \tag{7.25}$$

Contrast this result with the result previously obtained for a two-dimensional point source (i.e., corresponding to the radial flow through a cylindrical control volume with a line source along the axis of the cylinder). In the two-dimensional case, the radial-source velocity varies inversely as the radial distance from the source; in

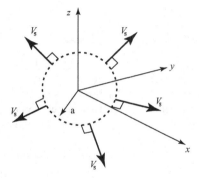

Figure 7.6. Mass flux through fixed control volume.

the three-dimensional case, the radial-source velocity varies inversely as the square of the radial distance from the source.

7.6 Incompressible Flow around a Sphere

The three-dimensional point-source flow described herein can be extended to the representation of a three-dimensional doublet by superposing a source–sink pair similarly to the analysis detailed for two-dimensional, planar, incompressible flow in Chapter 4. Superposing uniform flow and a doublet at the origin with its axis parallel to the freestream leads to the solution for the flow around a sphere (see Anderson, 1991 and Karamcheti, 1967). The radius of the sphere is given by:

$$\text{sphere radius} = \left(\frac{\Omega}{2\pi V_\infty}\right)^{1/3} \tag{7.26}$$

where Ω is the doublet strength.

In Chapter 4, a two-dimensional, planar, flow field generated by the superposition of a doublet and a uniform flow yielded the flow over a right-circular cylinder placed normal to the flow. The superposition of a three-dimensional doublet and a uniform flow gives the flow around a sphere. Both the radius of the cylinder and the radius of the sphere depend on the doublet strength. However, the radius of the cylinder varies as the doublet strength to the one-half power, whereas the radius of the sphere varies as the doublet strength to the one-third power.

According to the inviscid-flow model, the flow field about the right cylinder with its axis placed normal to the flow and also about the sphere is symmetrical fore and aft, meaning that there is a stagnation point on both the upstream and downstream surfaces of the bodies along the flow axis of symmetry. Boundary-layer separation profoundly influences the flow field on the downstream side of both the cylinder and the sphere in the case of a real (i.e., viscous) flow situation.

The maximum velocity on the surface of both of these bodies occurs 90° in polar angle away from the two stagnation points (i.e., at the top and bottom). The results are, for the same radius:

$$\text{cylinder:} \quad V_{\max} = 2.0 V_\infty.$$
$$\text{sphere:} \quad V_{\max} = 1.5 V_\infty.$$

Thus, the freestream does not accelerate to as high a velocity (or to as low a static pressure) around the surface of the sphere as it does around the surface of the right-circular cylinder of the same radius. This is evidence of a *three-dimensional relief effect*. The two-dimensional flow approaching a right-circular cylinder placed normal to the flow must *split* and pass either above or below the body. In the three-dimensional case, the flow passes *around* the sphere. Thus, the sphere causes less of a disturbance in a flow than the cylinder if both bodies are of the same radius. This same relief effect is observed when comparing supersonic flow around a two-dimensional wedge and an axisymmetric cone as developed in tests on compressible gas dynamics.

7.7 Elementary Solutions for the Stream Function in Axisymmetric Flow

We now develop solutions for axisymmetric flow analogous to those described in Chapter 4.

Uniform Flow

Consider a uniform flow, V_∞, in the x-direction in cylindrical coordinates. Then, from Eq. 7.20:

$$u_x = V_\infty = \frac{1}{r}\frac{\partial \psi}{\partial r} \quad \Rightarrow \quad \psi = \frac{V_\infty r^2}{2} + f(x)$$

$$u_r = -\frac{1}{r}\frac{\partial \psi}{\partial x} = -\frac{1}{r}f'(x) = 0 \quad \Rightarrow f(x) = \text{constant}.$$

Arbitrarily setting $\psi = 0$ at $r = 0$ (i.e., along the x-axis), the stream function for uniform flow is given by:

$$\psi = \frac{V_\infty r^2}{2}. \tag{7.27}$$

As expected, lines of constant stream function (i.e., streamlines) are parallel to the x-axis. Recall from Chapter 4 that each streamline may be assigned a value corresponding to the volume (or mass) flux between it and a reference streamline. For planar, two-dimensional flow, $\psi(x,y) \approx y$, $\Delta\psi(x,y) = \Delta y$, and equal increments in mass flux are represented by straight lines equally spaced, as shown in Fig. 7.7a. For axisymmetric flow, equal increments in mass flux again are represented by straight lines parallel to the axis of symmetry, but these lines are not equally spaced in the x-r plane because $\Delta\psi(x,r) = \Delta r^2$. Thus, for axisymmetric flow, lines of constant increment in stream function become closer together as the distance, r, from the axis of symmetry increases, as illustrated in Fig. 7.7b.

Source Flow

Consider a three-dimensional source at the origin of coordinates in three-space. One particular radial streamline from the source passes through any arbitrary point $P(x,r)$ in any plane containing the x-axis and making an angle θ with the y-axis, as shown in Fig. 7.8a. Now, we focus on one particular (x, r) plane (Fig. 7.8b) and recognize that

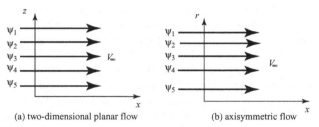

(a) two-dimensional planar flow   (b) axisymmetric flow

Figure 7.7. Constant increment in stream function, freestream flow.

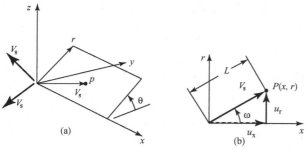

Figure 7.8. Three-dimensional source flow in axial plane of symmetry.

because of the symmetry of the source flow field, the flow behavior in any such plane is independent of ω—that is, the source flow has axial (as well as spherical) symmetry.

In Fig. 7.8b, is the magnitude of the radial velocity at the point (x,r) due to a three-dimensional source at the origin of coordinates. From Eq. 7.25 this may be written as:

$$V_s = \frac{K}{L^2}. \tag{7.28}$$

where K is a constant describing the source strength:

$$K \equiv \frac{\Lambda}{4\pi},$$

and L is the distance from the source to the point in question.
Then, from Fig. 7.8b:

$$L = \sqrt{x^2 + r^2}$$

$$u_x = V_s \cos\omega = \frac{K}{L^2}\frac{x}{L} = \frac{Kx}{L^3} = \frac{Kx}{\left(x^2 + r^2\right)^{3/2}}$$

$$u_r = V_s \sin\omega = \frac{Kr}{L^3} = \frac{Kr}{\left(x^2 + r^2\right)^{3/2}}.$$

To find the stream function defined in a plane of symmetry and due to a three-dimensional source, apply Eq. 7.20, with the result that:

$$\frac{\partial\psi}{\partial r} = ru_x = \frac{Krx}{\left(x^2 + r^2\right)^{3/2}} \tag{7.29}$$

Integrating:

$$\psi = -\frac{Kx}{\left(x^2 + r^2\right)^{1/2}} + f(x). \tag{7.30}$$

Again, from Eq. 7.20:

$$\frac{\partial\psi}{\partial x} = -ru_r = -\frac{Kr^2}{\left(x^2 + r^2\right)^{3/2}}. \tag{7.31}$$

Now, we take the derivative of Eq. 7.30 with respect to x, which yields:

$$\frac{\partial \psi}{\partial x} = -\frac{K}{\sqrt{x^2 + r^2}} + \frac{Kx^2}{\left(x^2 + r^2\right)^{3/2}} + f'(x). \tag{7.32}$$

Equating Eqs. 7.31 and 7.32 and solving for $f'(x)$:

$$f'(x) = 0, \tag{7.33}$$

which means that $f(x)$ is a constant. Reflecting this in Eq. 7.30 and again setting $\psi = 0$ along the x-axis ($r = 0$), it follows that:

$$f(x) = \frac{Kx}{\sqrt{x^2 + 0}} = K = \text{constant.}$$

Thus,

$$\psi = -\frac{Kx}{\sqrt{x^2 + r^2}} + K,$$

or, because, $K \equiv \Lambda / 4\pi$

$$\psi = \frac{L}{4\pi}\left(1 - \frac{x}{\sqrt{x^2 + r^2}}\right). \tag{7.34}$$

Eq. 7.34 represents the equation for the streamlines (ψ = constant) in any down-stream axial plane of symmetry due to the flow from a three-dimensional source at the origin of coordinates.

7.8 Superposition of Uniform Flow and a Source Flow

Similar to two-dimensional, planar flow (see Chapter 4), it may be shown by sub-stitution that the stream functions described by Eqs. 7.27 and 7.34 are both solu-tions of the defining equation for the stream function in axisymmetric flow, Eq. 7.23. Furthermore, because Eq. 7.23 is a linear equation, the sum of these two solu-tions is also a solution. In particular, the streamline $\psi = 0$ is the body surface with a stagnation point. Thus, a valid stream function for axisymmetric flow is given by:

$$\psi = \frac{V_\infty r^2}{2} + \frac{\Lambda}{4\pi}\left(1 - \frac{x}{\sqrt{x^2 + r^2}}\right). \tag{7.35}$$

Taking an arbitrary but fixed value of the source strength, Λ, and the freestream velocity, V_∞, the resulting streamlines may be found by taking various values of ψ = constant and finding pairs of points (x, r) that satisfy Eq. 7.35. The result is shown in Fig. 7.9.

The flow field is seen to represent the axisymmetric flow over an open-ended body of revolution at zero angle of attack. Figure 7.9 appears the same for any (x,r) plane; thus, for any angle θ in cylindrical coordinates. Notice that the stagnation point is upstream of the origin of coordinates, as expected. The fineness of the body of revolution may be altered by changing the values of Λ and/or V_∞.

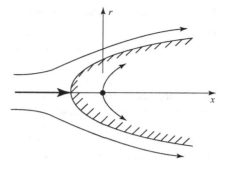

Figure 7.9. Superposition of a three-
dimensional source and uniform flow.

The velocity components u_x and u_r at any point in the flow field may be evaluated by
using Eq. 7.35 in Eq. 7.20.

7.9 Flow Past a Rankine Body

The superposition in Section 7.8 may be extended by combining a uniform axial flow with a three-dimensional source and sink of equal strengths
and located a distance, a, upstream and downstream of the origin, respectively. For this axisymmetric flow, the equation of the streamlines is given in
Karamcheti, 1967 as:

$$\psi = \frac{V_\infty r^2}{2} + \frac{\Lambda}{4\pi}\left[\frac{(x-a)}{\sqrt{(x-a)^2 + r^2}} - \frac{(x+a)}{\sqrt{(x+a)^2 + r^2}}\right]. \tag{7.36}$$

The resulting body shape in axisymmetric flow is a body of revolution called a
Rankine body. The two-dimensional, planar-flow counterpart of this body is the
Rankine oval. The shape of the two bodies and the streamlines associated with them
are illustrated in Fig. 7.10.

PROGRAM PSI. Run the same program as for the two-dimensional, planar
flow (see Chapter 4, but now choose the axisymmetric option. Explore what
happens to the axisymmetric body shape as the parameters are varied).

Figure 7.10. Comparison of
a two-dimensional Rankine
oval with a Rankine body
of revolution with the same
length.

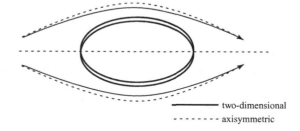

————— two-dimensional
- - - - - - - axisymmetric

7.10 Flow Past a General Body of Revolution

The approach outlined in Section 7.9 may be extended by distributing increasingly source–sink pairs of finite strength inside the body or along the body axis. The sum of the source and sink strengths must be zero if the body is to be closed. Simple body shapes may be treated in this way; however, solutions for body shapes of practical interest are necessarily numerical in character. If it is required to find the pressure distribution on a suitably slender body of revolution, then an approximate analytical solution may be found that uses source singularities distributed along the body axis. The velocity potential for a uniform flow superposed with a source distributed along the body axis from x_1 to x_2 is expressed as follows (see Karamcheti, 1967):

$$\phi(x,r) = V_\infty x - \frac{1}{4\pi} \int_{x_1}^{x_2} \frac{f(\xi)d\xi}{\sqrt{(x-\xi)^2 + r^2}}, \tag{7.37}$$

where $f(\xi)$ is the source strength per unit length, $f(\xi)d\xi$ is the infinitesimal source strength at $x = x$, and the point (x, r) is a field point in cylindrical coordinates as shown in Figure 7.11. The student may verify that this expression for the velocity potential satisfies the Laplace's Equation. Also, study the derivation of an analogous expression for the stream function that is carried out in Section 7.11.

Using Eq. 7.37 and applying the tangency-boundary condition at the surface of a given body leads to an integral equation for the unknown source-strength distribution, $f(\xi)$. Rather than directly solving this integral equation for the source-strength distribution, it is assumed that the slenderness ratio, ε, of the body is small, where $\varepsilon = [(\text{maximum body radius})/(\text{body length})]$. For small ε, the integral operator containing $f(\xi)$ can be expanded asymptotically in a power series (Wang, et al, 1985). It then may be shown that the source distribution, $f(x)$, as well as the values of x_1 and x_2, can be related to the axial distribution of the body cross-sectional area, $S(x)$. When these expressions are used in Eq. 7.37, the result is an asymptotic expansion for the velocity potential, which is expressed in terms of a simple geometric property, $S(x)$, of the body. The velocity components on the body surface and, hence, the inviscid-flow pressure distribution follow directly by definition. A similar approach for slender axisymmetric bodies in a supersonic flow is discussed in compressible flow text books.

Figure 7.12 shows the pressure coefficient on the surface of an ellipsoid as evaluated by an exact analytical solution and by a numerical solution. Also shown in this figure (i.e., the solid line) is the pressure coefficient as calculated from the approximate velocity potential obtained by using the asymptotic approach discussed previously. Notice that when the body is slender, $\varepsilon = 0.10$, the approximate solution agrees

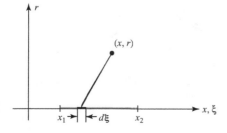

Fig. 7.11. Coordinate system and continuous source distribution.

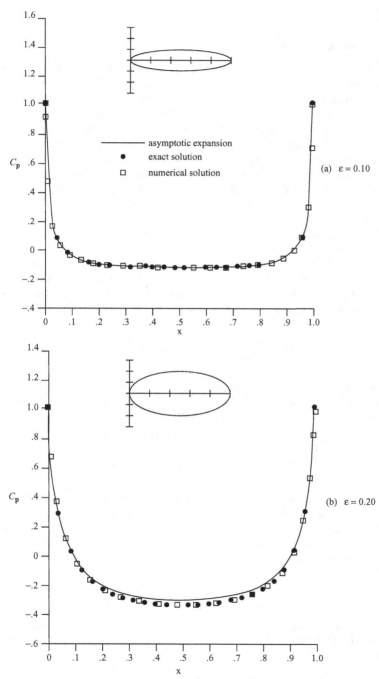

Figure 7.12. C_p for ellipsoid of revolution, $S(x) = 4x(1-x)$ (Wang et al., 1985).

well with the exact result, whereas for a body that is not so slender, $\varepsilon = 0.20$, the asymptotic solution does not predict the exact result as well.

7.11 Numerical Methods

There are two classes of axisymmetric flow problems that are treated numerically: (1) the direct problem of finding the flow about a body of given shape, which may be solved by using a distributed singularity along the body axis or a panel method; and (2) determining a body shape that results in low or minimum drag. This can be accomplished by using two different approaches. The first is to consider an inverse problem: Find the body shape that gives a desired pressure distribution on the body. The pressure distribution, in turn, is chosen so as to reduce or minimize the friction and pressure drag due to the boundary layer growing on the body surface. The second approach uses optimization algorithms and seeks to optimize the body shape for minimum drag subject to certain constraints. A brief discussion of each method follows.

1. The direct problem: What is the surface-pressure distribution for a body of specified shape?

    (a) *Distributed singularities along the body axis.* We first derive an expression for the stream function in axisymmetric uniform flow due to a distributed source along the x-axis. We begin by generalizing Eq. (7.35) to represent the superposition of uniform flow and a point source located at $x = x_0$. Thus,

$$\psi(x,r) = \frac{\Lambda}{4\pi}\left[1 - \frac{(x-x_0)}{\sqrt{(x-x_0)^2 + r^2}}\right] + \frac{V_\infty r^2}{2}, \tag{7.38}$$

    where Λ is the strength of the source. Next, we consider a source distributed along the x-axis between $x = 0$ and $x = L$. We rewrite Eq. 7.38 to describe a differential contribution to the stream function at a field point (x, r) due to a differential length, $d\xi$, of that distributed source located at $\xi = x$:

$$d\psi(x,r) = \frac{\mu(\xi)d\xi}{4\pi}\left[1 - \frac{(x-\xi)}{\sqrt{(x-\xi)^2 + r^2}}\right] + \frac{V_\infty r^2}{2}, \tag{7.39}$$

    where $\mu(\xi)$ is the local strength of the distributed source per unit length.

    Now, we sum the contributions to the stream function at (x, r) due to all of the differential lengths of distributed source between $x = 0$ and $x = L$. We let the distributed source represent a closed body of revolution, where L is the length of the body. Summing by integration:

$$\psi(x,r) = -\frac{1}{4\pi}\int_0^L \frac{\mu(\xi)(x-\xi)d\xi}{\sqrt{(x-\xi)^2 + r^2}} + \frac{V_\infty r^2}{2}. \tag{7.40}$$

    The first term in the square brackets in Eq. 7.39 vanishes on integration because if the body is closed, the net source–sink strength over the length of the body must be zero. Thus,

$$\int_0^L \mu(\xi)d\xi = 0.$$

Recall that by definition, Eq. 7.30, the expressions for the velocity components u_x and u_r follow from the expression for the stream function by differentiation namely:

$$u_x \equiv \frac{1}{r}\frac{\partial \psi}{\partial r}; \qquad u_r \equiv -\frac{1}{r}\frac{\partial \psi}{\partial x}. \tag{7.41}$$

In Eq. 7.40, the integration is relative to the variable ξ so that the integrand may be differentiated directly. If $\mu(x)$ is represented as a polynomial, then the integral expressions for ψ, Eq. 7.40, and for u_x and u_r, Eq. 7.41, may be evaluated in closed form.

Following Zedan and Dalton (1978), the source-strength distribution along the x-axis is split into N segments, each of length ΔL_j, and Eq. 7.40 is expressed as a summation. Zedan and Dalton, 1978 considers a linear-source distribution over each segment; this is generalized to a polynomial distribution in Zedan and Dalton, 1980. Figure 7.13 illustrates an assumed linear variation over each segment.

Notice that the source-strength variation is assumed and known but the magnitude of the strength is to be found. Referring to Fig. 7.13, let there be N source segments used to represent the body of revolution and let the strength distribution in each segment be of the following form:

$$\mu_\xi = (a + b\xi),$$

where a, b are constants to be determined. The parameter ξ is the local axial coordinate over each segment ΔL_j.

With two constants a,b to be found for each of N segments, there are 2N unknowns. We next require that the source distributions are continuous at the junctions of the segments (i.e., there cannot be any step in source strength from one segment to the next). If there are N segments, then there are (N-1) junctions. Thus, the demand that the strength distribution be continuous at each juncture results in (N-1) equations involving the 2N unknowns, leaving (N+1) equations to be generated if the a,b unknowns are to be found and the source-strength distribution along the axis determined.

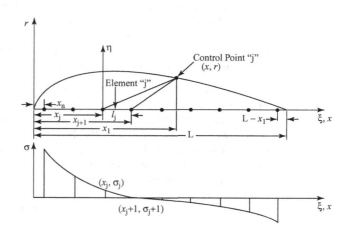

Figure 7.13. Representation of a body of revolution by discrete linear segments (Zedan and Dalton, 1978).

A total of N equations may be written by selecting N control points on the body surface. That is, let i assume integer values from 1 to N. The value of the stream function, ψ, at a control point i on the surface follows from Eq. 7.40 as:

$$\psi_{i(x,r)} = \frac{1}{4\pi} \sum_{i=1}^{N} \left[\sum_{j=1}^{N+1} A_{ij}\mu_j + V_\infty \frac{Ri^2}{2} \right], \qquad (7.42)$$

where A_{ij} is a matrix that is a function only of the known body geometry (see Zedan and Dalton, 1978), μ_j is the source intensity at the jth juncture, and $r = R$ on the surface. The closure condition, which states that the net efflux of all of the source–sink elements must be zero, supplies the final equation required; namely:

$$\sum_{j=1}^{N} Q_j = 0 \ \Rightarrow \sum_{j=1}^{N} \int_0^{\Delta Lj} \mu_\xi(\xi)d\xi = 0, \qquad (7.43)$$

where Q_j is the mass efflux over the jth source–sink segment.

As noted in previous discussions, the body surface must be a streamline, meaning that on the body surface ψ = constant and, for convenience, $\psi = 0$. Setting $\psi_i = 0$ in Eq. 7.42 at N control points on the body surface results in a system of N linear-algebraic equations in (N+1) unknowns μ_j. The addition of the closure equation, Eq. 7.43, provides the needed additional equation. Solving this set of equations provides the values of the source strength, μ_j, at the juncture points of the source distribution.

Now, using Eq. 7.41, the expression for the axial-velocity component at the ith control point follows from Eq. 7.42 as:

$$u_{x,i}(x,r) = \frac{1}{4\pi} \sum_{i=1}^{N} \left[\sum_{j=1}^{N+1} B_{ij}\mu_j \right] + V_\infty, \qquad (7.44)$$

where B_{ij} is a matrix that is a function only of the known body geometry (Zedan and Dalton, 1978). Because the source strength μ_j was found in the previous step, the axial-velocity component u_x is determined at each control point on the body surface.

Tangency requires that $u_x/u_r = dR/dx$ at the body surface, where $R(x)$ is known. Hence, the radial-velocity component, u_r, is determined at each control point. Lastly, $V_t^2 = u_x^2 + u_r^2$ and the tangential velocity, V_t at each control point is found. The surface pressure on the prescribed body follows directly from the Bernoulli Equation and the direct problem is solved.

(b) *The panel method.* The direct problem also may be analyzed numerically by distributing singularities over the surface of a given body. This is simply the panel method discussed in Chapter 6; again the surface singularities may be sources, doublets, or vortices. We restrict the discussion here to surface sources because we are considering bodies of revolution at zero angle of attack. As before, flat quadrilateral surface panels approximate the given body, as shown in Fig. 7.14. Notice in what follows that the panel method is a three-dimensional problem even for a body of revolution at zero angle of attack because each source panel affects the flow at all other panels present in three-dimensional space.

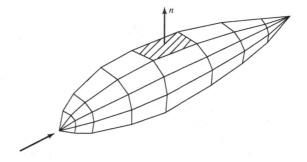

Figure 7.14. Representation
of a body of revolution with
flat panels.

In contrast, solution methods using source distributions along the body axis
are formulated in cylindrical coordinates and use axisymmetric-flow equations.
Thus, the panel method has an advantage over the distributed-source method
in that the former can be extended to fuselage shapes that are not bodies of
revolution.

The velocity-potential expression valid for a three-dimensional source may
be derived analogously in Cartesian coordinates to the way in which the stream
function was derived previously in this chapter. The velocity potential at Point
$P(x, y, z)$ due to a point source of strength L located at Point $Q(\xi, \eta, \zeta)$ is given by:

$$\phi(P) = -\frac{\Lambda}{4\pi} \frac{1}{[(x-\xi)^2 + (y-\eta)^2 + (z-\zeta)^2]^{1/2}} = -\frac{\Lambda}{4\pi} \frac{1}{\overline{PQ}}, \tag{7.45}$$

where $\overline{PQ}$ is the distance between Points P and Q. This solution of the potential
equation may be superposed with another solution representing a uniform flow
from infinity, $\phi(x, y, z) = V_\infty x$.

Now, we assume that instead of being a point source, the source has a strength
μ per unit area, which is distributed over a differential panel area $d\hat{A}$. The source
distribution may be assumed to be constant over the surface, $\hat{A}_J$, of a quadrilateral panel, or a higher-order distribution over the panel surface may be used.
Then, the velocity potential at Point P due to the superposition of a uniform flow
and a distributed source on Panel Q is given by:

$$\phi(x, y, z) = V_\infty x - \frac{\displaystyle\iint_{\hat{A}_j} \mu_j d\hat{A}_j}{4\pi} \frac{1}{\overline{PQ}}. \tag{7.46a}$$

Assuming that the source distribution per unit area is constant over any quadrilateral panel, $\hat{A}_J$, Eq. 7.46a simplifies to:

$$\phi(x, y, z) = V_\infty x - \frac{\mu_j A_j}{4\pi \overline{PQ}}. \tag{7.46b}$$

Next, we assign each panel a control point located at the middle of the panel and
let the body of revolution be represented by N panels. Then, the velocity potential
at the control point on the ith panel, $P_i(x, y, z)$, due to the influence of the constant-
strength sources on all of the N panels plus the influence of the freestream, is:

$$\phi_i(x,y,z) = V_\infty x - \frac{1}{4\pi} \sum_{j=1}^{N} \frac{\mu_j \hat{A}_j}{\overline{P_i Q_j}} + \phi\big|_{j=i}. \qquad (7.47)$$

Notice that the term $j = i$ is excluded from the summation. When $j = i$, the behavior of that term in the series is singular because the source panel Q is the same panel as that where Point P is located, and $\overline{PQ} = 0$ Thus, this term must be treated in a special way, as indicated by the separate term $\phi\big|_{j=1}$ in Eq. 7.47. Notice that all of the source panels contribute to the potential at a given control point, even the panels on the opposite side of the body from the control point. This is the same behavior that we observed in the case of vortex rings representing a finite wing. Also, we note that the effect of a source at a control point does not depend on the distance between the source panel and the control point as measured along the body surface; rather, the effect is dependent on the straight-line distance between the two, $\overline{PQ}$.

Taking $(\partial\phi / \partial n)_i$ in Eq. 7.47, where n is the unit outward normal at each control point, yields the velocity component normal to the panel at each control point. According to the surface-tangency-boundary condition, each normal component of velocity must be zero. Applying this boundary condition at each control point, in turn, yields a system of N linear-algebraic equations for the N unknowns, μ_j. When the source strengths are determined, the magnitude of the tangential velocity at each control point may be found and, hence, the surface pressure, after an appeal to the Bernoulli Equation. Thus, the pressure distribution over a body of revolution is determined. This pressure distribution may be integrated to obtain forces, if desired.

A comparison of results using axial and surface singularity (i.e., panel) methods for inviscid flow can be found in D'Sa and Dalton, 1986. Here, we conclude that the accuracy of the method using singularities distributed along the axis of symmetry and the accuracy of the panel method using singularities distributed over the body surface are comparable for problems involving smooth bodies. For bodies with sudden changes in curvature, the panel method is more accurate. A boundary-layer code must be appended to both of the solution methods for the direct problem to obtain the drag of the body of revolution.

2. The optimum body-shape problem: Find the body shape that has low or minimum drag in a given application.
(a) *The inverse method.* This method seeks to answer the "optimum" body-shape question by focusing on the behavior of the boundary layer because it causes the drag. Thus, if a pressure distribution is specified along a body surface with the aim of generating low drag due to viscous effects, what body shape is needed to obtain this pressure distribution? Tailoring the shape of a body of revolution to achieve a low-drag boundary layer is attractive because, as noted previously, the fuselage drag comprises a major fraction of the total drag of a flight vehicle. The viscous boundary layer is discussed in detail in Chapter 8; suffice it to say here that a laminar boundary layer is preferred to a turbulent layer because the former exhibits a much smaller skin-friction drag. The boundary-layer solution code used in conjunction with body-shape tailoring is, of necessity, complex. It must be

able to provide results for both laminar and turbulent boundary layers and it should be able to predict transition (i.e., the laminar boundary layer becomes turbulent) and separation (i.e., the boundary layer leaves the body surface). Either surface-panel or axial-singularity methods may be used in the inverse problem; both are iterative in nature. The axial-singularity method, discussed herein, often is used in practice because it is computationally more efficient than panel methods in this application. In any case, the goals are to (1) achieve a strong favorable pressure gradient $(\partial p / \partial x < 0)$ over the front portion of the body of revolution to delay transition, thus reducing frictional drag; and (2) to minimize the adverse pressure gradient $(\partial p / \partial x > 0)$ over the afterbody so as to delay boundary-layer separation and thereby reduce pressure drag.

Zedan and Dalton, 1978 and Zedan and Dalton, 1980 discuss an iterative inverse method for finding an axisymmetric body shape with low drag. Again, the basis of the method is a source singularity distributed in N segments along the body axis, (see Fig. 7.13). In Zedan, 1978, the distribution over each segment is linear, whereas in Zedan and Dalton, 1980, the distribution over each segment is represented by a polynomial. Zedan et al., 1994, treat the problem by using a doublet singularity distributed linearly over several segments along the body axis.

The solution method begins similar to the solution of the direct problem. Following (Zedan and Dalton, 1978) the linear-source-strength distribution over each segment shown in Fig. 7.13 is represented by $\mu_\xi = (a + b\xi)$.

With N segments of sources and two unknown constants describing the linear distribution of source strength over each segment, again there are 2N unknowns. By appealing to the continuity of source-strength distribution at each of (N-1) junctures between segments, the number of unknowns is reduced to (N+1). The (N+1) source strengths are found by writing a set of N linear-algebraic equations at N control points, Eq. 7.42, plus one closure condition, Eq. 7.43. This is the same methodology used previously in the axial-source-distribution method for the solution of the direct problem. Thus, rewriting Eq. 7.42 for the stream function at each control point i:

$$\psi_i(x,r) = \frac{1}{4\pi} \sum_{i=1}^{N} \left[\sum_{j=1}^{N+1} (A_{ij}\mu_j + V_\infty \frac{R_i^2}{2} \right] \qquad (7.48)$$

and rewriting the closure equation, Eq. (7.43):

$$\sum_{J=1}^{N} \int_0^{\Delta L_j} \mu_\xi(\xi)d\xi = 0. \qquad (7.49)$$

Again, the axial-velocity component at the ith control point is found by differentiation of Eq. 7.48; namely:

$$u_{x,i}(x,r) = \frac{1}{4\pi} \sum_{i=1}^{N} \left[\sum_{j=1}^{N+1} B_{ij}\mu_j \right] + V_\infty, \qquad (7.50)$$

where A_{ij} and B_{ij} are matrices that are a function only of body geometry (Zedan and Dalton, 1978).

The iterative solution to find the body shape corresponding to a prescribed pressure (i.e., velocity) distribution proceeds as follows:

(1) Make an initial guess at the body shape, $R°(x)$. This may be a body shape known to have a low drag for which improvement is sought or it may be a baseline ellipsoid shape.

(2) Calculate matrices A_{ij} and B_{ij} for this geometry.

(3) Select $i = N$ control points on $R°(x)$. Then, at each control point, calculate dR/dx.

(4) Impose a pressure distribution on the assumed initial body shape. For example, to delay boundary-layer transition, this might be a pressure distribution that has a stronger favorable pressure gradient over the forebody than that exhibited by the low-drag body shape that was used as a first guess. The imposed pressure distribution is changed easily to an imposed velocity distribution through use of the Bernoulli Equation. The velocity calculated in this way is an imposed tangential velocity at each control point.

(5) Require flow tangency at each control point on the body surface. Thus,

$$\frac{u_r}{u_x} = \frac{dR}{dx}.$$

Recalling that:

$$V_t^2 = u_x^2 + u_r^2 \Rightarrow V_t = u_x \sqrt{1 + \frac{u_r^2}{u_r^2}},$$

the tangency condition at each point i on the body surface may be written as:

$$u_x = \frac{V_t}{\sqrt{1 + (dR/dx)^2}}.$$

(6) With u_x corresponding to the imposed pressure distribution now known from Step 5 at each control point, write Eq. 7.50 as a set of (N) linear-algebraic equations in terms of (N+1) unknown source strengths, μ_j. The remaining equation for μ_j comes from closure, Eq. 7.49. Solving this system yields the source-strength distribution μ_j at the juncture points corresponding to the $R°$ body shape but with the imposed velocity distribution.

(7) Write Eq. 7.48 on the body surface by setting $\psi = \text{constant} = 0$ at each control point. With μ_j and A_{ij} known, a new body shape is found by solving the resulting equation for R_i; namely:

$$R_i = \sqrt{\frac{-1}{2\pi V_\infty} \sum_{j=1}^{N+1} A_{ij} u_j}.$$

Notice how the two different descriptions of the body surface—flow tangency and $\psi = \text{constant}$—are used to set up the iteration.

(8) The body radius from Step 7 represents an improved body shape, R^1. Use this in Step 1 and calculate the new matrices in Step 2. Then, repeat the iteration until satisfactory convergence is achieved. The final result is a body shape corresponding to the imposed pressure gradient.

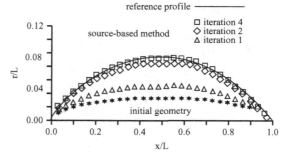

Figure 7.15. Body profile calculated by source-based inverse method (Zedan et al., 1994).

Figure 7.15 shows a body generated by the source-based inverse method discussed previously for the case of a shape with a known solution. The converged shape closely matches the reference profile after only a few iterations. Zedan, 1978, point out that this method fails for a body geometry that has an inflection point. An alternative plan is presented for this special case.

The drag of the body shape given by the iterative solution is found by performing a viscous-flow solution for a boundary layer growing on a surface in the presence of the prescribed pressure gradient. Running a few solutions provides a good physical idea of what is happening and a physical basis for a drag-reduction strategy.

(b) *Mathematical optimization techniques.* These techniques require a large number of numerical calculations. Parsons et al., 1974 present a computer-based optimization procedure for generating low-drag hull shapes useful for hydrodynamic applications. The method incorporates a parametric body description, a drag computation, hydrodynamic constraints, and an optimization scheme in the form of a computer program. Lutz and Wagner, 1998, describe an optimization scheme that uses a segmented linear distribution of sources along the body axis, much like the inverse method discussed herein. However, the source strength is used as a variable for the optimization process. This indirect method is combined with an integral boundary-layer method and an optimization algorithm. The focus of their work is on airship applications. The objective is to minimize the drag of the hull for maximum hull volume (i.e., lifting capacity) at different speeds (i.e., different Re number values). The student is referred to the literature for details of these and other optimization methods.

7.12 Numerical Methods for the Complete Airplane

Panel methods also are used to solve for the flow around a complete airplane. Such a solution furnishes the inviscid-flow pressure distributions on the various airplane components and predicts the forces and moments on the vehicle in an inviscid flow. Thus, the results are most useful for flight vehicles in cruise or without large separated-flow regions. The interference effect between a wing and a fuselage and in a wing/fuselage juncture also may be studied by using these numerical solutions. That is, the flow around a vehicle component in isolation is

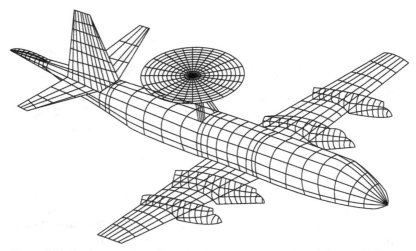

Figure. 7.16. Surface-panel configuration for a complete airplane (Tinoco, 1990).

not identical to the flow around that same component when the component is in proximity to another body in the flow. The aerodynamic characteristics of an airplane, then, are not simply the sum of the aerodynamic characteristics of all of its parts. In numerical solutions for a complete airplane, source or doublet panels usually are used to represent the fuselage, and the lifting surfaces are analyzed by the VPM or the VLM.

A panel representation of a complete airplane is shown in Fig. 7.16. The objective of this computer simulation was to guide the positioning of the large radome so that the aerodynamic center of the aircraft was unchanged.

Singh et al., 1989 describe a calculation method for finding the potential flow around a complete aircraft configuration. The fuselage is represented by sources distributed along the wetted surface, whereas the wing and wing-like components are modeled by using source and vorticity distributions on their respective mean-camber surfaces. Boundary-layer effects are not included.

The aerodynamic behavior of a complete subsonic transport and partial-model results for the same transport (i.e., results for wing alone, tail on/off) are found by using a panel method and discussed by Troeger and Selby, 1998. Code VSAERO, which uses piecewise constant source and doublet singularities distributed on flat, quadrilateral panels, is used for the computations. The results are compared with experimental data, and it is found that discrepancies between analysis and experiment are caused primarily by viscous effects, which are ignored in the panel method. The model of the complete transport uses 3,655 surface panels for the vehicle and 2,133 wake panels.

As mentioned previously, CFD is being used more frequently in the field of aerodynamics and is of great value as a wind-tunnel partner in the solution of many complex flow problems. Recently, CFD has made increasingly larger contributions to the aircraft-design process; this evolving role of CFD is discussed by Tinoco, 1998 and Dodbele et al., 1987. The student is encouraged to read the literature on this fascinating application of numerical methods.

PROBLEMS

7.1 Show by direct subsitution that the stream functions obtained for a uniform flow, a source, and a doublet satisfy Eq. 7.23, the differential equation for the stream function in an axisymmetric flow.

7.2 Derive the equation for the axisymmetric stream function in spherical coordinates, where r is the distance from the origin, θ is the angle between the radial line to a point in the field and the reference axis x, and ϕ is the azimuthal angle. For the irrotational case, show that the stream function must satisfy:

$$r^2 \frac{\partial^2}{\partial r^2} + \sin\theta \frac{\partial}{\partial\theta}\left(\frac{1}{\sin\theta}\frac{\partial\psi}{\partial\theta}\right) = 0.$$

Is this the same as Eq. 7.23? Explain.

7.3 Look for separable solutions to the stream function equation derived in Problem 7.2. That is, write $\psi(r,\theta) = R(r)\Theta(\theta)$. Show that the finite solutions for $R(r)$ are:

$$R_n(r) = A_n r^{-n}$$

and that the equation for $\Theta(\theta)$ is:

$$(1-\eta^2)\frac{\partial^2\Theta}{\partial\eta^2} + n(n+1)\Theta = 0,$$

where $\eta \equiv \cos\theta$.

REFERENCES AND SUGGESTED READING

Anderson, J. D., *Fundamental of Aerodynamics*, McGraw Hill, New York, 2nd ed., 1991.

Dodbele, S. S., Van Dam, C. P., Vijgen, P. M., and Holmes, B. J., "Shaping of Airplane Fuselages for Minimum Drag," AIAA Journal of Aircraft, Vol. 24, No. 5, (pp. 298–304), July 1987.

D'Sa, J. M., and Dalton, C., "Body of Revolution Comparisons for Axial and Surface-Singularity Distributions," AIAA Journal of Aircraft, Vol. 23, No. 8, (pp. 669–672), August 1986.

Jameson, A., "Re-Engineering the Design Process Through Computation," AIAA Journal of Aircraft, Vol. 36, No. 1 (pp. 36–50), Jan.–Feb. 1999.

Karamcheti, K., *Principles of Ideal-Fluid Aerodynamics*, John Wiley and Sons, New York, 1967.

Lutz, T., and Wagner, S., "Drag Reduction and Shape Optimization of Airship Bodies," AIAA Journal of Aircraft, Vol. 20, No. 35, (pp. 345–51), May–June 1998.

Parsons, J. S., Goodson, R. E., and Goldschmied, F. R., "Shaping of Axisymmetric Bodies for Minimum Drag in Incompressible Flow," AIAA Journal of Hydronautics, Vol. 8, No. 3, (pp. 100–07), July 1974.

Singh, N., Aikat, S., and Basu, B. C., "Incompressible Potential Flow About Complete Aircraft Configurations," Aeronautical Journal, Vol. 93, No. 929, (p. 335), November 1989.

Tinoco, E. N., "CFD Applications to Complex Configurations: A Survey, " *Applied Computational Aerodynamics*, Progress in Astronautics and Aeronautics, Vol. 125, P. A. Henne, Editor, Chapter 15, (pp. 559–615), AIAA, Washington, DC, 1990.

Tinoco, E. N., "The Impact of Computational Fluid Dynamics in Aircraft Design," CASI Canadian Aeronautics and Space Journal, Vol. 44, No. 3., (pp. 132–144), September. 1998.

Troeger, L. P., and Selby, G. V., "Computation of the Aerodynamic Characteristics of a Subsonic Transport," AIAA Journal of Aircraft, Vol. 35, No. 2, (pp. 183–190) March–April 1998.

Wang, T-C., Liu, C. H., and Geer, J., "Comparison of Uniform Perturbation and Numerical Solutions for Some Potential Flows Past Slender Bodies," *Computers and Fluids*, Vol. 13, No. 3, (pp. 271–83), March 1985.

Zedan, M. F., and Dalton, C., "Potential Flow Around Axisymmetric Bodies: Direct and Inverse Problems," AIAA Journal, Vol. 16, No. 3, (pp. 242–50), March 1978.

Zedan, M. F., and Dalton, C., "Higher-Order Axial Singularity Distributions for Potential Flow About Bodies of Revolution," *Computer Methods in Applied Mechanics and Engineering,* Vol. 21 (pp. 295–314), Elsevier, 1980.

Zedan, M. F., Seif, A. A., and Al-Moufadi, S., "Drag Reduction of Airplane Fuselages Through Shaping by the Inverse Method," AIAA Journal of Aircraft, Vol. 31, No. 2, (pp. 279–87), March–April 1994.

8 Viscous Incompressible Flow

> External aerodynamics was a disturbingly mysterious subject before Prandtl solved
> the mystery with his work on boundary layer theory from 1904 onwards.
>
> L. Rosenhead
> *Laminar Boundary Layers*, Oxford 1963

8.1 Introduction

This chapter examines the role of viscosity in the flow of fluids and gases. Although
the viscosity of air is small, it must be included in a flow model if we are to explain
wing stall and frictional drag, for example. The four preceding chapters are concerned with the analysis of airfoils, wings, and bodies of revolution based on an
assumption of inviscid flow (i.e., negligible viscous effects). The inviscid-flow model
allowed analytical solutions to be developed for predicting, with satisfactory accuracy, the pressure distribution on bodies of small-thickness ratio at a modest (or
zero) angle of attack. However, the inviscid-flow model leads to results that are at
odds with experience, such as the prediction that the drag of two-dimensional airfoils and right-circular cylinders is zero. This contradiction is resolved by realizing
that actual flows exhibit viscous effects.

Viscosity is discussed from a physical viewpoint in Chapter 2. In Chapters 5, 6,
and 7, the existence of viscosity is acknowledged when it is necessary to advance
an analytical derivation for an inviscid flow. Also, viscous effects are called on,
with words like *viscous drag* and *separation*, when comparing the predicted and
observed behavior of airfoils and wings. However, no analysis in this textbook has
been developed thus far that provides the required detailed physical basis for these
effects.

The focus of this chapter is a detailed study of the role of viscosity in an incompressible flow, particularly regarding modifications to the behavior of airfoils and
wings that was predicted based on an inviscid-flow model. This study answers questions such as: Why does a wing stall? How can stall be predicted or prevented? How
is frictional drag on a vehicle calculated?

When we include detailed viscous (and, later, heat-transfer) terms in the conservation equations of Chapter 3, it leads to a set of equations that defy analytical

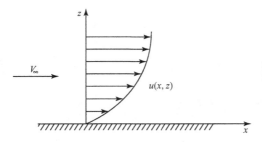

Figure 8.1. Typical velocity profile in a boundary layer.

solution—even for incompressible flows—except in special cases. In 1904, Prandtl[*] showed that a very thin "boundary layer" exists on the surface of bodies immersed in a flowing medium that has a small coefficient of viscosity (e.g., air). Within the boundary layer, flow-velocity gradients are large, whereas outside of the boundary layer, they usually are small (Fig. 8.1). Recall from Chapter 2 the argument at the molecular level that the velocity at the wall must be zero. This is termed the *no-slip boundary condition*.

Recall also from Chapter 2 that the velocity changes from the freestream value are the net result of mixing at the molecular level. This mixing leads to a shear stress in the fluid. Shear stress in a flow is proportional to velocity gradient, as discussed later. Thus, within the thin boundary layer, shear forces dominate, whereas outside of the boundary layer, inertial forces (due to directed motion of the fluid particles) dominate. This allows the problem of flow around a vehicle to be decomposed into two parts, both of which are treated simply by approximate equations appropriate to either an external inviscid flow or a thin viscous boundary layer.

Because the boundary layer is thin, the problem of flow around a vehicle first is treated as if there were no boundary layer present (i.e., inviscid-flow model). Problems in which the viscous layer is not thin (e.g., airfoils beyond the stall) are excluded. The boundary-layer problem is solved next, and then the two solutions are synthesized (i.e., combined) to describe the entire flow field. The pressure distribution arising from the external-flow solution is used as a known in the boundary-layer solution, as shown later. Iterations are carried out to account for the small modification to the body shape caused by the presence of the boundary layer.

Alternatively, numerical finite-difference (i.e., CFD) solutions have been developed that can treat the entire flow field in one unified manner and the external flow and the boundary-layer flow need not be decomposed. Such CFD solutions are complex and make strong demands on computer capability. The CFD methods are numerous and the mathematical theory is beyond the scope of this textbook, but we use the simpler concepts to demonstrate the power of such methods and to obtain useful solutions. We begin here by examining many simpler analyses to introduce several concepts that are applicable to all viscous flows.

For streamlined bodies in an incompressible flow, boundary-layer behavior depends on the value of a nondimensional parameter Re, the Reynolds number, which was introduced in Chapter 2:

$$\mathrm{Re} \equiv \frac{\rho V L}{\mu}, \qquad (8.1)$$

---

[*] Ludwig Prandtl (1875–1953) was a German professor of applied mechanics whose concept of the boundary layer was a breakthrough in the study of viscous flow.

where ρ is the density, V is the velocity, μ is the coefficient of viscosity, and L is a characteristic length such as a wing chord. This ratio was described in a discussion of similarity parameters, Eq. 2.20. The Re number usually is evaluated at a convenient reference condition such as the freestream, and it is the most important parameter that influences the characteristics of any viscous flow. If the flow problem involves compressibility, then an additional parameter—the Mach number—must be accounted for in the mathematical modeling.

If all of the streamlines in the flow within the boundary layer are parallel to one another, the boundary-layer flow is said to be in *laminas*, or to be *laminar*. In this case, momentum is transported normal to the lamina only by molecular diffusion. If the flow within the boundary layer has a random structure and contains large eddies, then the boundary layer is said to be *turbulent*. In the turbulent case, momentum is carried across streamlines by convective effects resulting from random fluctuations of the fluid. Laminar and turbulent boundary layers each have distinctive properties and methods of solution. Both types are discussed in this chapter.

Both analytical and numerical solutions to incompressible viscous-flow problems are discussed herein. These solutions primarily apply to boundary-layer flows and lead to methods for predicting the following characteristics of an incompressible boundary layer:

1. *Profile.* The shape of the velocity profile within the boundary layer, from the surface of the body to the edge of the layer, must be determined. Knowledge of this profile shape then allows the prediction of skin friction and, thus, frictional drag of a flight vehicle.
2. *Growth.* The boundary-layer thickness increases with increasing downstream distance along a vehicle surface. The rate of growth of the boundary layer must be found if, for example, it is desired to locate an engine inlet near a vehicle surface but out of the boundary layer so as to avoid ingesting nonuniform boundary-layer flow into the engine. This "growth" may be negative in regions of highly accelerated flow.
3. *Separation.* Under certain circumstances, the flow in the boundary layer leaves the body, or separates. When this occurs, the apparent shape of the body, as seen by the oncoming flow, changes dramatically and the oncoming flow no longer perceives a streamlined body. This may have drastic consequences (e.g., wing stall); thus, it is important to determine which parameters influence separation and how separation effects can be minimized or avoided. Separation occurs primarily when the flow is decelerating.
4. *Transition.** A boundary layer usually starts to develop at the leading edge of a body as a laminar boundary layer. Farther downstream, it may begin to "transition" into a turbulent profile. This does not occur instantaneously but rather over a streamwise distance. Because turbulent boundary layers cause higher skin friction (and heat transfer) than laminar boundary layers in the same external conditions, it is important that designers have knowledge that allows prediction of the transition location, as well as an understanding of how to delay or prevent transition.

---

* This term is often confused with *separation*. It is important that the student clearly understands the meaning of these two terms.

The chapter begins by detailing the viscous terms in the Conservation of Momentum equations (see Chapter 3), which were written in symbolic form when the equations were derived originally. The momentum equations, with all of the viscous terms specified, are called the *Navier–Stokes equations**. Following a discussion of the no-slip boundary condition, exact solutions of the Navier–Stokes equations for incompressible flow are reviewed. Unfortunately, the practical application of these solutions is severely limited, so approximate equations (i.e., the boundary-layer equations) are derived next.

Analytical and numerical (i.e., CFD) solutions to these laminar boundary-layer equations are carried out assuming incompressible flow. These solutions then are examined to interpret the observed behavior of airfoils and wings. Finally, transition of the boundary layer from laminar to turbulent is discussed, and turbulence is introduced. The steady flow of an incompressible fluid with negligible body forces is assumed throughout the discussion.

There is no need to discuss anything further regarding the continuity equation because it contains no force (i.e., viscous) terms. Thus, the continuity equation is the same whether the flow is assumed to be inviscid or viscous. The differential continuity equation was developed in Cartesian coordinates as Eq. 3.53, which for a steady, incompressible flow is:

$$\frac{\partial u}{\partial x} + \frac{\partial v}{\partial y} + \frac{\partial w}{\partial z} = 0, \tag{8.2}$$

where the local velocity components are u, v, w.

In an incompressible viscous or inviscid flow, the variations in density and temperature are so small that the energy (and state) equations may be set aside (i.e., uncoupled). Thus, solutions are sought here for a set of partial-differential equations consisting only of the continuity and momentum equations.

8.2 Navier–Stokes Equations

The differential-momentum equations that include viscous terms are called the Navier–Stokes equations. Recall from Chapter 3 that the differential-momentum equation can be derived by applying the Conservation of Momentum principle (in the form of Newton's Second Law) to a moving fluid particle. The result for two-dimensional viscous flow was Eq. 3.74. Here again, we begin by applying Newton's Second Law to a fluid particle in Cartesian coordinates; however, the viscous forces now must be accounted for in detail. For convenience, the fluid particle is assumed to be a cube with volume $(dxdydz)$. Because the fluid particle is of constant mass $\rho(dxdydz)$:

$$\frac{D}{Dt}[\rho(dxdydz)V] = \rho(dxdydz)\frac{DV}{Dt} = F, \tag{8.3}$$

where DV/Dt for steady flow is the convective acceleration of the particle and F is the net force acting on the fluid particle.

---

* Named after Navier (France) and Stokes (England), who independently developed the equations in the early 1800s.

Writing the Eulerian derivative for steady flow in Cartesian coordinates:

$$\rho(dxdydz)\,(\mathbf{V}\cdot\nabla)\mathbf{V} = F$$

$$\rho(dxdydz)\left[u\frac{\partial V}{\partial x}+v\frac{\partial V}{\partial y}+w\frac{\partial V}{\partial z}\right] = F. \tag{8.4}$$

This is a vector equation. Writing Eq. 8.4 in components:

$$\rho(dxdydz)\left[\rho u\frac{\partial u}{\partial x}+\rho v\frac{\partial u}{\partial y}+\rho w\frac{\partial u}{\partial z}\right] = dF_x$$

$$\rho(dxdydz)\left[\rho u\frac{\partial v}{\partial x}+\rho v\frac{\partial v}{\partial y}+\rho w\frac{\partial v}{\partial z}\right] = dF_y \tag{8.5a}$$

$$\rho(dxdydz)\left[\rho u\frac{\partial w}{\partial x}+\rho v\frac{\partial w}{\partial y}+\rho w\frac{\partial w}{\partial z}\right] = dF_z,$$

where the right side is written as differential-force components because we are dealing with an infinitesimal fluid particle. It remains to detail the force terms on the right side of this equation. In the absence of body forces, these terms express forces (or stresses) exerted on the *outside* surfaces of the fluid particle by the surroundings. Recall that in addition to normal pressure forces (i.e., the only forces present in an inviscid flow), the force terms in Eq. 8.5 must include all of the shear and normal forces due to the presence of viscosity.

For convenience, the fluid-particle cube with volume $dxdydz$ is assumed to be at the origin of coordinates in a snapshot of the moving particle taken at an instant of time, as shown in Fig. 8.2. In general, the surface stress, τ, is at an angle with respect to each surface of the cube, as illustrated in the figure for the surface passing through the origin and perpendicular to the x-axis. Now, we decompose this general surface stress, τ, into a component normal to the face, τ_n, and a component tangential to the face, τ_t. The component τ_t can be decomposed further into two components in the coordinate directions, as shown in Fig. 8.3, where $\tau_{xx} = \tau_n$.

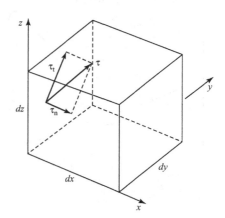

Figure 8.2. Differential element and general stress.

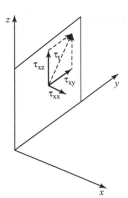

Figure 8.3. Notation for stresses.

The following notation is used:

1. The first subscript on the symbol for stress indicates the axis *normal* to the plane on which the stress acts.
2. The second subscript corresponds to the direction in which the stress acts.

Thus, τ_{xz} indicates the stress on a plane normal to the x-axis and acting in the z-direction. Figures similar to Fig. 8.3 may be drawn for the five other planes that define the fluid particle. Fig. 8.4 indicates all of the stresses acting in the x-direction on the six faces of the fluid particle. For convenience, baseline values of the stresses are assigned to the faces in the coordinate planes. Thus, the tangential stress on the rear surface is given by:

$$\tau_{yx} + \frac{\partial \tau_{yx}}{\partial y} dy.$$

As a matter of convention, the stresses on a plane are taken to be the force per unit area acting on the side of the plane that faces in the *positive* coordinate direction. Stresses acting on the *outside* surfaces of the element are shown with a dot ● on the arrow in Fig. 8.4. Thus, the stress on the upper surface of the fluid element acts on the *outside* of the element in Fig. 8.4, whereas the shear stress on the lower surface acts on the *inside* of the surface. Likewise, τ_{xx} is a normal stress acting on the *inside* of

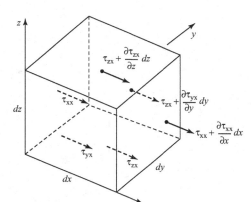

Figure 8.4. Stresses acting on a fluid element.

the left face of the element in Fig. 8.4; hence, $-\tau_{xx}$ is a normal stress acting on the *outside* of the left face of the element. Notice how this convention returns the familiar result in the case of inviscid flow. In that case, $\tau_{xx} = -p$ (see Eq. 8.13), and the normal pressure force acts toward the right on the outside of the left face and toward the left on the outside of the right face of the element. Thus, the pressure force acts *on* the fluid particle, as our formulation of the momentum-balance equation requires.

Using this notation and convention and taking a force (i.e., stress times area) to be positive in the positive coordinate direction, the net force acting on the element in the x-direction (for example) can be deduced. Recall that Newton's Second Law requires this to be the net force acting *on* the element. Remember that a force acting on the inside of a surface represents the force of the element acting *on* the surroundings, so that the direction of this force must be reversed before it is entered into the expression for Newton's Second Law. The net force acting on the outside surfaces of the particle, Fig. 8.4, in the x-direction is then as follows:

$$dF_x = \left[\left(\tau_{xx} + \frac{\partial \tau_{xx}}{\partial x} dx\right) - \tau_{xx}\right] dydz + \left[\left(\tau_{yx} + \frac{\partial \tau_{yx}}{\partial y} dy\right) - \tau_{yx}\right] dxdz +$$

$$\left[\left(\tau_{zx} + \frac{\partial \tau_{zx}}{\partial z} dz\right) - \tau_{zx}\right] dxdy = 0,$$

(8.5b)

or, canceling:

$$dF_x = \left[\frac{\partial \tau_{xx}}{\partial x} + \frac{\partial \tau_{yx}}{\partial y} + \frac{\partial \tau_{zx}}{\partial z}\right] dxdydz.$$

(8.6a)

similarly, the net forces on this element in the y- and z-directions can be derived as:

$$dF_y = \left[\frac{\partial \tau_{xy}}{\partial x} + \frac{\partial \tau_{yy}}{\partial y} + \frac{\partial \tau_{zy}}{\partial z}\right] dxdydz$$

(8.6b)

and

$$dF_z = \left[\frac{\partial \tau_{xz}}{\partial x} + \frac{\partial \tau_{yz}}{\partial y} + \frac{\partial \tau_{zz}}{\partial z}\right] dxdydz.$$

(8.6c)

Notice that Eqs. 8.6b and 8.6c can be obtained directly from Eq. 8.6a by permuting the second subscript because this is the direction subscript. Also notice that $dxdydz$ is simply the volume of the fluid particle.

Finally, we substitute Eq. 8.6 for the force terms on the right side of Eq. 8.5. Then, we cancel the coefficient $(dxdydz)$ on both sides of the equation. This implies that the resulting equation is independent of the volume of the fluid particle, which is expected. The result is:

$$\rho\left[\rho u \frac{\partial u}{\partial x} + \rho v \frac{\partial u}{\partial y} + \rho w \frac{\partial u}{\partial z}\right] = \frac{\partial \tau_{xx}}{\partial x} + \frac{\partial \tau_{yx}}{\partial y} + \frac{\partial \tau_{zx}}{\partial z}$$

$$\rho\left[\rho u \frac{\partial v}{\partial x} + \rho v \frac{\partial v}{\partial y} + \rho w \frac{\partial v}{\partial z}\right] = \frac{\partial \tau_{xy}}{\partial x} + \frac{\partial \tau_{yy}}{\partial y} + \frac{\partial \tau_{zy}}{\partial z}$$

$$\rho\left[\rho u \frac{\partial w}{\partial x} + \rho v \frac{\partial w}{\partial y} + \rho w \frac{\partial w}{\partial z}\right] = \frac{\partial \tau_{xz}}{\partial x} + \frac{\partial \tau_{yz}}{\partial y} + \frac{\partial \tau_{zz}}{\partial z}.$$

(8.7)

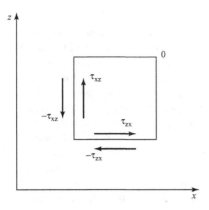

Figure 8.5. Tangential stresses on a fluid element.

These are the components of the momentum equation in three-dimensional Cartesian coordinates. Notice that even with the help of the continuity equation, there are many more unknowns than there are equations. The fluid stresses on the right side of Eq. 8.7 must be studied further, with the aim of expressing them in terms of velocity-component derivatives. Recall that in these equations, the normal stresses (with the double subscript) consist of both pressure and viscous stresses. The mixed subscripts xy, zx, and so on denote shear (i.e., tangential) stresses. The problem of the large number of unknowns in Eq. 8.7 is alleviated by realizing that the shear stresses are not independent. Consider the tangential stresses acting on the two-dimensional fluid element in Fig. 8.5.

The tangential stresses acting on the outside faces of the fluid element are shown. If the torque about the Point 0 is not zero, the nonzero torque causes an infinite angular acceleration of the fluid element because it has infinitesimal mass. Because this cannot occur in nature, the tangential stresses must be in equilibrium such that:

$$\tau_{xz} = \tau_{zx}, \tag{8.8a}$$

and, by a similar argument:

$$\tau_{xy} = \tau_{yx}; \ \tau_{yz} = \tau_{zy}, \tag{8.8b}$$

This reduces the number of unknown stresses to six: three shear stresses and three normal stresses. All of these viscous stresses can be related to velocity-component derivatives. How this is accomplished is discussed in two parts.

1. *Shear stress.* The molecular origins of shear stress are discussed in Chapter 2. Recall that in solid mechanics, the shear stress on a solid is proportional to the strain or angular distortion (i.e., Hooke's Law). For liquid or gases, it is assumed that the shear stress is proportional to the strain rate, or to the rate of change of angular distortion of the fluid element. For air and water, this assumption has been well validated by experiment. The proportionality between shear stress and strain rate is linear for air and water; hence, they are termed *Newtonian fluids*. If there is a nonlinear dependence between shear stress and strain rate (e.g., as in blood), then the fluid is termed *non-Newtonian*. Only Newtonian fluids are considered here.

   In Chapter 4, the concept of the strain of a fluid element is discussed. The rate of strain, ε_{xy}, of a two-dimensional, fluid particle in the x-y plane was defined in terms of the rate of change of angular distortion of the particle and then in terms of velocity-component derivatives, Eq. 4.3. This result in Chapter 4 was set aside for later use and is used now. Rewriting Eq. 4.3:

$$\varepsilon_{xy} = \frac{\partial v}{\partial x} + \frac{\partial u}{\partial y}. \tag{8.9}$$

For a Newtonian fluid, it is assumed that $\tau_{xy} \approx \varepsilon_{xy}$ or $\tau_{xy} = \mu\varepsilon_{xy}$, where m is a constant of proportionality. Thus,

$$\tau_{xy} = \mu\varepsilon_{xy} = \mu\left[\frac{\partial v}{\partial x} + \frac{\partial u}{\partial y}\right] = \tau_{yx} \tag{8.10a}$$

Writing expression for the strain in the xz- and yz-planes (i.e., for τ_{xz} and τ_{yz}) by paralleling the derivations of Eq. 4.3,

$$\tau_{x\varepsilon} = \mu\varepsilon_{x\varepsilon} = \mu\left[\frac{\partial w}{\partial x} + \frac{\partial u}{\partial z}\right] = \tau_{ex}$$

$$\tau_{y\varepsilon} = \mu\varepsilon_{y\varepsilon} = \mu\left[\frac{\partial w}{\partial y} + \frac{\partial v}{\partial z}\right] = \tau_{ex} \tag{8.10b,c}$$

The constant of proportionality, μ, in Eqs. (8.10a,b,c) is called the coefficient of viscosity, and is termed a *transport property* of the fluid. It has different values for different fluids. The magnitude of the coefficient of viscosity may be thought of physically as being a measure of the ability of a gas/liquid to resist shear deformations. Since shear stress in a gas/liquid arises from molecular interactions, as discussed in Chapter 2, it may be shown from kinetic theory that the coefficient of viscosity is a function of pressure and temperature. For typical aerodynamic problems the pressure levels are such that the small pressure dependence may be ignored. Thus, viscosity may be assumed to be a function of temperature only. For incompressible flow, where any temperature excursion in the flow is very small, the coefficient of viscosity may be taken to be a constant. For air at standard temperature,

$$\mu = 3.72{\cdot}10^{-7}\ \text{slug/fts} \qquad (\mu = 1.78{\cdot}10^{-5}\ \text{kg/ms}). \tag{8.11}$$

Because the ratio μ/ρ appears often in the analysis, it is given a special symbol and name, $v = \mu/\rho$, the kinematic viscosity. At standard conditions, the kinematic viscosity of air has the value:

$$v = 1.55{\cdot}10^{-4}\ \text{ft}^2/\text{s} \qquad (1.44{\cdot}10^{-5}\ \text{m}^2/\text{s}).$$

In some applications, involving flows with large temperature variations may be necessary to account for the variations of the viscosity with temperature. For such cases, we use results from kinetic theory, such as Sutherland's Law:

$$\mu = \mu_b\left(\frac{T}{T_0}\right)^{3/2}\frac{T_0 + S}{T + S}, \tag{8.12}$$

where for air:

$$\begin{cases} T_0 = 491.76°\ R & (T_0 = 273.2°\ K) \\ \mu_b = 3.585{\cdot}10^{-7}\ \text{slug/fts} & (\mu_b = 1.716{\cdot}10^{-5}\ \text{kg/ms}) \\ S = 199.8°\ R & (S = 111°\ K). \end{cases}$$

For temperatures between 378 and 3,420° R, this expression gives values within 2 percent of the experimentally measured values. The viscosity coefficient and other physical parameters also may vary with pressure. Values of the viscosity coefficients and parameters for other gases and liquids are in Thwaites, 1949 and Hiemenz, 1911. For problems involving incompressible flow, constant values for the viscosity coefficient (e.g., those given in Eq. 8.11) are appropriate.

2. *Normal Stress.* In an inviscid flow, the only normal stress on a fluid particle is the pressure. The molecular origin of this pressure is discussed in Chapter 2. However, in a viscous flow, there are normal stresses present in addition to the familiar pressure stresses. This fact is emphasized by introducing the following notation:

$$\tau_{xx} = -p + \sigma_{xx}$$
$$\tau_{yy} = -p + \sigma_{yy} \qquad (8.13)$$
$$\tau_{zz} = -p + \sigma_{zz}.$$

This representation indicates that the normal stress is the sum of two parts: the part due to pressure, p (the negative sign indicates that it acts inward *on* the particle surface), and a second part, σ, which represents normal stresses caused by viscosity. These viscous normal stresses can be significant if there are large velocity gradients $\partial u/\partial x$, $\partial u/\partial y$, and $\partial u/\partial z$ at the faces of the fluid particle. These effects often are neglected in aerodynamic applications because the shearing stresses are far more important.

Previously, from a macroscopic viewpoint, we described a fluid particle as being of fixed identity. However, at the molecular level, the actual particles that comprise the fluid particle are changing continually. Because of the random motion of the molecules, some are leaving the fluid particle and others are entering. Thus, there is a continual diffusion of mass and momentum (at the molecular level) in and out through the surfaces of the particle. If there are rapid inward or outward movements of the faces of the fluid particle, this results in an unbalance of the molecular diffusion of momentum across the particle boundaries in the normal directions. By Newton's Second Law, there must be a corresponding unbalanced force. It is this force that is denoted as σ_{xx}, σ_{yy}, and σ_{zz}, the normal stresses due to viscosity. Recall that the existence of a shear stress also is explained at the molecular level (see Chapter 2).

The derivation of the defining equations for the viscous normal stresses is presented in detail in Ref. 5. Here, we simply display the results for completeness:

$$\sigma_{xx} = 2\mu \frac{\partial u}{\partial x} + \lambda \left[\frac{\partial u}{\partial x} + \frac{\partial v}{\partial y} + \frac{\partial w}{\partial z} \right]$$
$$\sigma_{yy} = -2\mu \frac{\partial v}{\partial y} + \lambda \left[\frac{\partial u}{\partial x} + \frac{\partial v}{\partial y} + \frac{\partial w}{\partial z} \right] \qquad (8.14)$$
$$\sigma_{zz} = 2\mu \frac{\partial w}{\partial z} + \lambda \left[\frac{\partial u}{\partial x} + \frac{\partial v}{\partial y} + \frac{\partial w}{\partial z} \right].$$

These viscous, normal-stress expressions contain two constants of proportionality. The constant μ is the same coefficient of viscosity observed before in the shear-stress expressions. The second constant, λ, is called the *coefficient of bulk viscosity* because it is associated with a change in particle volume. For an incompressible flow, the quantity in square brackets in Eq. 8.14 is zero by virtue of continuity, so that λ does not enter into consideration of an incompressible-flow problem.

It remains to substitute the expressions for the shear stresses in Eq. 8.10 and the normal stresses in Eqs. 8.13 and 8.14 into the appropriate force terms in the momentum expression, Eq. 8.7. The result is:

$$
\begin{cases}
\rho u \dfrac{\partial u}{\partial x} + \rho v \dfrac{\partial u}{\partial y} + \rho w \dfrac{\partial u}{\partial z} = -\dfrac{\partial p}{\partial x} + \dfrac{\partial}{\partial x}\left[2\mu\dfrac{\partial u}{\partial x} + \lambda\left(\dfrac{\partial u}{\partial x} + \dfrac{\partial v}{\partial y} + \dfrac{\partial w}{\partial z}\right)\right] + \\[2mm]
\qquad + \dfrac{\partial}{\partial y}\left[\mu\left(\dfrac{\partial v}{\partial x} + \dfrac{\partial u}{\partial y}\right)\right] + \dfrac{\partial}{\partial z}\left[\mu\left(\dfrac{\partial u}{\partial z} + \dfrac{\partial w}{\partial x}\right)\right] \\[4mm]
\rho u \dfrac{\partial v}{\partial x} + \rho v \dfrac{\partial v}{\partial y} + \rho w \dfrac{\partial v}{\partial z} = -\dfrac{\partial p}{\partial y} + \dfrac{\partial}{\partial y}\left[2\mu\dfrac{\partial v}{\partial y} + \lambda\left(\dfrac{\partial u}{\partial x} + \dfrac{\partial v}{\partial y} + \dfrac{\partial w}{\partial z}\right)\right] + \\[2mm]
\qquad + \dfrac{\partial}{\partial x}\left[\mu\left(\dfrac{\partial v}{\partial x} + \dfrac{\partial u}{\partial y}\right)\right] + \dfrac{\partial}{\partial z}\left[\mu\left(\dfrac{\partial w}{\partial y} + \dfrac{\partial v}{\partial z}\right)\right] \\[4mm]
\rho u \dfrac{\partial w}{\partial x} + \rho v \dfrac{\partial w}{\partial y} + \rho w \dfrac{\partial w}{\partial z} = -\dfrac{\partial p}{\partial z} + \dfrac{\partial}{\partial z}\left[2\mu\dfrac{\partial w}{\partial z} + \lambda\left(\dfrac{\partial u}{\partial x} + \dfrac{\partial v}{\partial y} + \dfrac{\partial w}{\partial z}\right)\right] + \\[2mm]
\qquad + \dfrac{\partial}{\partial x}\left[\mu\left(\dfrac{\partial u}{\partial z} + \dfrac{\partial w}{\partial x}\right)\right] + \dfrac{\partial}{\partial y}\left[\mu\left(\dfrac{\partial w}{\partial y} + \dfrac{\partial v}{\partial z}\right)\right].
\end{cases}
\tag{8.15}
$$

These are the Navier–Stokes equations for a steady, compressible, three-dimensional, viscous flow.

Notice in Eq. 8.15 that the viscosity coefficients are located *inside* the spatial derivative. In a compressible flow, the temperature may change greatly over the flow field; hence, the temperature-dependent viscosity coefficients cannot be treated as constants. Recall that derivatives of density (i.e., ρ, mass per unit volume) do not appear in Eq. 8.15 even for a compressible flow. This is because the mass of the fluid particle, $\rho\,dx\,dy\,dz$, is constant so that the mass term passes through the Eulerian derivative in the statement of Newton's Second Law. The volume of the fluid particle then divides out, leaving the density as a coefficient of the substantial derivative.

Navier–Stokes Equations for Incompressible Flow

The Navier–Stokes equations for an incompressible, viscous flow may be written by simplifying Eq. 8.15. As mentioned herein, the terms containing the bulk viscosity, λ, drop out when considering an incompressible flow because the continuity equation requires that:

$$
\frac{\partial u}{\partial x} + \frac{\partial v}{\partial y} + \frac{\partial w}{\partial z} = 0.
\tag{8.16}
$$

Also, the coefficient of viscosity, μ, may be treated as a constant because temperature variations throughout the flow are very small. Thus,

$$
\begin{cases}
\rho u \dfrac{\partial u}{\partial x} + \rho v \dfrac{\partial u}{\partial y} + \rho w \dfrac{\partial u}{\partial z} = -\dfrac{\partial p}{\partial x} + \mu\left(\dfrac{\partial^2 u}{\partial x^2} + \dfrac{\partial^2 u}{\partial y^2} + \dfrac{\partial^2 u}{\partial z^2}\right) \\[2mm]
\rho u \dfrac{\partial v}{\partial x} + \rho v \dfrac{\partial v}{\partial y} + \rho w \dfrac{\partial v}{\partial z} = -\dfrac{\partial p}{\partial y} + \mu\left(\dfrac{\partial^2 v}{\partial x^2} + \dfrac{\partial^2 v}{\partial y^2} + \dfrac{\partial^2 y}{\partial z^2}\right) \\[2mm]
\rho u \dfrac{\partial w}{\partial x} + \rho v \dfrac{\partial w}{\partial y} + \rho w \dfrac{\partial w}{\partial z} = -\dfrac{\partial p}{\partial z} + \mu\left(\dfrac{\partial^2 w}{\partial x^2} + \dfrac{\partial^2 w}{\partial y^2} + \dfrac{\partial^2 w}{\partial z^2}\right).
\end{cases}
\tag{8.17}
$$

These are the Navier–Stokes equations for steady, incompressible, viscous, three-dimensional flow. Notice that there are now four equations: the continuity equation, Eq. 8.16, written for an incompressible flow; and the three momentum equations in Eq. 8.17. Thus, there are four unknowns: u, v, w, and p. The density, ρ, is a known constant and the coefficient is viscosity, μ, is a known constant for a particular fluid.

Eq. 8.16 now is written for two-dimensional flow in the x-z (or airfoil) plane. Thus, $v = \partial/\partial y = 0$ and Eq. 8.17 reduces to:

$$
\begin{cases}
\rho u \dfrac{\partial u}{\partial x} + \rho w \dfrac{\partial u}{\partial z} = -\dfrac{\partial p}{\partial x} + \mu\left(\dfrac{\partial^2 u}{\partial x^2} + \dfrac{\partial^2 u}{\partial z^2}\right) = -\dfrac{\partial p}{\partial x} + \mu\nabla^2 u \\[2mm]
\rho u \dfrac{\partial w}{\partial x} + \rho w \dfrac{\partial w}{\partial z} = -\dfrac{\partial p}{\partial z} + \mu\left(\dfrac{\partial^2 w}{\partial x^2} + \dfrac{\partial^2 w}{\partial z^2}\right) = -\dfrac{\partial p}{\partial z} + \mu\nabla^2 w
\end{cases}
\tag{8.18}
$$

where ∇^2 is the Laplacian operator. These are the Navier–Stokes equations for a steady, two-dimensional, incompressible, viscous flow. Rewriting the continuity equation, Eq. 8.16, for two-dimensional incompressible flow:

$$
\frac{\partial u}{\partial x} + \frac{\partial w}{\partial z} = 0.
\tag{8.19}
$$

This provides the third equation needed to determine the three unknowns, u, w, and p. Carefully examine Eq. 8.18. Note that even after the considerable simplification of assuming an incompressible flow, the Navier–Stokes equations still are *nonlinear* due to the convective acceleration terms on the left side involving products of the velocity components and their derivatives. We cannot use the powerful superposition principle that is a benefit in linear problems. There exist no general methods for integrating these equations, although exact analytical solutions exist for a few special cases that we examine carefully in subsequent sections. This suggests that simplifying assumptions must be applied if solutions to viscous problems are to be found. An important simplification takes advantage of the *thinness* of the usual viscous region, or *boundary layer*. The Navier–Stokes equations are found to be considerably simpler for the case of a thin boundary layer, but the resulting boundary-layer equations are still nonlinear. Before the boundary-layer equations are developed, it is useful to examine the proper boundary conditions for a viscous-flow problem.

Boundary Conditions for Viscous Flows

Boundary conditions are the mechanism by which a solution of a differential equation is related to a specific physical problem. Whether or not a flow is viscous, there can be no velocity component normal to the solid surface of a body. If the body surface is not solid and there is suction or blowing through the surface, then the normal component of velocity is not zero at the wall but rather proportional to the mass flow through the surface. The focus here is on solid surfaces so that one boundary condition at our disposal is that there is no flow through the solid surface (i.e., the normal component of velocity at the surface is zero).

For a viscous flow, the physically observed fact is that the tangential velocity of the flow at the body surface is the same as that of the body. If the body is in motion, the fluid "sticks" to the surface and is carried along by the body. If the body is at rest (i.e., the usual point of view for problem solving), then the tangential velocity at the body surface is zero. This is the so-called no-slip boundary condition that must be imposed on the tangential velocity at a surface in a viscous flow.

The reason that the fluid sticks to the surface is explained at the molecular level in Chapter 2 in terms of the momentum transfer between the molecules and the molecular structure of the solid surface. In effect, all momentum parallel to the surface is destroyed by inelastic collisions on the rough interface.

Notice the fundamental difference in the boundary condition that is imposed on the tangential-velocity component at the surface of a body in inviscid and viscous flows. If the flow is inviscid, the velocity vector at the surface simply must be tangent to the body without any magnitude specified (i.e., a repetition of the previous "no-flow-through-the-surface" boundary condition). However, if the flow is viscous, then the tangential-velocity component at a solid surface at rest must be zero (i.e., the no-slip condition).

8.3 Exact Solutions of the Navier–Stokes Equations

If sufficient simplifying assumptions are available in framing a viscous-flow problem, it is possible to find an exact solution to the resulting Navier–Stokes equations. Three examples are presented in this section. In all of these examples, the Navier–Stokes equations become linear and analytical solutions are then possible. Other examples of steady and unsteady problems that permit an exact solution of the Navier–Stokes equations include the impulsively started flat plate, the stagnation-point flow, and the flow between two concentric, rotating cylinders. Detailed treatment of these problems is in White, 1986 and Schlichting, 2003. In the first two examples that follow, two-dimensional airfoil coordinates (x, z) are used rather than the usual two-dimensional coordinates (x, y) to keep our focus on the airfoil/wing viscous-flow problem.

Steady, Parallel, Incompressible Flow Through a Straight Channel

It is assumed that there is a two-dimensional steady flow through an infinitely long channel of unit width and of height H, as shown in Fig. 8.6a. It is assumed further that all flow streamlines are parallel to the solid wall (i.e., $w = 0$ everywhere). This limits the Re number of the problem to a small enough value so that the flow is laminar. Such a flow is sometimes called *Poiseuille flow* (Schlichting, 2003).

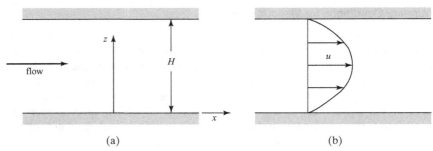

Figure 8.6. Poiseuille flow.

Because $w = 0$, the continuity equation, Eq. 8.19, requires that:

$$\frac{\partial u}{\partial x} = 0 \;\rightarrow\; u = f(z) \;\text{ only.} \tag{8.20}$$

It follows that the Navier–Stokes Eq. 8.18 reduces to:

$$\left\{ \begin{aligned} 0 + 0 &= -\frac{\partial p}{\partial x} + \mu \frac{\partial^2 u}{\partial z^2} && \text{(8.21a)} \\[2mm] 0 + 0 &= -\frac{\partial p}{\partial z} + 0. && \text{(8.21b)} \end{aligned} \right.$$

Notice how the assumption of parallel flow eliminated the nonlinear *convective acceleration* terms on the left side of the equation. The resulting linear equations are solved easily. The z-momentum equation, Eq. 8.21b, indicates that p depends on the axial position, $p = p(x)$ only. Eq. 8.21a then states:

$$\frac{dp}{dx} = \mu \frac{d^2 u}{dz^2}, \;\text{ or}$$

$$\begin{pmatrix} \text{function of} \\ x \text{ only} \end{pmatrix} = \begin{pmatrix} \text{function of} \\ z \text{ only} \end{pmatrix}.$$

If the left side of the equation, which is independent of z, is equal to the right side, which is independent of x, then both sides must be independent of both x and z; namely, they both must be equal to a constant. Thus, Eq. 8.21a may be written:

$$\mu \frac{d^2 u}{dz^2} = \frac{dp}{dx} = \text{constant, or}$$

$$\frac{dp}{dx} = \text{constant;} \quad \mu \frac{d^2 u}{dz^2} = \text{constant.} \tag{8.22}$$

Notice that the pressure gradient in the streamwise direction must be constant. This is the pressure (supplied by a pump or compressor, for example) that pushes the flow in the channel from left to right. The expression, $d^2u/dz^2 = $ constant, may be integrated twice. This process leads to two constants of integration that are evaluated by applying the no-slip boundary condition at both of the solid walls; namely:

$$u|_{z=0} = 0 = u|_{z=H}$$

The result of the integration and the application of the boundary condition is:

$$u = \frac{1}{2\mu}\frac{dp}{dx}(z^2 - Hz),\qquad(8.23)$$

where H is the height of the channel. The student should verify this result. This is the equation for a parabolic profile, as plotted in Fig. 8.6b.

Steady, Incompressible Flow Between Two Semi-Infinite Parallel Plates

We consider two long parallel plates with a viscous fluid between them, as shown in Fig. 8.7. The fluid initially is at rest. At some time, the top plate is set in motion to the right with a constant velocity U. We are interested in the steady-flow problem—that is, after the plate has been in motion for a long time. This flow problem is called *Couette flow* (see Schlichting, 2003).

With parallel flow again assumed, the defining equation is the same as Eq. 8.21. However, in the case of Couette flow, the problem is easier because no pressure difference is imposed on the flow between the two plates to make the fluid move. The fluid is moving as a consequence of the viscous force at the upper surface—it is being pulled along by the moving upper plate. Thus, there is no pressure difference driving the flow, $dp/dx = 0$, and the defining equation for Couette flow reduces to:

$$\frac{\partial^2 u}{\partial z^2} = \frac{\partial^2 u}{\partial x^2} = 0,$$

subject to the boundary conditions:

$$u|_{z=0} = 0;\ u|_{z=H} = U.$$

Integrating twice and evaluating the constants of integration using these boundary conditions, the result is a linear-velocity profile given by:

$$u = \frac{U}{H}z,\qquad(8.24)$$

as shown in Fig. 8.7b. The student should verify this result. If a left-to-right-driving pressure difference is applied to the fluid between the two plates in Fig. 8.7, the solution to the resulting linear problem is a superposition of the

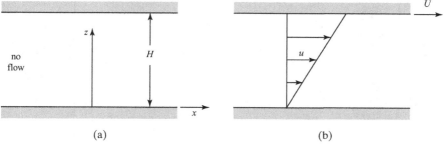

(a)   (b)

Figure 8.7. Couette flow.

linear-velocity profile of Couette flow and the parabolic profile found previously in Poiseuille flow.

Decay of a Potential Vortex

Consider a two-dimensional potential vortex at the origin of coordinates in an incompressible flow, as shown in Fig. 8.8a and discussed in Chapter 4. The fluid is inviscid and the velocity field is given by $V = \Gamma/2\pi r$, where Γ is the strength of the vortex.

    The potential vortex continues the behavior shown in Fig. 8.8a indefinitely in the absence of viscosity. Now, imagine that at a time $t = 0$, a viscosity switch is turned on so that the fluid instantaneously becomes viscous, whereas nothing is done to sustain the strength of the potential vortex. A prediction of the resulting flow field requires the solution of a linear, time-dependent form of the Navier–Stokes equations.

    The continuity and momentum (Navier–Stokes) equations in polar coordinates become:

$$
\left\{
\begin{aligned}
&\frac{\partial}{\partial r}(ru_r) + \frac{\partial u_\theta}{\partial \theta} = 0 \\[2mm]
&\frac{\partial u_r}{\partial t} + u_r\frac{\partial u_r}{\partial r} - \frac{u_\theta^2}{r} + \frac{u_\theta}{r}\frac{\partial u_r}{\partial \theta} = -\frac{1}{\rho}\frac{\partial p}{\partial r} + v\left[\frac{1}{r}\frac{\partial}{\partial r}\left(r\frac{\partial u_r}{\partial r}\right) - \frac{u_r}{r^2} + \frac{1}{r^2}\frac{\partial^2 u_r}{\partial \theta^2} - \frac{2}{r^2}\frac{\partial u_\theta}{\partial \theta}\right] \\[2mm]
&\frac{\partial u_\theta}{\partial t} + u_r\frac{\partial u_\theta}{\partial r} + \frac{u_r u_\theta}{r} + \frac{u_\theta}{r}\frac{\partial u_\theta}{\partial \theta} = -\frac{1}{\rho r}\frac{\partial p}{\partial \theta} + v\left[\frac{1}{r}\frac{\partial}{\partial r}\left(r\frac{\partial u_\theta}{\partial r}\right) - \frac{u_\theta}{r^2} + \frac{1}{r^2}\frac{\partial^2 u_\theta}{\partial \theta^2} + \frac{2}{r^2}\frac{\partial u_r}{\partial \theta}\right].
\end{aligned}
\right.
$$

The student should verify these equations. (Examine the equation following Eq. 3.54 and Example 3.10 but recall that both the time dependence and the viscous stresses must be retained in the present situation.) Notice that the momentum balance was divided through by the (constant) density so that the kinematic viscosity appears in the viscous terms. For the conditions specified, it is assumed that streamlines are circular (as they are in the potential vortex); thus, the radial velocity component, u_r, is everywhere zero. The continuity equation reveals that the circumferential velocity, u_θ, cannot be a function of θ and therefore is a function only of time and radial

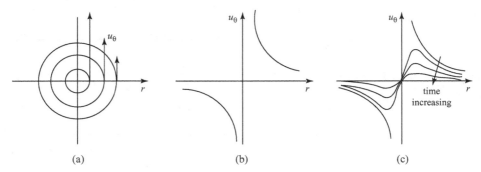

         (a)   (b)   (c)

Figure 8.8. Decay of a potential vortex.

position. It also is apparent that none of the variables can depend on the angular position by symmetry. The θ-momentum equation then reduces to:

$$\frac{\partial u_\theta}{\partial t} = v\left[\frac{1}{r}\frac{\partial}{\partial r}\left(r\frac{\partial u_\theta}{\partial r}\right) - \frac{u_\theta}{r^2}\right].$$

The solution of this linear equation is found by standard methods (see Exercise 8.26) to be:

$$u_\theta(r,t) = \frac{\Gamma}{2\pi r}\left[1 - e^{(r^2/4vt)}\right] \tag{8.25}$$

and is shown with solid lines in Fig. 8.8c for increasing values of time. The constant of integration is expressed in the form of the vorticity, Γ, for a potential vortex. Notice that when viscosity is introduced into the flow model, the infinity in flow velocity at the origin (i.e., a physical impossibility) disappears. At the origin, the viscous vortical flow exhibits solid-body rotation. As time progresses, the vortex decays and, ultimately, the signature of the vortex in the flow field disappears. This is the observed behavior of the vortices trailing behind the wings of an aircraft, and the disturbance eventually is smeared out by the action of viscosity. The kinetic energy originally contained in the potential vortex is dissipated by viscosity and changed into internal (i.e., thermal) energy, resulting in a (slight) increase in air temperature. In addition to being of interest because it represents an exact solution to the Navier–Stokes equations, the solution for a decaying vortex (sometimes called a *Taylor*, or a *Lamb*, vortex) is useful as an exact benchmark solution against which to validate certain CFD codes.

Notice that it was not necessary to use the radial-momentum equation to find the velocity solution. However, the radial momentum yields additional information of interest; namely, the pressure distribution through the vortex as a function of time. The r-momentum equation reduces to:

$$\rho\frac{u_\theta^2}{r} = \frac{\partial p}{\partial r}, \tag{8.26}$$

which can be integrated to yield the pressure from the known solution for the velocity, u_θ. The result (refer to Exercise 8.26) shows that the pressure has minimum value at the center of the vortex (zero pressure in a potential vortex). Hence, a particle trapped in the center of a vortex tends to stay there until it can overcome the radial pressure gradient.

8.4 Role of the Reynolds Number

The Re number was introduced previously (see Chapter 2) as an important viscous-flow similarity parameter that strongly controls the viscous boundary-layer behavior. We now show how this similarity parameter appears as a major governing parameter in the incompressible viscous-flow equations. Steady, two-dimensional Cartesian flow is discussed for convenience, but all results are readily extended to time-dependent flows in three dimensions and to other coordinate systems. Because the flow is assumed to be incompressible, ρ = constant. The momentum equation, Eq. 8.18, and the continuity equation, Eq. 8.19, describe the flow field. Dividing Eq. 8.18 by the density, ρ (constant), the governing equations become:

$$\begin{cases} \dfrac{\partial u}{\partial x} + \dfrac{\partial w}{\partial z} = 0 \\[2mm] u\dfrac{\partial u}{\partial x} + w\dfrac{\partial u}{\partial z} = -\dfrac{1}{\rho}\dfrac{\partial p}{\partial x} + \dfrac{\mu}{\rho}\left(\dfrac{\partial^2 u}{\partial x^2} + \dfrac{\partial^2 u}{\partial z^2}\right) \\[2mm] u\dfrac{\partial w}{\partial x} + w\dfrac{\partial w}{\partial z} = -\dfrac{1}{\rho}\dfrac{\partial p}{\partial z} + \dfrac{\mu}{\rho}\left(\dfrac{\partial^2 w}{\partial x^2} + \dfrac{\partial^2 w}{\partial z^2}\right). \end{cases} \tag{8.27}$$

We proceed now to nondimensionalize these three equations to identify the similarity parameters they may contain. All velocity components are divided by a reference velocity, V_∞; all coordinate distances are divided by a reference length, L (perhaps the chord of an airfoil or the length of a flat plate); and the pressure term is divided by the reference quantity (ρV_∞^2), which has the units of pressure (this is clearly twice the dynamic pressure).

Multiply both sides of the continuity equation by (L/V_∞) and denote dimensionless quantities by overbars. Thus, continuity becomes:

$$\frac{\partial \bar{u}}{\partial \bar{x}} + \frac{\partial \bar{w}}{\partial \bar{z}} = 0,$$

where, for example:

$$\left(\frac{L}{V_\infty}\right)\frac{\partial u}{\partial x} = \frac{\partial(u/V_\infty)}{\partial(x/L)} = \frac{\partial \bar{u}}{\partial \bar{x}}.$$

Next, multiplying through the momentum equations in Eq. (8.27) by (L/V_∞^2) and again denoting dimensionless quantities by overbars, the result is:

$$\begin{cases} \bar{u}\dfrac{\partial \bar{u}}{\partial \bar{x}} + \bar{w}\dfrac{\partial \bar{u}}{\partial \bar{z}} = -\dfrac{\partial \bar{p}}{\partial \bar{x}} + \dfrac{\mu}{\rho V_\infty L}\left(\dfrac{\partial^2 \bar{u}}{\partial \bar{x}^2} + \dfrac{\partial^2 \bar{u}}{\partial \bar{z}^2}\right) \\[2mm] \bar{u}\dfrac{\partial \bar{w}}{\partial \bar{x}} + \bar{w}\dfrac{\partial \bar{w}}{\partial \bar{z}} = -\dfrac{\partial \bar{p}}{\partial \bar{z}} + \dfrac{\mu}{\rho V_\infty L}\left(\dfrac{\partial^2 \bar{w}}{\partial \bar{x}^2} + \dfrac{\partial^2 \bar{w}}{\partial \bar{z}^2}\right), \end{cases} \tag{8.28}$$

where, for example:

$$\left(\frac{L}{\rho V_\infty^2}\right)\frac{\partial p}{\partial x} = \frac{\partial(p/\rho V_\infty^2)}{\partial(x/L)} = \frac{\partial \bar{p}}{\partial \bar{x}}$$

and:

$$\left(\frac{L}{\rho V_\infty^2}\right)\mu\frac{\partial^2 w}{\partial z^2} = \mu\left(\frac{1}{\rho V_\infty}\right)\frac{\partial}{\partial(z/L)}\left[\frac{\partial(w/V_\infty)}{\partial(z/L)}\cdot\frac{1}{L}\right] = \frac{\mu}{\rho V_\infty L}\frac{\partial^2 \bar{w}}{\partial \bar{z}^2}.$$

The only parameter that appears in the resulting set of dimensionless equations:

$$\frac{\partial \bar{u}}{\partial \bar{x}} + \frac{\partial \bar{w}}{\partial \bar{z}} = 0 \tag{8.29}$$

$$\bar{u}\frac{\partial \bar{u}}{\partial \bar{x}} + \bar{w}\frac{\partial \bar{u}}{\partial \bar{z}} = -\frac{\partial \bar{p}}{\partial \bar{x}} + \frac{1}{\text{Re}}\left(\frac{\partial^2 \bar{u}}{\partial \bar{x}^2} + \frac{\partial^2 \bar{u}}{\partial \bar{z}^2}\right) \tag{8.30}$$

$$\bar{u}\frac{\partial \bar{w}}{\partial \bar{x}} + \bar{w}\frac{\partial \bar{w}}{\partial \bar{z}} = -\frac{\partial \bar{p}}{\partial \bar{z}} + \frac{1}{\text{Re}}\left(\frac{\partial^2 \bar{w}}{\partial \bar{x}^2} + \frac{\partial^2 \bar{w}}{\partial \bar{z}^2}\right) \tag{8.31}$$

is the coefficient $(\rho V_\infty L/\mu) = \text{Re}$, the Reynolds number. The student should confirm that this also is true for the Navier–Stokes equations that describe three-dimensional flow. Additional dimensionless groups appear if the flow is unsteady or compressible. The Re number is a *similarity parameter* (see Chapter 2).

Suppose that a viscous-flow problem involving a body of given shape and size has a certain numerical value of the Re number and that the Navier–Stokes equations for this problem are written in dimensionless form and solved. Now, consider a second flow problem involving a body of geometrically similar shape (i.e., same shape and orientation to the flow but different size). If the Re number for this second case has the same magnitude as the first (even though the individual magnitudes of velocity, viscosity, and reference length are different in the two problems), then the solution to the first problem (in dimensionless form) is also the solution to the second problem. That is, any solution to Eq. 8.29 is also the solution for any other problem with geometrical similarity and the same Re number. This demonstrates one of many powerful features associated with the similarity concept.

We saw previously how exact solutions to the Navier–Stokes equations may be found for certain problems by using restrictive assumptions. The Navier–Stokes equations also may be simplified in two limiting cases distinguished by very small and by very large values of the Re number:

1. *Flows with Re << 1.0.* Such flows are called *Stokes* or *Oseen* flows. They correspond to very slow *creeping* fluid motion. The continuity and Navier–Stokes equations can be combined into a single linear equation, which is the starting point for the theory of lubrication and other important applications.
2. *Flows with Re >> 1.0.* This situation usually arises from a small coefficient of viscosity (as in most aerodynamics applications with atmospheric air as the medium) and leads to the *Prandtl boundary-layer problem.*

In the next section, we study this second case, which represents the important aerodynamics problem of viscous flow at high Re number over a surface such as an airplane wing. In such problems, the Re number is typically in the range of 10^6 to 10^7. Notice that the effects of viscosity in the dimensionless governing equation (Eq. 8.29) are represented by terms divided by this very large parameter. This suggests that the viscous effects are mathematically unimportant. In fact, it is this feature that enabled us to solve external aerodynamics problems of the types intro-duced in preceding chapters without regard for the effect of viscous effects or fric-tional losses.

However, it is clear that viscous effects must be important near the surface of the body where the velocity must tend to zero. This paradox was resolved by the work of Prandtl in 1904. The answer is, of course, that one of the derivative terms multiplied by the inverse of the Re number must be very large so that the product is as important as other terms in the momentum balance. From the viewpoint of differential-equation theory, this gives rise to an important class of mathematical problems known as *singular perturbation theory*. The Prandtl boundary-layer problem was the first of this type to be solved; the central role of the Re number in its formulation and solution are the subject of the next section.

8.5 The Prandtl Boundary-Layer Equations

The derivation of the Prandtl boundary-layer equations for a two-dimensional, incompressible flow is carried out in detail in this section because it is the basis for many practical aerodynamics applications. The resulting equations describe a *laminar boundary layer*. These equations are modified in later sections to represent turbulent boundary layers.

The shape of the velocity distribution in a boundary layer is deduced from physical reasoning in Chapter 2, as shown in Fig. 2.3, which is repeated here. In Fig. 8.9, the boundary-layer thickness is given the symbol δ and represents the dis-tance from the wall at which the velocity in the boundary layer approaches that of the flow outside the boundary layer. It must be emphasized that this symbol, as used throughout this chapter, always denotes a length in physical units (e.g., inches or centimeters). A more precise definition of this thickness is established after the solu-tion for the velocity profile is obtained in Section 8.6 and interpreted in Section 8.7.

Two assumptions are made in deriving the boundary-layer equations, as follows.

1. The boundary-layer thickness, δ, is very small compared to a streamwise distance, x, along a body; thus, $\delta/x \ll 1.0$.
2. The Re number is very large; Re $\gg$ 1.

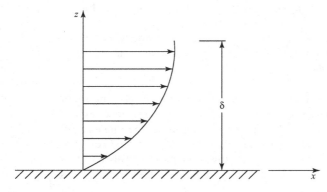

Figure 8.9. Velocity distribution in a boundary layer.

The assumption that a typical boundary layer is very thin is confirmed by experimental data. For example, at a distance of 1 foot back from the leading edge of a thin airfoil at zero angle of attack in a 50-ft/s freestream at standard sea-level conditions, the boundary-layer thickness is $\delta = 0.009$ ft (0.11 in). The value of the Re number for this example (based on a characteristic length of 1 foot) is Re = 322,500. Of course, near the leading edge, where x approaches zero, the assumption $\delta/x \ll 1.0$ fails. However, this occurs only over a length that is a minute fraction of an airfoil chord length and therefore is of little concern from a practical point of view.

Derivation of the Boundary-Layer Equations

Consider a thin flat plate extending to downstream infinity and aligned with a steady, incompressible flow, as shown in Fig. 8.10. The characteristic length L introduced previously now represents a fixed distance downstream of the plate leading edge. The flat plate is the simplest possible geometry. This simplifies the initial discussion of the boundary layer because no streamwise pressure gradients require attention. Later, we present the effects of pressure gradients that are introduced by the presence of the body in the flow or by an externally impressed pressure variation. The velocity at the edge of the boundary layer is essentially that of the oncoming freestream, V_∞. In what follows, the symbol U_e is used to represent the edge velocity instead of V_∞ and the pressure-gradient terms are retained in the defining equations. This allows for subsequent generalization to flows with local streamwise pressure gradients and, consequently, with $U_e = U_e(x)$ rather than $U_e = V_\infty = $ constant.

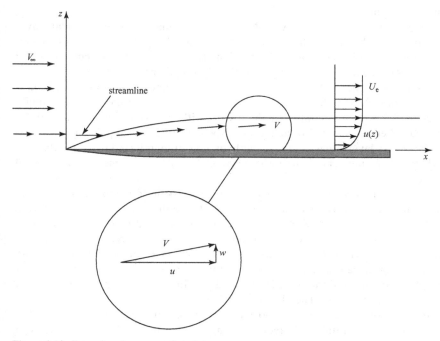

Figure 8.10. Boundary layer on a flat plate.

At the surface of the plate, the velocity is zero (no-slip). At the edge of the thin boundary layer the velocity is U_e. The velocity rapidly increases from zero to U_e within a very short distance, so that within the boundary layer the viscous shear stress ($\mu \, \partial u/\partial z$) must be very large. This region of significant shear stress defines the boundary layer thickness, $\delta(x)$, which varies along the surface as suggested in Figure 8.10. Outside of this *region of nonuniformity* the velocity gradients are small and the shear stress is negligible, so that the region may be taken to be potential flow field (frictionless). Experiments confirm that not only is $\delta \ll x$ but also $d\delta/dx \ll 1$. That is, the boundary layer thickness increases slowly so that within the boundary layer the velocity normal to the surface is very small compared to the streamwise value, that is, $w \ll u$, as shown in Fig. 8.10. The boundary layer thickness, δ, increases continuously with x because the streamwise momentum of the flow particles located further and further away from the surface is being reduced by the continuing shear force at the surface. Since the boundary layer thickness increases with distance downstream, continuity requires that the mass flow within the boundary layer must also be increasing. This means that streamlines pass into the boundary layer from the freestream flow as suggested in Fig. 8.10.

Equations (8.29) govern the flow described in Figure 8.10. However, this is a set of nonlinear partial differential equations in spite of the simplifications introduced already. There are no general solutions to this set. Clearly, one must either find additional valid simplifications or resort to a strictly numerical approach. The latter technique (CFD) is now in vogue due to the ready availablity of high-speed computers. However, valuable physical insight is lost if we yield to this "easy way out" at this juncture.

We proceed now to simplify the Navier-Stokes equations to the boundary layer equations, which are valid for $\delta/x \ll 1$ and Re $\gg 1$. This will be done by first estimating the size of each term in the nondimensional Navier-Stokes equations and then dropping some terms as being negligibly small compared to those retained. The same approach will be used later in deriving linearized compressible flow theory, although here the development will be a bit more formal. The opportunity is taken at this point to familiarize the student with the modern analytical methods that have been developed to handle the type of mathematical problem arising in the boundary layer situation. These methods, now often referred to as singular perturbation techniques, apply to a wide variety of similar problems and yield a vital physical insight.

Because it is necessary in the following discussion to compare nonambiguously the sizes of various quantities, it is useful to introduce a special notation to accomplish this. We work here with properly nondimensionalized quantities, as introduced in Section 8.5. The thickness of the boundary layer is a convenient length scale to use within the boundary layer because there is no natural length scale available in the boundary-layer problem. (In most cases, the physical length of the body is essentially infinite compared to the thickness of the boundary layer.) The chord length of an airfoil, or the length of the flat plate in a flow experiment, is too large to represent a useful measure of the properties of the thin fluid layer near the surface in which rapid changes in velocity take place.

Orders of Magnitude

We expect that the thickness of the boundary layer has a major impact on the problem formulation. Recall the fact that δ can represent a length smaller than even the thickness of the paper on which this page is written up to perhaps 1 centimeter or more, depending on the distance from the leading edge of the boundary layer. In most cases, δ is considerably smaller than the length of the aerodynamic surface.

The method used here for deducing the appropriate set of equations is the same as that used in the early 1900s by Ludwig Prandtl to achieve his astonishing insight into viscous boundary-layer flow. We must examine the various terms in the governing equations to determine which ones control the behavior. Some terms must be retained whereas others that have little effect can be ignored. This is accomplished by comparing sizes of terms on the basis of their orders of magnitude, using the boundary-layer thickness as a primary scaling variable. The arguments here closely follow those used by Prandtl.

To illustrate the approach in the simplest possible way, consider the continuity equation, Eq. 8.29:

$$\frac{\partial \bar{u}}{\partial \bar{x}} + \frac{\partial \bar{w}}{\partial \bar{z}} = 0.$$

It is clear that because u is of the same order of magnitude as the velocity outside the boundary layer (V_∞), then it is proper to assume that:

$$u \sim V_\infty,$$

which is often expressed mathematically as $u = O(V_\infty)$, where O is the *order symbol*. This is read as "u is of the order of V_∞." It expresses the fact that the streamwise velocity anywhere in the region of interest is comparable to that outside of the boundary layer. Therefore, the dimensionless velocity is:

$$\bar{u} \equiv \frac{u}{V_\infty} \sim 1 \qquad (\text{or}, \bar{u} = O(1)).$$

Similarly, if we move downstream from the origin a distance $x \sim L$ (much larger than the thickness of the boundary layer), then the dimensionless length in the x-direction is:

$$\bar{x} \equiv \frac{x}{L} \sim 1 \qquad (\text{or}, \bar{x} = O(1)).$$

Similarly, if we take changes in the x-position, Δx, and changes in the x-velocity, Δu also to be comparable in size to L and V_∞, respectively, then $\Delta \bar{x} \sim 1$ and $\Delta \bar{u} \sim 1$ Therefore, the derivative of u with respect to x that appears in the continuity equation is on the order of magnitude of unity; that is:

$$\frac{\partial \bar{u}}{\partial \bar{x}} \approx \frac{\Delta \bar{u}}{\Delta \bar{x}} \sim 1 \qquad \left(\text{or}, \frac{\partial \bar{u}}{\partial \bar{x}} = O(1)\right).$$

Now, changes in the z-direction in the boundary layer must be much smaller than L; that is, they are comparable to the dimensional boundary-layer thickness, δ, which is the appropriate length scale in the boundary layer. It therefore is useful to introduce the dimensionless boundary-layer thickness:

$$\bar{\delta} = \frac{\delta}{L} \ll 1,$$

which helps in scaling properties of the boundary layer normal to the surface. That is, it is appropriate to write:

$$\Delta \bar{z} \sim \bar{\delta} \qquad (\text{or, } \Delta \bar{z} = O(\bar{\delta})),$$

which expresses the fact that changes in distance in the boundary layer normal to the surface are much smaller than in the streamwise direction. So that the continuity equation, Eq. 8.29, is satisfied, it is clear that the velocity and changes in the velocity normal to the surface also must be on the order of $\bar{\delta}$. Then:

$$\bar{w} \sim \bar{\delta} \text{ and } \Delta \bar{w} \sim \bar{\delta}$$

so that the derivative:

$$\frac{\partial \bar{w}}{\partial \bar{z}} \approx \frac{\Delta \bar{w}}{\Delta \bar{z}} \sim 1,$$

as required for continuity. This justifies the previous observation that:

$$w \ll V_\infty \qquad (\bar{w} \ll 1).$$

This ordering process now can be applied to the full set of equations, Eqs. 8.29–30, so that any negligibly small terms can be identified and eliminated, if possible. The result is:

$$\frac{\partial \bar{u}}{\partial \bar{x}} + \frac{\partial \bar{w}}{\partial \bar{z}} = 0 \tag{8.29}$$

$$\quad 1 \qquad 1$$

$$\bar{u}\frac{\partial \bar{u}}{\partial \bar{x}} + \bar{w}\frac{\partial \bar{u}}{\partial \bar{z}} = -\frac{\partial \bar{p}}{\partial \bar{x}} + \frac{1}{Re}\left(\frac{\partial^2 \bar{u}}{\partial \bar{x}^2} + \frac{\partial^2 \bar{u}}{\partial \bar{z}^2}\right) \tag{8.30}$$

$$\quad 1\ 1 \quad \bar{\delta}\ \frac{1}{\bar{\delta}} \qquad\qquad 1 \qquad \frac{1}{\bar{\delta}^2}$$

$$\bar{u}\frac{\partial \bar{w}}{\partial \bar{x}} + \bar{w}\frac{\partial \bar{w}}{\partial \bar{z}} = -\frac{\partial \bar{p}}{\partial \bar{x}} + \frac{1}{Re}\left(\frac{\partial^2 \bar{w}}{\partial \bar{x}^2} + \frac{\partial^2 \bar{w}}{\partial \bar{z}^2}\right) \tag{8.31}$$

$$\quad 1\ \bar{\delta} \quad \bar{\delta}\ 1 \qquad\qquad \bar{\delta} \quad \frac{1}{\bar{\delta}},$$

where the order of each term is written under its position in the equations. As already noted, continuity is properly satisfied. It is important to notice that only the correct choice of scaling of the normal velocity and displacement allows this; otherwise, the two terms would not cancel to satisfy the equation.

Consider the x-momentum equation, Eq. 8.30. Any term that is to be retained must be of the order $O(1)$. Thus, both parts of the convective acceleration on the left must stay; unfortunately, the problem is still nonlinear! The axial pressure gradient is taken to be of the order $O(1)$ and is dependent only on conditions outside of the boundary layer, as we demonstrate herein. The viscous force on the right of Eq. 8.30 is dominated by the large Re number in the denominator. If it is not similar in size to the other terms in Eq. 8.30, then there is no mechanism present to slow the fluid particles down to zero speed at the surface. In other words, it then would not be possible to satisfy the no-slip condition. If viscous effects are not to disappear, it is necessary that:

$$\bar{\delta}^2 \sim \frac{1}{\text{Re}}; \qquad (8.32)$$

then, the derivative $\partial^2 \bar{u} / \partial \bar{x}^2$ represents a negligible term. At last, something drops out! Because the Re number is very large, the small factor $1/\text{Re}$ in a sense "cancels" the very large factor of the order of the inverse of the square of the dimensionless boundary-layer thickness. The result is a product that is of first order, $O(1)$. For emphasis, the viscous term can be retained only in the momentum balance if:

$$\bar{\delta}^2 = O\left[\frac{1}{\text{Re}}\right]. \qquad (8.33)$$

Thus, all of the terms in Eq. 8.32 are of the same order (namely, $O(1)$), which is consistent with the terms in the continuity equation.

Rewriting Eq. 8.33 in dimensional form:

$$u\frac{\partial u}{\partial x} + w\frac{\partial u}{\partial z} = -\frac{\partial p}{\partial x} + v\frac{\partial^2 u}{\partial z^2}. \qquad (8.34)$$

This equation often is called the *boundary-layer equation*. Notice that unlike most of the other problems solved in this chapter, it is not a linear-differential equation. The convective terms involving products of the variables and their derivatives still appear on the left side. Nevertheless, it is considerably simpler than the x-component of the Navier–Stokes equation.

The z-momentum equation shows that the pressure gradient normal to the surface:

$$\frac{\partial \bar{p}}{\partial \bar{z}} \sim \bar{\delta} \rightarrow \frac{\partial p}{\partial z} \rightarrow 0 \qquad (8.35)$$

is very small because $\bar{\delta} \ll 1$. It is on this basis that we can assume that the pressure through the boundary layer is controlled by conditions outside of the layer. Corrections to the pressure distribution are not much affected by the presence of the layer, as Eq. 8.33 shows. For these reasons, in the limit of a very small boundary-layer thickness, we assume that the pressure gradient normal to the surface is essentially zero. This is a key element of the Prandtl boundary-layer theory, which states that:

The static pressure gradient through a boundary layer is negligibly small.

Thus, if a static-pressure sensor is inserted into a wind-tunnel wall such that it reads the pressure very close to the surface, the reading also accurately represents the static pressure at the edge of the boundary layer. No correction is required to obtain the value of freestream static pressure. Eq. 8.36 is valid provided that the surface on which the boundary layer is growing is flat or nearly so. If the surface has appreciable curvature, then centripetal forces on the fluid particles may become significant; in that case, there must be a pressure gradient, $\partial p/\partial z$, normal to the surface to maintain the particles in equilibrium. Such a correction is estimated easily from the geometry, whenever necessary.

To summarize, the Prandtl boundary-layer equations rewritten in dimensional form are:

$$\begin{cases} \dfrac{\partial u}{\partial x} + \dfrac{\partial w}{\partial z} = 0 \\[2mm] u\dfrac{\partial u}{\partial x} + w\dfrac{\partial w}{\partial z} = -\dfrac{1}{\rho}\dfrac{\partial p}{\partial x} + v\dfrac{\partial^2 u}{\partial z^2}. \end{cases} \tag{8.36}$$

The continuity and x-momentum balance control the flow. The z-momentum equation provides the information that $p = p(x)$. Then, Eq. 8.37 represents two equations in the three unknowns, u, w, p. This difficulty is resolved by treating $p = p(x)$ as a given quantity. In many practical cases of practical importance, $p = p(x)$ is the pressure-distribution term provided by the external-flow (i.e., potential-flow) solution because p does not change in the direction perpendicular to the surface. The inviscid-flow solutions for the pressure distribution on airfoils, wings, and bodies, as discussed in Chapters 5 through 7, provide the required information describing the pressure distribution $p = p(x)$, which is inserted into Eq. 8.37 to give the $\partial p/\partial x$ term in the x-momentum equation. As a result, there are two unknowns, u and w, to be determined by the solution of the two boundary-layer equations. The potential flow-pressure field is said to be "impressed" on the boundary layer. The boundary-layer equations, Eq. 8.36, must be solved subject to boundary conditions appropriate for a given physical and geometrical situation.

8.6 Incompressible Boundary-Layer Theory

We now carry out a detailed solution for one of the most important viscous-flow problems in aerodynamics—namely, the high-Re-number flow over a flat plate. The analytical technique demonstrated for the flat-plate problem is the key to understanding a wide variety of other engineering solutions. Clearly, something special is needed to handle the nonlinear (i.e., convective acceleration) terms still present in the momentum balance. This means that the methods of differential-equation theory do not apply. Separation of variables and superposition are no longer applicable tools. A method of great utility in such situations is introduced, which often is called the *similarity method*.

By a flat plate, we mean a surface on which there is no impressed pressure. The governing equations are those shown in Eq. 8.37 without the pressure gradient in the x-direction. That is, both components of the pressure gradient are now zero:

$$\begin{cases} \dfrac{\partial u}{\partial x} + \dfrac{\partial w}{\partial z} = 0 \\[2mm] u\dfrac{\partial u}{\partial x} + w\dfrac{\partial w}{\partial z} = v\dfrac{\partial^2 u}{\partial z^2}. \end{cases} \qquad (8.37)$$

This special case was first solved in 1907 by Blasius, a graduate student of Prandtl, at Göttingen University. A solution to Eq. 8.38 is sought that satisfies the no-slip requirement at the surface and matches the flow field at great distance from it. In mathematical form, we require that:

$$\begin{cases} u = 0 \\ w = 0 \end{cases} \quad \text{for } x > 0,\ z = 0, \text{ and}$$

$$\begin{cases} u = U_e(x) & \text{for } x > 0,\ z \to \infty \\ u = V_\infty & \text{for } x = 0,\ z > 0. \end{cases} \qquad (8.38)$$

A major difficulty in carrying out this solution is that the momentum equation is nonlinear. Students undoubtedly have noticed that the nonlinearity is associated with terms arising in the convective acceleration part of the momentum balance. Classical methods for solving partial-differential equations (e.g., a separation of variables) are not applicable in this problem. Also, we lose the ability to superimpose simple linear solutions in representing more complex cases, a technique exploited many times in preceding chapters. Therefore, we must approach this problem differently than in the more conventional situations that arose in the Couette and Poiseiulle flows and in the potential flow problems addressed previously; these were linear problems.

A useful step is to reduce the number of dependent variables by using a stream function to represent the velocity field. The continuity equation is satisfied exactly if we write:

$$u = \frac{\partial \psi}{\partial z}, \quad w = -\frac{\partial \psi}{\partial x}.$$

Then, the momentum equation becomes:

$$\frac{\partial \psi}{\partial z}\frac{\partial^2 \psi}{\partial x \partial z} - \frac{\partial \psi}{\partial x}\frac{\partial^2 \psi}{\partial z^2} = v\frac{\partial^3 \psi}{\partial z^3}. \qquad (8.39)$$

The corresponding boundary conditions are:

$$\begin{cases} \dfrac{\partial \psi}{\partial z} = 0 \\[2mm] \dfrac{\partial \psi}{\partial x} = 0 \end{cases} \text{for } x > 0,\ z = 0 \text{ and} \begin{cases} \dfrac{\partial \psi}{\partial z} = U_e(x) & \text{for } x > 0,\ z \to \infty \\[2mm] \dfrac{\partial \psi}{\partial z} = V_\infty & \text{for } x = 0,\ z > 0. \end{cases} \qquad (8.40)$$

Modern computational techniques (i.e., CFD methods) are used routinely to solve nonlinear problems of this type. However, we want to demonstrate a powerful technique that uses mathematical finesse rather than computational brute force to give the required answers. Recall that much physical insight can be gained by using this approach in place of the "black-box" computations in difficult problems. The price of these benefits is mastery of special mathematical methods not often taught in undergraduate programs (and seldom even in graduate engineering programs). The method introduced often is called the similarity method. It has widespread application in the solution of nonlinear problems and is strongly couched in simple geometrical ideas.

Application of the Similarity Method

The word *similarity* is used in several contexts throughout this book. There is a significant probability that the student is becoming confused from the frequent appearance of this terminology in seemingly different uses of the same word. Therefore, the several meanings are reviewed briefly, as follows:

1. *Geometric Similarity*. This describes two or more body shapes that are exact photographic images (i.e., enlargement or reduction) of one another.
2. *Similarity Parameters*. These are dimensionless quantities that can be deduced from dimensional-analysis considerations or from experiment—for example, the Mach number, Re number, and lift coefficient. These are especially useful in presenting results or when planning experiments.
3. *Similarity Relationships*. These are functional relationships between similarity parameters. They may be explicit when equation solutions may be found, or functional in form if the defining equation is difficult to solve, as in nonlinear-flow problems.
4. *Dynamic Similarity*. If two geometrically similar bodies are in two flows that have the same values of the pertinent similarity parameters, then the two flows are dynamically similar. The pertinent similarity parameters may be found by dimensional analysis or by writing the defining equations for that class of problem in nondimensional form. Similarity parameters appear as coefficients in the equations. For example, if the laminar boundary-layer equations for incompressible flow are written in nondimensional form, the only parameter that appears is the Re number. Hence, a known solution to the equations for one flow problem also is a solution (in nondimensional form) to a second flow problem, provided that the two flows have the same Re number. For compressible flow, additional pertinent similarity parameters usually are the Mach number, Prandtl number, and ratio of specific heats, γ. Dynamically similar flows have equal-force coefficients.
5. *Similarity Rules*. Two flow problems can be called similar in the sense that they are related by a coordinate transformation connecting the two defining equations and the two sets of appropriate boundary conditions. Certain scale factors arise that relate to the pressure and force coefficients and the shape of the two bodies in the two flows.
6. *Similarity Solutions*. These are methods of solution in which mathematical advantage may be taken if the solution at one station in a flow is geometrically similar to solutions at other stations. For example, in certain boundary layers, the velocity profiles are of the same mathematical family. When this happens, it becomes possible to construct a new independent variable—which is a

combination of the original dependent variables—such that the derivatives in the governing equation(s) all can be written as derivatives of this new independent variable. For two-dimensional, viscous-flow problems, partial-differential equations become ordinary differential equations when similar solutions are assumed. Similar solutions in viscous flow involve "self-similar" velocity profiles, meaning that the profiles collapse to a single curve when graphed in appropriate coordinates. An example is the velocity profile in the incompressible boundary layer growing along a flat plate.

7. *Similitude.* Two flow problems have similitude if they have identical defining equations and boundary conditions when written in dimensionless form.

It is in the context of the sixth meaning that we are working herein. Referring to Fig. 8.10, notice that we expect the flat-plate boundary layer to increase in thickness as the distance from the leading edge is increased. It is certainly plausible that the velocity distribution across this thickness would be "similar" at any station. We now examine the differential equations to see if this is a possible outcome, which is accomplished by checking whether the equation and the boundary conditions are unaffected by an *affine transformation*. Such a transformation is one in which a parameter—say, λ—or its powers can be used to scale each variable without changing the form of the mathematical problem. That is, the problem is *invariant* to the transformation.

If the differential equation for the stream function (Eq. 8.40) and the four boundary conditions (Eq. 8.41) are transformed in this manner by setting:

$$\begin{cases} \psi = \lambda \psi' \\ x = \lambda^n x' \\ z = \lambda^m z' \end{cases} \tag{8.41}$$

where the primes denote the "stretched" variables, we find that the problem is now governed by:

$$\frac{\partial \psi'}{\partial z'}\frac{\partial^2 \psi'}{\partial x' \partial z'}(\lambda^{2-n-2m}) - \frac{\partial \psi'}{\partial x'}\frac{\partial^2 \psi'}{\partial z'^2}(\lambda^{2-n-2m}) = v\frac{\partial^3 \psi'}{\partial z'^3}(\lambda^{1-3m})$$

$$\begin{cases} \dfrac{\partial \psi'}{\partial z'}(\lambda^{1-m}) = 0 & \text{for } x'(\lambda^n) > 0,\ z'(\lambda^m) = 0 \\[2mm] \dfrac{\partial \psi'}{\partial x'}(\lambda^{1-n}) = 0 & \text{for } x'(\lambda^n) > 0,\ z'(\lambda^m) = 0 \\[2mm] \dfrac{\partial \psi'}{\partial z'}(\lambda^{1-m}) = V_\infty & \text{for } x'(\lambda^n) > 0,\ z'(\lambda^m) \to \infty \\[2mm] \dfrac{\partial \psi'}{\partial z'}(\lambda^{1-m}) = V_\infty & \text{for } x'(\lambda^n) = 0,\ z'(\lambda^m) > 0 \end{cases} \tag{8.42}$$

For a similarity solution to be available, it is necessary for all occurrences of the parameter λ to disappear in the transformed equations. This is so only if:

$$\lambda^{2-n-2m} = \lambda^{1-3m}$$

$$\text{and } \lambda^{1-m} = 1. \tag{8.43}$$

Notice that the boundary conditions requiring zero values are unaffected by the transformation. Therefore, for a similarity solution, it is necessary only that $m = 1$ and $n = 2$. Then, Eq. 8.43 reduces to the same set before the transformation. The real benefit from the transformation process begun here is that it suggests that we can combine the original independent variables such that the number of variables required in the problem can be reduced. For instance, the combination:

$$\eta \approx \frac{z}{\sqrt{x}} = \frac{\lambda z'}{\sqrt{\lambda^2 x'}} = \frac{z'}{\sqrt{x'}} \tag{8.44}$$

yields such a result. Notice that the new combined "length" variable does not have the dimensions of length but rather the square root of length. It is often better to work with dimensionless variables, so we seek to introduce appropriate scaling factors from the natural parameters in the problem to render the transformed problem in dimensionless form. The velocity of the uniform parallel flow outside the boundary layer, $U_e = V_\infty$ can used as a reference velocity. Since there is no natural length scale of the sort present in the Couette and Poiseuille flows, it is necessry to use the parameters in the differential equation to form an equivalent reference length. Notice that the combination:

$$\frac{\nu}{U_e} \rightarrow \frac{[\text{ft}^2/\text{sec}]}{[\text{ft/sec}]}$$

has the dimensions of length, so all the dimensional length variables like x and y or x' and y' can be rendered dimensionless by dividing by ν/U_e. Therefore, if we replace x and y and x' and y' by their dimensionless equivalents in Eq. (8.45) we can define a dimensionless transformed coordinate, η as:

$$\eta = \frac{z\dfrac{U_e}{\nu}}{\sqrt{x\dfrac{U_e}{\nu}}} = z\sqrt{\frac{U_e}{\nu x}} \tag{8.45}$$

Combining the original dependent variable, ψ, with the original independent variables can result in a new dependent variable that is independent of the parameter λ. Call the new (dimensionless) dependent variable f, which we assume will be a function only of the new single dependent variable η. To eliminate the λ in the transformed stream function, we note that $x = \lambda^2 x'$ so $\lambda = \sqrt{x/x'}$ and f can then be defined in dimensionless form as:

$$f = \frac{\psi}{\sqrt{x}}\left[\frac{U_e}{U_e \nu}\sqrt{\frac{\nu}{U_e}}\right] = \frac{\psi}{\sqrt{\nu U_e x}} = \frac{\lambda \psi'}{\sqrt{\nu U_e x \lambda^2 x'}} = \frac{\psi'}{\sqrt{\nu U_e x x'}}$$

which satisfies the criterion that parameter λ does not appear explicitly. The parameters within the square brackets represent the dimensions of the stream function and the inverse square root of the characteristic length. Since f is dimensionless by definition, we can then write the (dimensional) stream function as:

$$\psi(x, \eta) = \sqrt{\nu U_e x}\ f(\eta) \tag{8.46}$$

The detailed steps we have shown here are not usually presented in undergraduate textbooks. What is often done is to introduce the results shown in Eqs. (8.46) and (8.47) as assumptions with little justification other than that "it works." Although you may have trouble understanding fully the significance of the similarity transformation process in a single reading, you are invited to study the steps again when confronted in the future by problems that may require a similiarity solution.

Now, to see if this new set of variables represents any kind of an advantage, we attempt to rewrite Eqs. 8.40 and 8.41 in terms of η and f. Several derivatives must be evaluated, so we note first that:

$$\frac{\partial \eta}{\partial x} = z\sqrt{\frac{U_e}{\nu}} \frac{d}{dx} x^{-\frac{1}{2}} = -\frac{z}{2}\sqrt{\frac{U_e}{\nu}} x^{-\frac{3}{2}} = -\frac{\eta}{2x}$$

$$\frac{\partial \eta}{\partial z} = \sqrt{\frac{U_e}{\nu x}}$$

Please keep in mind as we use these results that we are transforming the set of independent variables (x, z) to the new set of independent variables (x, η). Care must be taken in the differentiation process. The derivatives of any function with respect to the original variables must now be written as:

$$\begin{cases} \dfrac{\partial}{\partial x} \rightarrow \dfrac{\partial}{\partial x} + \dfrac{\partial \eta}{\partial x}\dfrac{\partial}{\partial \eta} = \dfrac{\partial}{\partial x} - \dfrac{\eta}{2x}\dfrac{\partial}{\partial \eta} \\[3mm] \dfrac{\partial}{\partial z} \rightarrow \dfrac{\partial \eta}{\partial z}\dfrac{\partial}{\partial \eta} = \sqrt{\dfrac{U_e}{\nu x}}\dfrac{\partial}{\partial \eta} \end{cases} \qquad (8.47)$$

so that the derivatives needed in Eq. (8.40) are:

$$\begin{cases} \dfrac{\partial \psi}{\partial x} = \dfrac{\partial}{\partial x}\sqrt{\nu U_e x}\, f - \dfrac{\eta}{2x}\sqrt{\nu U_e x}\, \dfrac{\partial f}{\partial \eta} = \dfrac{1}{2}\sqrt{\dfrac{\nu U_e}{x}}(f - \eta f') \\[3mm] \dfrac{\partial \psi}{\partial z} = \sqrt{\nu U_e x}\sqrt{\dfrac{U_e}{\nu x}}\dfrac{\partial f}{\partial \eta} = U_e f' \\[3mm] \dfrac{\partial^2 \psi}{\partial x \partial z} = \dfrac{1}{2}\sqrt{\dfrac{\nu U_e}{x}}\sqrt{\dfrac{U_e}{\nu x}}\dfrac{\partial}{\partial \eta}(f - \eta f') = -\dfrac{1}{2}\dfrac{U_e}{x}\eta f'' \\[3mm] \dfrac{\partial^2 \psi}{\partial z^2} = U_e\sqrt{\dfrac{U_e}{\nu x}}\dfrac{\partial f'}{\partial \eta} = U_e\sqrt{\dfrac{U_e}{\nu x}}f'' \\[3mm] \dfrac{\partial^3 \psi}{\partial z^3} = U_e\sqrt{\dfrac{U_e}{\nu x}}\sqrt{\dfrac{U_e}{\nu x}}\dfrac{\partial}{\partial \eta}f'' = \dfrac{U_e^2}{\nu x}f''' \end{cases} \qquad (8.48)$$

The Blasius Flat Plate Boundary-Layer Equation

Inserting these results directly into Eq. (8.40), and denoting derivatives with respect to η with primes, results in the third-order equation

$$\frac{1}{2}U_e f'\left(-\frac{U_e}{x}\eta f''\right) - \frac{1}{2}\sqrt{\frac{\eta U_e}{x}}(f - \eta f')\left(U_e\sqrt{\frac{U_e}{\nu x}}f''\right) = \nu\frac{U_e^2}{\nu x}f'''$$

and after simplifying,

$$2f''' + ff'' = 0 \tag{8.49}$$

which is often called the Blasius boundary layer equation. The boundary conditions in like fashion simplify to

$$\begin{cases} f'(0) = 0 \\ f(0) = 0 \\ f'(\infty) = 1 \end{cases} \tag{8.50}$$

since f is a function of η only. The original complex partial differential equation reduces to a simple (but nonlinear!) ordinary equation. Since Eq. 8.50 is nonlinear it is necessary to apply numerical methods for its solution. Another feature requiring special attention is that the boundary conditions must be satisfied at two widely separated locations ($\eta = 0$, and $\eta = 0$). This means that we must solve a *two-point boundary value problem* by numerical means. Many approaches have been developed in the years since Blasius presented his Ph.D. thesis work. Many of these were developed before the ready availability of fast digital computers, so they necessarily employed a variety of mathematical stratagems that are no longer needed and are therefore not discussed here. The problem can now readily be handled on your personal computer or even a good programmable hand calculator.

For example, one can easily solve the problem using a simple fourth-order *Runge-Kutta* integrator and a simple *shooting method* to deal with the two-point boundary value features. Equation (8.50) is first broken down into three simultaneous first-order ordinary differential equations. For example, one can write

$$\begin{cases} \dfrac{df}{d\eta} = g \\[2mm] \dfrac{dg}{d\eta} = h \\[2mm] \dfrac{dh}{d\eta} = -\dfrac{1}{2}fh \end{cases} \quad \text{with} \quad \begin{cases} g = 0 \\ f = 0 \end{cases} \quad \eta = 0 \ \text{ and } \quad g = 1 \quad \eta \to \infty \tag{8.51}$$

where the first two equations simply define the derivatives of the variable f. Since there is no boundary condition on h, (equivalent to the second derivative of f with respect to η, then we can adjust this value until $g = 1$ at a large value of the independent variable η. Newton's method or a similar technique can be used to adjust an initial guess for h at $\eta = 0$. It is not necessary to carry the integration to a large value of η. Experience shows that once a value of about 10 is reached, this is effectively infinite on the scale of η. The detailed calculation is left to the student in a problem at the end of this chapter.

Numerical results from this procedure are shown in Table 8.1 and Fig. 8.11a. Notice that in Fig. 8.11a, the plot for f' versus h is a representation for the velocity distribution through the boundary layer if f' is proportional to the x-velocity, as indicated in Eq. 8.47. That is, the x-velocity component through the layers is:

Table 8.1. *Function f and its derivatives*

η	f	f'	f''
0.0	0.00000	0.00000	0.33206
0.5	0.04149	0.16589	0.33091
1.0	0.16557	0.32978	0.32301
1.5	0.37014	0.48679	0.30258
2.0	0.65003	0.62977	0.26675
2.5	0.99631	0.75126	0.21741
3.0	1.39681	0.84605	0.16136
3.5	1.83770	0.91304	0.10777
4.0	2.30575	0.95552	0.06423
4.5	2.79014	0.97952	0.03398
5.0	3.28328	0.99154	0.01591
5.5	3.78058	0.99688	0.00658
6.0	4.27963	0.99897	0.00240
6.5	4.77933	0.99970	0.00077
7.0	5.27925	0.99992	0.00022
7.5	5.77923	0.99998	0.00006
8.0	6.27922	1.00000	0.00001
8.5	6.77922	1.00000	0.00000
9.0	7.27922	1.00000	0.00000
9.5	7.77922	1.00000	0.00000
10.0	8.27922	1.00000	0.00000

$$u = \frac{\partial \psi}{\partial z} = \frac{\partial}{\partial z}\left(\sqrt{\nu U_e x} f(\eta)\right) = \sqrt{\frac{U_e}{\nu x}} \frac{\partial}{\partial \eta}\left[\sqrt{\nu U_e x} f(\eta)\right] = U_e f', \quad (8.52)$$

where Eq. 8.48 for the z-derivative is used. Thus, the solid curve in Fig. 8.11 is the famous Blasius laminar-velocity profile for a flat-plate boundary layer. Carefully measured velocity profiles show essentially exact agreement with the theoretical predictions from the Blasius analysis shown in Fig. 8.11a.

Comparison of Theoretical Solution to Experimental Data

It is of interest at this point to compare the analytical solution to actual experimental data for the laminar boundary-layer results. The problem just solved may seem rather abstract because of the several stages required in its definition. It was necessary first to derive the basic equations describing the fluid dynamics. The equations then were modified using Prandtl's physical insights to properly incorporate the viscous terms. This led to stretching the coordinate normal to the wall and the introduction of the boundary-layer coordinate, η. A powerful mathematical technique, the similarity method, then was used to find an analytical solution. After all of this manipulation of the problem, the student may wonder if there is anything left that properly represents reality. As usual, the answers can be found

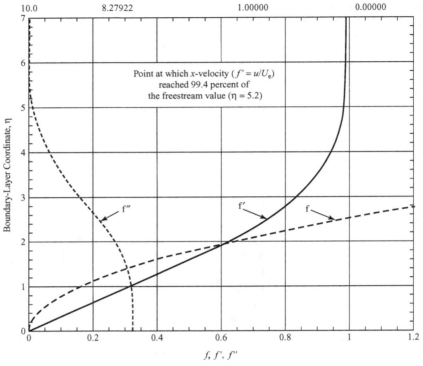

Figure 8.11a. Plots of the Blasius function, f, and its derivatives.

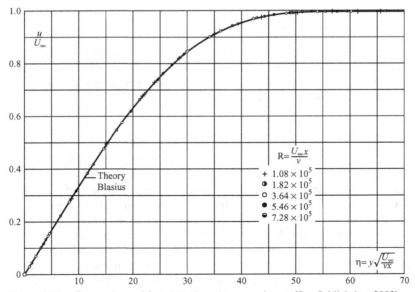

Figure 8.11b. Comparison of flat-plate theory to experiment (See Schlichting, 2003).

only by comparing analytical results to experimental measurements. This was done by many investigators, but a particularly careful set of measurements was carried out by Nikuradse in 1942 (See Thwaites, 1949). His measurements were made at five locations along the flat plate to study the effect of the Re number (in the range $1.1.10^5 < R_e < 7.3.10^5$) on the velocity profile. One item that he wanted to verify was the similarity of the velocity profiles. That is, the shape of the velocity distribution should be the same at any axial location. Figure 8.11b shows his measurements. The horizontal axis is the boundary-layer coordinate as we defined it in Eq. 8.46. The vertical axis in this plot is the value of the dimensionless x-velocity, $f' = u/U_e$. There can be no question that the agreement between the Blasius flat-plate theory and the experimental data is virtually perfect. This should help to build the student's confidence in the powerful mathematical methods introduced herein.

The z-Velocity Component in the Boundary Layer

Notice that although we have not required anything from the z-component of the momentum equation other than information regarding the pressure gradient normal to the surface, it does not mean that there is no velocity component in the z-direction. This is a misunderstood feature of boundary-layer theory. Because there is a momentum deficit near the surface (i.e., the flow speed in the parallel direction must decrease to zero to satisfy the no-slip requirement), the result is a velocity component normal to the surface. That it is required by continuity is seen easily because from the first of Eq. (8.37):

$$\frac{\partial w}{\partial z} = -\frac{\partial u}{\partial x} = -U_e \frac{\partial f'}{\partial x},$$

(8.53)

which indicates clearly that w cannot be zero. That is, the boundary-layer flow is not one-dimensional although the flow upstream of the flat plate is assumed to be one-dimensional. To determine w, we could integrate Eq. 8.54 after first transforming to the boundary-layer coordinate system (x, η). However, it is simpler to use the solution we already found for the stream function. Recall that by definition, the existence of the stream function guarantees that continuity is satisfied. Then, we can use the definition:

$$w = -\frac{\partial \psi}{\partial x}$$

to determine the normal-velocity component. From Eq. 8.47, we find:

$$w = -\frac{\partial}{\partial x}\sqrt{\nu U_e x}\, f + \frac{\eta}{2x}\sqrt{\nu U_e x}\frac{\partial f}{\partial \eta} = -\frac{1}{2}\sqrt{\frac{\nu U_e}{x}}\left(f - \eta\frac{\partial f}{\partial \eta}\right),$$

(8.54)

which shows that the normal velocity is zero at the surface (as is required by the solid-wall-boundary condition) and increases to a maximum as the edge of the boundary layer is approached. This result also shows that the normal velocity is very small because it is proportional to the square root of the product of the freestream velocity and the viscosity coefficient.

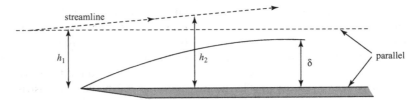

Figure 8.12. Upward deflection of streamlines due to the boundary layer.

Figure 8.12 shows the behavior of the flow near the leading edge of the plate and indicates the upward deflection of the streamlines resulting from the formation of the boundary layer.

The magnitude of the z-velocity decreases as distance along the plate increases; that is, the first term in Eq. 8.54 decreases as the inverse of the square root of x whereas the second term decreases faster because it is inversely dependent on x. Far downstream, there is no normal velocity. Equation 8.54 also demonstrates a fault with the theoretical results. Notice that at the leading edge of the plate, the simple theory predicts that the normal velocity must be infinite because $x = 0$ at that point. This is a result of failure of our assumptions regarding relative sizes of the various quantities in proximity to the leading edge. That is, the simplified equations do not hold there. This defect can be corrected by application of powerful singular-perturbation methods that are based on the simple analysis we used. It is beyond the scope of this book to discuss these so-called leading-edge corrections. It also is possible to resort to a complete two-dimensional, CFD numerical solution in which derivatives in both the x- and z-directions are represented with finite difference formulas. If care is taken to use a sufficiently fine grid near the leading edge, the apparent mathematical difficulty (usually referred to as a *singularity*) can be resolved. The CFD solution method is introduced in the next subsection. However, it displays the same defect at the leading edge that we just discovered because it attempts to solve the simple Blasius equation rather than the complete two-dimensional, Navier–Stokes equations. In other words, the defect is in the simplified differential equation rather than in the method of solution.

Numerical Solution of the Blasius Equation by Finite Differences

Because computational methods are emphasized throughout this textbook, we now describe an alternate numerical approach to the flat-plate boundary-layer solution that introduces CFD methods. Of course, we cannot undertake an in-depth discussion of all of the subtleties of these methods. However, the present problem represents an excellent opportunity to introduce a powerful procedure for obtaining solutions to otherwise intractable boundary-value problems.

We introduce the subject with no pretense of completeness and let the interested student seek further. Here, we primarily are interested in applying an approximate solution procedure to a nonlinear, ordinary differential equation and using these results in our boundary-layer analysis. Thus, the subject is developed as it applies to the Blasius equation given by:

$$2f''' + f f'' = 0 \tag{8.55}$$

with boundary conditions:

$$f(0) = f'(0) = 0, \ f'(\infty) = 1$$

as before. For simplicity we introduce:

$$g = f'$$

so that the Blasius equation becomes:

$$2g'' + f \, g' = 0 \tag{8.56}$$

with boundary conditions given as:

$$f(0) = g(0) = 0, g(\infty) = 1. \tag{8.57}$$

In what follows, we attempt to represent the derivatives by their finite-difference equivalents. Only the derivative normal to the surface must be accounted for in this simple problem. In effect, the integration in the direction of the flat plate already was accomplished. That is, we are dealing only with an ordinary differential equation rather than the partial-differential equation that must be confronted in a more general problem. The point to remember is that the same approach could be applied in the case of more independent variables, which we do in later applications.

Note that Eq. 8.56 is linear in g. However, it is still a nonlinear equation and we set up an iterative solution procedure at selected points in the domain. Here, the domain is the boundary layer. Thus, we divide the domain into a set of discrete points and number them from 1 to N (with N being the number of points that we are using). We choose to place the first point on the wall and the last point at the edge of the boundary layer. The next step is to create an approximate solution method whereby we find the values of the variables at these points. We do so by using these values and Taylor-series expansions to replace the derivatives in the equation with finite differences.

We note from the general Taylor series for a function $f(x)$ evaluated at a nearby point, $x + \Delta x$, that:

$$f(x \pm \Delta x) = f'(x) + f'(x)\Delta x + f''(x)\frac{\Delta x^2}{2!} \pm f'''(x)\frac{\Delta x^3}{3!} + f''''(x)\frac{\Delta x^4}{4!} \pm \ldots \tag{8.58}$$

Thus, in terms of our independent variable, η, we have:

$$f(\eta \pm \Delta\eta) = f'(\eta) \pm f''(\eta + \Delta\eta) + f''\frac{\Delta\eta^2}{2!} \pm f'''(\eta)\frac{\Delta\eta^3}{3!} + f''''(\eta)\frac{\Delta\eta^4}{4!} \pm \ldots \tag{8.59}$$

Now, with $\Delta\eta$ being the spacing between points in the domain, we can see that:

if $\qquad\qquad\qquad\qquad f_i = f(\eta),$

then $\qquad\qquad\qquad\qquad f_{i+1} = f(\eta + \Delta\eta), \tag{8.60}$

and $\qquad\qquad\qquad\qquad f_{i-1} = f(\eta - \Delta\eta).$

By adding or subtracting Eqs. 8.58 and 8.59, it is now possible to find the expressions:

$$f'(\eta) = \frac{f_{i+1} - f_{i-1}}{2\Delta\eta} - \frac{1}{2}f'''(\eta)\frac{\Delta\eta^2}{3!} + \theta(\Delta\eta^3) \tag{8.61}$$

$$f''(\eta) = \frac{f_{i+1} - 2f_i + f_{i-1}}{\Delta\eta^2} + f''''(\eta)\frac{\Delta\eta^2}{4!} + \Theta(\Delta\eta^4). \qquad (8.62)$$

We note that the second and remaining terms on the right-hand side of Eqs. 8.61 and 8.62 are of much smaller size (small $\Delta\eta$ and derivatives not too large) than the first and we choose to drop them. This gives an algebraic expression for the first and second derivatives.

Next, we apply Eqs. 8.61 and 8.62 to Eq. 8.56 for $g(h)$. This results in:

$$\frac{g_{i+1} - 2g_i + g_{i-1}}{\Delta\eta^2} + f_i\frac{g_{i+1} - g_{i-1}}{2\Delta\eta} = 0, \qquad (8.63)$$

which is a second-order approximation to Eq. 8.56. Eq. 8.63 can be rewritten by collecting the coefficients of the different g's to obtain:

$$\left(\frac{1}{\Delta\eta^2} - \frac{f_i}{2\Delta\eta}\right)g_{i-1} + \left(\frac{-2}{\Delta\eta^2}\right)g_i + \left(\frac{1}{\Delta\eta^2} + \frac{f_i}{2\Delta\eta}\right)g_{i+1} = 0,$$

or $$c_i g_{i=1} + b_i g_i + a_i g_{i+1} = 0, \qquad (8.64)$$

where:

$$c_i = \left(\frac{1}{\Delta\eta^2} - \frac{f_i}{2\Delta\eta}\right) \qquad b_i = \left(\frac{-2}{\Delta\eta^2}\right) \qquad a_i = \left(\frac{1}{\Delta\eta^2} + \frac{f_i}{2\Delta\eta}\right).$$

The subscript i now refers to the fact that Eq. 8.64 represents the equation arising from applying finite differencing to the Blasius equation at the ith grid point. Also notice that the coefficients are functions of f_i and, hence, vary throughout the domain.

We note that Eq. 8.64 is applicable to every point in the domain from $i = 2$ to $i = N-1$ (we already know the values at $i = 1$ and $i = N$; these are the boundary conditions). Thus, there are N-2 equations for the N-2 unknown values of g. (We address the unknown value of f_i after setting up a procedure for g.) Applying Eq. 8.64 at each point between $i = 2$ and $i = N-1$ yields a linear system of the following form:

$$\begin{bmatrix} b_2 & a_2 & & & & \\ c_3 & b_3 & a_3 & & & \\ & c_4 & b_4 & a_4 & & \\ & & & c_{N-2} & b_{N-2} & a_{N-2} \\ & & & & c_{N-1} & b_{N-1} \end{bmatrix} \begin{bmatrix} g_2 \\ g_3 \\ g_4 \\ g_{N-2} \\ g_{N-1} \end{bmatrix} = \begin{bmatrix} -c_2 g_1 \\ 0 \\ 0 \\ 0 \\ -a_{N-1} g_N \end{bmatrix}. \qquad (8.65)$$

This shows how to calculate the N-2 values of g. We would be finished were it not for the fact that we have not yet found the f_i's. To do this, we recall that f and g are related through Eq. 8.57 and we can find the values of f_i once we have found g_i by a simple numerical quadrature. However, we still need values for f to obtain the values of g from Eq. 8.65. Thus, we make an initial guess for the g's and compute the f's from this guess. Once we have found new values of g, we then can recompute f. This iterative process repeats until the solution no longer changes.

When the solution no longer changes, we say that the process has *converged*. Generally, the way this is checked is to see whether the maximum change in one variable is no larger than a small number. We call this number the *convergence criterion*. In our process, we check for the maximum change in g to be no larger than some reasonably small number—say, $|\Delta g_i|_{max} < 10^{-6}$.

We now return to obtaining the values of f. Recall that $f' = g$; hence:

$$f(\eta) = \int_0^\eta g\,d\eta'. \tag{8.66}$$

Because we only know g at discrete points, we must approximate this integral. We do this using the trapezoidal rule. Thus, Eq. 8.66 becomes:

$$f_i = \sum_{j=2}^i \frac{1}{2}(g_j + g_{j-1})(\eta_j - \eta_{j-1}). \tag{8.67}$$

The summation starts at 2, recalling that f_1 is at the wall and has a value of zero.

Now, we can set up the entire numerical procedure as follows:

1. Select N and the height of the domain (maximum value of h).
2. Make an initial guess for the value of g at the points in the domain (linear between the boundary values suffices).
3. Find values of f_i in the domain from Eq. 8.67.
4. Solve Eq. 8.65 for new values of g_i in the domain.
5. Check whether the process converged.
6. If not converged, repeat from Step 3.
7. When converged, both g and f now are known at all points in the domain and the boundary-layer properties may be computed from them.

Program **BLASIUS**, a program on the website, implements this scheme. The student is invited to run this program now to explore the behavior of the boundary-layer numerical solution. The output is presented in graphical form showing the variation of f and its derivatives with η. The next section is a detailed description of the solution and its interpretation. In particular, it is important to determine the shearing stress and, hence, the drag force on the surface created by the viscous effects that lead to the presence of the boundary layer.

8.7 Results from the Solution of the Blasius Equation

Certain parameters describing the laminar boundary layer on a flat plate (i.e., zero pressure gradient) may be evaluated using the solution of the Blasius equation. The definitions of the parameters are valid for either a laminar or a turbulent boundary layer, with or without a pressure gradient. After the key parameters are defined, their values are expressed for a laminar boundary layer (i.e., Blasius profile) on a flat plate. Although the freestream flow is little disturbed by the presence of the flat plate, the velocity at the outer edge of the boundary layer is denoted by $U_e(x)$ rather than a constant value such as V_∞ implied in the flat-plate solution. This allows for later generalization. In particular, the former notation implies that the velocity outside the boundary layer is allowed to change in the x-direction as required by an imposed pressure gradient or as a result of curvature of the surface.

Boundary-Layer Thickness, δ

The boundary-layer thickness is defined previously (see Fig. 8.9) as the distance from the wall at which the velocity is "essentially" that of the external stream. The Blasius solution indicates that the boundary-layer velocity profile, $u(z)$, approaches the velocity at the edge of the boundary layer, U_e, asymptotically. The boundary-layer thickness, then, may be defined precisely (but arbitrarily) as the distance from the surface at which the velocity, u, is an agreed-on percentage of the external flow velocity. An often-used definition is $f' = u/U_e = 0.994$ as the edge of the boundary layer (i.e., the velocity, u, in the boundary layer is 99.4 percent of the external freestream velocity, U_e), as shown in Fig. 8.11. Then, from Table 8.1:

$$\eta \equiv z\sqrt{\frac{U_e}{\nu x}} \quad \rightarrow \quad 5.2 \equiv z_{BL}\sqrt{\frac{U_e}{\nu x}}$$

$$\delta = z_{BL} = 5.2\sqrt{\frac{\nu x}{U_e}} = \frac{5.2x}{\sqrt{Re_x}}. \tag{8.68}$$

In this equation:

$$Re_x \equiv \frac{U_e x}{\nu}$$

and the subscript x on the Re number signifies that it is to be based on a length x from the plate leading edge to the downstream station in question.

Eq. 8.68 provides an explicit expression for the magnitude of the boundary-layer thickness, but it is neither useful nor unique. For instance, if it were decided to define the magnitude of δ as the value of z where the velocity ratio u/U_e is 0.999 (i.e., u is within 0.1 percent of the external velocity), then Table 8.1 indicates that the numerical factor in Eq. 8.51 is now about 6.01; there is a larger constant in the expression for δ in Eq. (8.51). However, no matter which value of u/U_e that is chosen to define δ(x), it is always true that the boundary-layer thickness on a flat plate varies inversely as the square root of the Re number. Considering the additional linear dependence on x in the numerator, it can be seen that the thickness grows as the square root of the distance downstream from the leading edge.

Thus far, we have found the velocity profile for the flat-plate boundary layer and how the boundary layer grows. Now we define two other thickness properties of the boundary layer that have certain advantages over δ. In particular, they yield unique values for the thickness in any given situation.

Boundary-Layer Displacement Thickness, δ^*

We consider an external flow streamline well away from the edge of a boundary layer growing on a flat plate, as depicted in Fig. 8.12. Because the flow within the boundary layer is moving slower than the freestream due to the action of viscosity, continuity requires that $h_2 > h_1$ and the external streamline is displaced (slightly) from the horizontal. (The same effect was shown in Example 3.6 when considering the decelerated flow in the wake of an airfoil, with the control surface defined as a streamline.) The displacement thickness is a measure of how far the external flow streamlines are displaced away from the plate surface due to the presence of the boundary layer. The displacement thickness is defined by the expression:

$$\rho U_e \delta^* = \int_0^\infty \rho U_e\, dz - \int_0^\infty \rho u\, dz \qquad (8.69)$$
$$[1] \qquad [2] \qquad [3]$$

or, because the density is constant:

$$\delta^* = \int_0^\infty \left(1 - \frac{u}{U_e}\right) dz. \qquad (8.70)$$

In Eq. 8.69, the left side is set equal to the difference between the mass flux above the plate if there were no boundary layer, Term [2], and the mass flux above the plate with the boundary layer present, Term [3]. Thus, Term [1] represents the mass flux *deficit* (or *defect*) in the boundary layer caused by viscous action retarding the local-flow velocity.

To interpret the definition, Eq. 8.70, physically, we consider Fig. 8.13. On the left is a Blasius velocity profile. On the right is a step profile that has $u = 0$ up to a height $z = \delta^*$ and then an instantaneous change in velocity to the full external velocity U_e. Both profiles have the same mass-flow-rate (i.e., mass-flux) deficit. Thus, the mass-flux deficit due to the presence of the boundary layer, which causes the displacing of the flow streamlines outward, can be represented as the outward displacement of the solid surface by an amount δ^*. A new effective body shape results, shown in Fig. 8.14, consisting of the flat plate plus an additional curved solid shape $\delta^*(x)$.

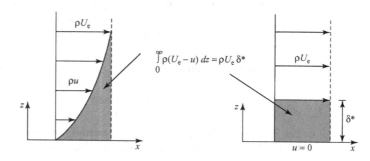

Figure 8.13. Two velocity profiles with the same mass-flux deficit.

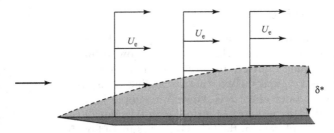

Figure 8.14. Equivalent body shape in a viscous flow.

This discussion may be generalized to consider an airfoil in a viscous flow operating below the stall. The pressure distribution on the airfoil is calculated by using an inviscid-flow model. Then, the boundary-layer equations are solved for the velocity profile within the boundary layer (by methods discussed later) and, hence, δ^* may be evaluated from Eq. 8.70, which is valid for laminar or turbulent boundary-layer flow. This displacement thickness then is added to the airfoil shape (as δ^* was added to the plate) and a first iteration is carried out by calculating the pressure distribution on this equivalent airfoil shape using potential flow methods. A new δ^* follows and the process is repeated until a satisfactory convergence is achieved. Because δ^* normally is small, only a few iterations are required.

The displacement thickness for a laminar boundary layer on a flat plate can be evaluated from the Blasius solution. In this case, Eq. 8.70 becomes:

$$\delta^* = \int_0^\infty \left(1 - \frac{u}{U_e}\right) dz = \int_0^\infty \left(1 - \frac{f'}{2}\right)\left(2\sqrt{\frac{vx}{U_e}}\, d\eta\right). \tag{8.71}$$

This integral can be evaluated numerically to yield:

$$\frac{\delta^*}{x} = \frac{1.7208}{\sqrt{R_{ex}}}, \tag{8.72}$$

where, again, x is the distance from the leading edge of the flat plate. This is a useful result in that it accounts for the presence of the boundary layer, and the expression is unique.

Boundary-Layer Momentum Thickness, θ

This parameter is defined in terms of the *momentum* deficit in the boundary layer rather than the *mass* deficit used in defining the displacement thickness. Thus,

$$\rho U_e^2 \theta \equiv \int_0^\infty \rho u U_e\, dz - \int_0^\infty \rho u^2\, dz, \tag{8.73}$$

or, for incompressible flow:

$$\theta = \int_0^\infty \frac{u}{U_e}\left(1 - \frac{u}{U_e}\right) dz. \tag{8.74}$$

In Eq. 8.73, Term [1] represents the momentum of flow within the boundary layer if the mass in the boundary layer has a velocity U_e, whereas Term [2] represents the momentum of the boundary-layer flow if the boundary-layer mass has a velocity u. Thus, by this definition, the momentum thickness, θ, represents a thickness of the freestream flow that has a momentum equal to the momentum deficit in the boundary layer. The quantity cannot be understood physically as easily as δ^*, but it is useful in later discussions.

Evaluating Eq. 8.74 from the Blasius boundary-layer solution yields a unique expression:

$$\frac{\theta}{x} = \frac{0.664}{\sqrt{R_{ex}}}. \tag{8.75}$$

The ratio of the displacement thickness to the momentum thickness of a boundary layer is called the *shape factor, H*; that is:

$$H \equiv \frac{\delta^*}{\theta} = \int_0^\infty \left(1 - \frac{u}{U_e}\right) dz \bigg/ \int_0^\infty \frac{u}{U_e}\left(1 - \frac{u}{U_e}\right) dz. \tag{8.76}$$

Shear Stress on the Wall, τ_w

An important feature of the formation of a boundary layer on a surface is that there is a *surface traction* (i.e., a force parallel to the surface) in addition to the normal force due to the pressure. This can be an important source of drag on the body. Recall from Eq. 8.10 that in the boundary layer, the shear stress at any point in the flow is proportional to the rate at which the parallel velocity changes normal to the wall, $\partial u/\partial z$. In particular, at a wall, the shear stress is:

$$\tau_w = \mu \frac{\partial u}{\partial z}\bigg|_{z=0}. \tag{8.77}$$

This shear stress is a viscous (friction) force per unit area acting in a tangential direction on the surface. Therefore, it is a *local boundary-layer property*, dependent on x. Differentiating the Blasius solution for $u(z)$, we find:

$$\tau_w = 0.332 \mu U_e \sqrt{\frac{U_e}{\nu x}}. \tag{8.78}$$

The average shear stress on the wall over a length interval $x = 0$ to $x = L$ then is given by:

$$\overline{\tau}_w = \frac{1}{L} \int_0^L \tau_w \, dx. \tag{8.79}$$

Notice that the factor $1/L$ is required in the definition so that if an average value is assumed for τ_w over the interval, then the definition is consistent. Applying Eq. 8.79 to a Blasius profile:

$$\overline{\tau}_w = (0.664) \frac{\rho U_e^2}{\sqrt{R_{eL}}}, \tag{8.80}$$

where the Re number is formed with the length L. This average wall-shear stress acts on the surface area of a flat plate of length L (per unit width).

Friction Coefficient, C_f

The friction coefficient is a *dimensionless* expression for the wall shear stress, defined as:

$$C_f \equiv \frac{\tau_w}{\frac{1}{2}\rho U_e^2} \tag{8.81}$$

so that, for the Blasius profile:

$$C_f = \frac{(0.664)}{\sqrt{R_{ex}}}. \tag{8.82}$$

The average friction coefficient for a surface over the length $x = 0$ to $x = L$ is given for the Blasius profile by:

$$\bar{C}_f = \frac{\bar{\tau}_w}{\frac{1}{2}\rho U_e^2} = \frac{1.328}{\sqrt{R_{eL}}}, \tag{8.83}$$

where the Re number is based on the characteristic length L.

Study the following example to see how these several definitions can be useful in the estimation of the viscous drag on a body.

EXAMPLE 8.1 *Given*: A thin-wing model at a small angle of attack (approximate by a flat plate) has a chord length of 2 feet and a span of 10 feet. The wing is tested in an air stream of 100 ft/s at standard conditions. Assume a Blasius solution for the wing boundary layer and assume two-dimensional flow over the entire wing span.

Required: Find the frictional drag force on the wing (both sides). Give the units of the answer.

Approach: Calculate the drag by first evaluating the average wall shear stress from Eq. 8.74 or Eq. 8.76.

Solution: Evaluating Eq. 8.76 using standard atmosphere values of ρ and ν:

$$\bar{\tau}_w = \frac{(0.664)(0.00238)(100)^2}{\sqrt{(2)(100)(0.00238)/(3.72\cdot10^{-7})}} = 0.093\frac{\text{lbf}}{\text{ft}^2}.$$

An estimate of the viscous drag of the plate is then:

$$D = \bar{\tau}_w(L)(\text{width})\times(2 \text{ sides}) = 40\bar{\tau}_w = 3.72\,\text{lb}_f.$$

Appraisal: In calculating the drag, the average wall-shear stress must be used because the local stress is a "point" property. Check the units of the answer. Also, note that drag force is viscous stress times area. Accounting for the two sides of the wing is equivalent to stating that this is the frictional drag of the wing "wetted" area—that is, the total wing area exposed to the flow.

The experimental value of the local-friction coefficient for a flat-plate laminar boundary is about 0.001 at a Re number of 500,000. This experimental result and the prediction of Eq. 8.82 agree to within 6 percent. The experimental measurement is difficult (Liepmann and Dhawan, 1951), involving the direct measurement of the shear force acting on a small section of the flat plate. When the turbulent boundary layer is discussed later in this chapter, it is found that the turbulent-velocity profile is fuller near the wall than the laminar profile; that is, near the wall, $u_{turb} > u_{lam}$. This implies that the wall shear stress and, hence, the frictional drag, is larger for a turbulent boundary layer than for a laminar boundary layer.

8.8 Boundary Layer with a Streamwise Pressure Gradient

The presence of a streamwise pressure gradient has a significant effect on the boundary-layer profile and growth. Thus, a flat-plate solution should not be used (except as an approximation) to describe the boundary layer on an airfoil or wing surface, where the results of Chapters 5 and 6 show significant variations in chordwise surface-pressure distribution, especially with increasing angle of attack.

The effect of a streamwise pressure gradient first is discussed from a qualitative physical viewpoint to fix ideas. Then, similar solutions are explained. These solutions are restricted to pressure gradients that have a particular dependence on streamwise distance; namely, x^m. The topic of streamwise pressure gradients concludes with a discussion of integral and numerical methods, both of which can be used for boundary-layer profiles in the presence of arbitrary streamwise pressure distributions.

Pressure Gradients and Boundary-Layer Profiles: A Qualitative Assessment

To discuss the effect of a streamwise pressure gradient on the behavior of a boundary layer, it is necessary to examine the x-momentum boundary-layer equation, Eq. 8.37, with the pressure-gradient term included; namely:

$$u\frac{\partial u}{\partial x} + w\frac{\partial w}{\partial z} = -\frac{\partial p}{\partial x} + v\frac{\partial^2 u}{\partial z^2}. \qquad (8.84)$$

Recalling that $\partial p/\partial z = 0$, $\partial p/\partial x$ previously could have been written as dp/dx (i.e., as the total derivative) because p only changes in the direction parallel to the surface.

For convenience, we consider a flat plate installed in a curved nozzle, shown in Fig. 8.15, as a simplified model of a flow along a curved airfoil surface that is experiencing a pressure gradient.

For an incompressible flow, continuity demands that as the nozzle area decreases, the freestream velocity must increase (see Fig. 8.14a). The Bernoulli Equation may be written for the freestream flow to relate velocity changes to static-pressure changes. Recalling that the freestream impresses a pressure distribution on the boundary

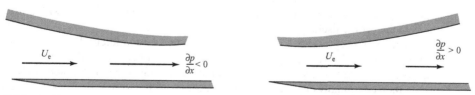

(a) favorable pressure gradient   (b) adverse pressure gradient

Figure 8.15. Flat-plate boundary layer with streamwise pressure gradient.

layer, the Bernoulli Equation may be thought of as written for a streamline only at the outer edge of the boundary layer, where the velocity is U_e. Thus,

$$p + \frac{1}{2}\rho U_e^2 = \text{constant.} \tag{8.85}$$

Differentiating in the x-direction along a streamline:

$$\frac{dp}{dx} + \rho U_e \frac{dU_e}{dx} = 0. \tag{8.86}$$

Thus, in discussing a boundary layer, we may speak interchangeably about a streamwise pressure gradient or an external-flow velocity gradient. In particular, the x-momentum equation, Eq. 8.83, may be written in an alternate form by replacing dp/dx with the expression from Eq. 8.86; namely:

$$u\frac{\partial u}{\partial x} + w\frac{\partial w}{\partial z} = U_e \frac{dU_e}{dx} + \nu \frac{\partial^2 u}{\partial z^2}. \tag{8.87}$$

We continue the discussion in terms of a pressure gradient because it is more meaningful physically. The velocity-gradient substitution is used as a convenience because the pressure gradient is usually known in terms of the velocity distribution in the external flow field.

Considering Fig. 8.15a, the boundary layer on this plate is experiencing a decrease in static pressure in the streamwise direction ($\partial p/\partial x < 0$). This is called a *favorable pressure gradient*. Conversely, the boundary layer in Fig. 8.15b is experiencing an increasing static-pressure field with downstream distance, or ($\partial p/\partial x > 0$). This is called an *adverse pressure gradient*.

Without solving anything, we determine which conclusions we can draw regarding the behavior of a boundary-layer velocity profile with a pressure gradient. First, we write Eq. 8.70 at the surface where, by virtue of the boundary conditions, $u = w = 0$. At the wall, then:

$$\mu\frac{\partial^2 u}{\partial z^2} = \mu\frac{\partial}{\partial z}\left(\frac{\partial u}{\partial z}\right) = \frac{dp}{dx}. \tag{8.88}$$

Consider three different streamwise pressure gradients, first making the right side of Eq. 8.84 zero, then less than zero, and, finally, greater than zero. Examine the qualitative behavior of the variation in the slope of the velocity profile, $\partial u/\partial z$, with distance, z, from the surface, as shown in Fig. 8.16. Recognize that for large z, the slope of the velocity profile must go to zero as the boundary-layer profile merges with the constant freestream velocity, U_e, at the edge of the boundary layer. Be careful: The graphs in Fig. 8.16 present the variation of the slope of the velocity profile with z, *not* the velocity profile itself.

Figure 8.16a represents a flat plate with zero pressure gradient. Eq. 8.74 states that at $z = 0$, the derivative of $\partial u/\partial z$ is zero. The second derivative of u being zero implies that the velocity profile $u(z)$ has an inflection point at $z = 0$. The slope of the profile then continually decreases and approaches zero for large z, that is, at large z, $u \approx$ constant and, hence, $\partial u/\partial z \approx 0$. This corresponds to the behavior of the Blasius boundary-layer (i.e., zero pressure gradient) profile, discussed previously.

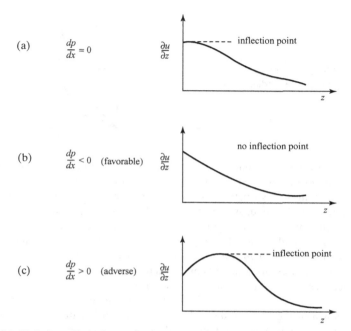

(a) $\dfrac{dp}{dx} = 0$ $\dfrac{\partial u}{\partial z}$ ----- inflection point

(b) $\dfrac{dp}{dx} < 0$ (favorable) $\dfrac{\partial u}{\partial z}$ no inflection point

(c) $\dfrac{dp}{dx} > 0$ (adverse) $\dfrac{\partial u}{\partial z}$ ------ inflection point

Figure 8.16. Variation of velocity-profile slope with pressure gradient.

Figure 8.16b corresponds to a favorable pressure gradient. Then, $\partial^2 u/\partial z^2 < 0$ at the wall, from Eq. 8.74, the slope of the velocity profile continually decreases with increasing z, and there is no inflection point in the profile.

Figure 8.16c corresponds to an adverse-pressure gradient. Here, the second derivative of the profile, $\partial^2 u/\partial z^2$, must be positive at the wall, Eq. 8.88 and also $\partial^2 u/\partial z^2$ must approach zero far from the wall, as discussed previously. However, this says that there must be an inflection point in the profile somewhere in between (i.e., at some z, $\partial^2 u/\partial z^2 = 0$) and there is a reversal of profile curvature there. The argument is too elementary to be able to specify the value of z at which this happens, but a change in curvature in the profile must occur. The character of the pressure gradient thus has a profound effect on the boundary-layer velocity profile.

We examine the effect of a streamwise pressure gradient from another physical viewpoint. Consider a fluid particle moving streamwise in the boundary layer. By virtue of its movement, it possesses a certain momentum. If the pressure acting on the back of the particle is greater than that on the front (a favorable pressure gradient), then the unbalanced force accelerates the particle and increases its momentum. This increase in momentum has a more pronounced effect near the wall, where the particle momentum initially is small. A favorable pressure gradient thus makes the boundary-layer velocity profile more full near the wall, as illustrated in Fig. 8.17. This fullness increases for a stronger favorable pressure gradient. There is no inflection point in the profile, as expected from the previous discussion. Be careful when thinking about the slope of a boundary-layer velocity profile, $\partial u/\partial z$! The student is used to thinking about slope expressions where u is the ordinate and z the abscissa, whereas boundary-layer convention plots z as ordinate and u as abscissa.

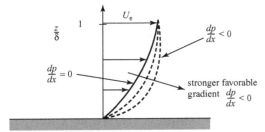

Figure 8.17. Velocity profile change
with favorable pressure gradient.

Now, we consider the same fluid particle in an adverse-pressure gradient. Here,
the particle is working against an unbalanced force directed upstream and the particle loses momentum. Again, the effect is most noticeable near the wall, where the
particle momentum is small, as shown in Fig. 8.18. Thus, the profile changes with an
inflection point appearing, as previously discussed. If the adverse-pressure gradient
becomes stronger, and as the adverse-pressure gradient persists, the velocity profile
increasingly distorts and $\partial u/\partial z$ near the wall becomes increasingly smaller.

If the adverse-pressure gradient continues to a certain point, then ultimately the
fluid particles very near the wall lose all of their momentum and a point is reached
where, at the wall, both $z = 0$ and $\partial u/\partial z = 0$. This is called the *incipient separation
point*, as shown in Fig. 8.19. A separation streamline originates at the separation
point and is a line of demarcation between the external flow and the "dead-air" flow
in the separation region. Thus, if there is separation on a surface, the oncoming flow
experiences a large apparent change in the geometry of the surface and it no longer
proceeds in a direction parallel to the surface, as in an inviscid flow. Within the separated region, there is a flow reversal and vortices are formed. Notice that within the
separated region, the point on the velocity profile where $u = 0$ must lie beneath the
separation streamline so that the reversed flow can circulate within the region.

Within the separated region, the velocity-component ratio w/u is no longer small
and the boundary-layer equations no longer hold. Likewise, $\partial p/\partial z$ is no longer negligibly small. Depending on the pressure field, the separation streamline may bend
back and reattach to the wall farther downstream, forming a separation bubble, or
it may trail downstream to form a wake. Recall the inviscid-pressure distribution on
the surface of a right-circular cylinder from Chapter 4. At 90° from the oncoming
stream, the local velocity is $2V_\infty$ and the pressure is a minimum. Beyond that point,
the surface pressure increases to a second stagnation point at 180°. However, in
a viscous flow, the boundary layer on a cylinder separates soon after 90° because

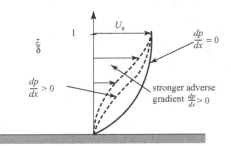

Figure 8.18. Velocity-profile change with
adverse-pressure gradient.

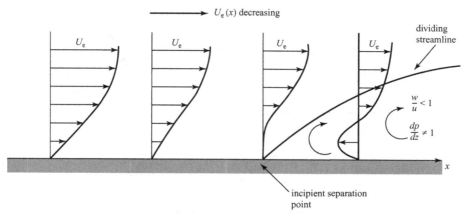

Figure 8.19. Separated boundary layer in an adverse-pressure gradient.

it experiences an adverse-pressure gradient and a wake forms downstream of the cylinder. The same general findings pertain to any bluff object.

It should be clear from this discussion that boundary-layer separation does not occur in the presence of a favorable pressure gradient. Separation is caused by an adverse pressure gradient. Furthermore, separation is to be avoided if possible because the dividing streamline greatly modifies the apparent shape of a carefully designed surface. That is, the intended pressure distribution is not achieved and the aerodynamic performance may be impaired. This usually means greatly increased drag and sometimes serious degradation of control and handling properties of a flight vehicle.

This physical description of boundary-layer behavior in the presence of an adverse pressure gradient may be applied to the performance of an airfoil.

We run Program **AIRFOIL** (in inviscid mode) and examine the chordwise pressure distributions on selected airfoil shapes at zero angle of attack. Typical shapes exhibit a suction peak near the leading edge and then the pressure coefficient decreases (i.e., surface static pressure increases) continuously toward the trailing edge, with a sharp increase in static pressure beginning a few percent chord ahead of the trailing edge as the pressure rises to atmospheric pressure. The separated flow observed near the trailing edge of a typical airfoil is a result of the presence of these adverse pressure gradients. Now, we select an airfoil and observe the inviscid-flow chordwise pressure distribution with increasing angle of attack. Notice how the magnitude of the adverse pressure gradient downstream of the suction peak increases as the angle of attack becomes larger. As a result of this increasing adverse pressure gradient, the boundary-layer separation point moves upstream and the lift-curve slope rapidly becomes nonlinear (see, for example, Fig. 5.3). Ultimately, the adverse pressure gradient becomes so large that the separation point on the airfoil moves rapidly upstream and the airfoil stalls.

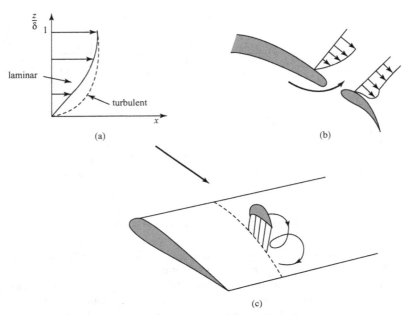

Figure 8.20. Methods for inducing turbulent boundary-layer flow.

It is clearly advantageous to delay or avoid separation. Consideration of the physical cause of separation in an adverse pressure gradient suggests that it is useful to make the boundary-layer profile "fuller," which allows more of a momentum loss near the surface before incipient separation occurs. There are several ways to accomplish this as follows:

1. *Make the boundary layer turbulent.* As mentioned previously and as shown later in this chapter, the turbulent boundary layer has a fuller profile than the laminar one, as shown in Fig. 8.20a. Thus, separation is less likely but at the cost of greater frictional drag. In low-speed applications (e.g., high-performance sailplane design) it is common to use *turbulators* to purposely introduce transition to turbulence. This avoids the formation of "laminar separation bubbles" that represent a serious loss of aerodynamic efficiency. The turbulators often take the form of "zig-zag" tape, which is bonded to the surface in areas where it is desired to induce turbulent transition (e.g., near the aileron or flap hinge line). The mechanism by which the tape "trips" the boundary layer is not well understood, but it is clear that the irregular protrusion into the airflow creates vortices or eddies—*vortical waves*—that trigger the required transition to turbulent flow downstream of the tape.
2. *Blow higher-energy air tangential to the surface prior to separation.* This creates a distorted profile, shown in Fig. 8.20b, with much higher momentum near the surface. A blown flap is illustrated where air from the higher-pressure underside of the flap flows through a slot and energizes the boundary layer on the upper, lower-pressure side. This is a form of *boundary-layer control*. Generally,

considerable design complexity and, hence, cost are introduced by using tangential blowing because it implies a source for the high-energy flow along with the attendant ducting, injection surfaces, and control systems.

3. *Use a vortex generator.* This generator is a miniature wing or vane projecting vertically from the upper surface of a wing and at an angle of attack to the oncoming flow, as shown in Fig. 8.20c. The mechanism here is similar to that described in Method 1; that is, the formation of vortices to stimulate the creation of a fully turbulent unsteady boundary layer. A vortex is continuously shed at the tip of the vortex generator for precisely the same reason that it forms at the tip of a three-dimensional wing. This vortex sweeps higher-momentum air downward from the outer edge of the boundary layer to "energize" the low-momentum region of the boundary layer near the surface. Rows of such generators are sometimes seen on the upper surfaces of wings. For example, they are a familiar feature of the wing surfaces on certain commercial jet aircraft, such as the Boeing 757. Their main purpose in such cases usually is to ensure that flow separation does not affect the behavior of the flaps or other aerodynamic controls. Any measure that aids in the prevention of large-scale separation is an advantage; however, as we will see, a turbulent boundary layer causes increased skin-friction drag.

The possibility of flow separation also is incorporated into the design of flow devices. For example, the diffuser of a subsonic wind tunnel, which slows the flow after it passes through the test section, usually has a slowly increasing cross section (making for a long length) that as small an adverse pressure gradient as possible is imposed on the diffuser flow.

It is obviously important to study in more detail the features of viscous-boundary-layer flows with pressure gradients. Qualitative discussions of a few experimental findings hardly can provide enough information. It is useful to first attempt to extend the analysis used for the flat plate (i.e., the zero-pressure-gradient problem) to situations with a pressure gradient. In a famous paper, Falkner and Skan showed that the similarity method can be used for a restricted class of freestream velocities of the following form:

$$U_e(x) = U_1 x^m,$$

where m can be 0, < 1, or >1 and U_1 is a constant velocity-scaling parameter. This special case is discussed briefly in the next subsection. An approximate integral treatment for the case of a more general pressure gradient follows.

Similar Boundary-Layer Solutions with Arbitrary Pressure Gradients

Following the same general approach used for the Blasius problem, we introduce a dimensionless stream function, Ψ, and similarity parameter, η, of the following form:

$$\psi(x,\eta) = \sqrt{\left(\frac{2}{m+1}\right) v U_1 x^{(m+1)/2}} f(\eta)$$

$$\eta = z\sqrt{\left(\frac{m+1}{2}\right)\frac{U_1}{vx}},$$

(8.89)

where the notation follows closely that used previously. If these expressions are used to rewrite the boundary-layer equation (Eq. 8.84), we find (as in the flat-plate problem) that the motion is now governed by an ordinary differential equation for the function $f(\eta)$. This is the Falkner–Skan equation:

$$f''' + ff'' + \frac{2m}{m+1}(1 - f'^2) = 0, \tag{8.90}$$

where primes denote differentiation with respect to the boundary-layer coordinate, η, and solutions are possible only for certain ranges of values of m as already suggested. Note that Eq. 8.89 reduces to the Blasius equation for $m = 0$. Because the equation is nonlinear, it is clear that we might encounter problems of convergence or numerical instability. In particular, it is required that the pressure-gradient parameter:

$$\beta = \frac{2m}{m+1}$$

must lie in a range for which there is no flow separation. It was found that flow separation occurs if $\beta < -0.1988388$, and numerical difficulties (e.g., failure to converge) are encountered in that pressure-gradient range.

The finite-difference numerical method introduced for solution of the Blasius equation can be used for the Falkner–Skan problem. Reducing the order by using $g = f'$, the equation can be rewritten as:

$$g'' + fg' + \beta(1 + g^2) = 0. \tag{8.91}$$

As before, the second-order central difference equations can be used to represent the derivatives, and the result is linear in the unknown value of g. The nonlinear term, g^2, can be converted to linear form by using a Taylor-series expansion. Expanding in the k_{th} iteration from the known (or initial) value of g to the unknown or $(k + 1)_{st}$ level, we write:

$$\left(g_i^{k+1}\right)^2 = 2g_i^k g_i^{k+1} - \left(g_i^k\right)^2.$$

Then, the finite-difference representation for the Falkner–Skan equation is:

$$\frac{g_{i+1}^{k+1} - 2g_i^{k+1} + g_{i-1}^{k+1}}{\Delta\eta^2} + f_i^k \frac{g_{i+1}^{k+1} - g_{i-1}^{k+1}}{\Delta\eta} + \frac{2m}{m+1}\left[1 - 2g_i^k g_i^{k+1} + \left(g_i^k\right)^2\right] = 0.$$

As before, this can be rearranged into a convenient tridiagonal system:

$$\left(\frac{1}{\Delta\eta^2} + \frac{f_i}{2\Delta\eta}\right)g_{i-1}^{k+1} + \left(\frac{-2}{\Delta\eta^2} - \frac{4m}{m+1}g_i^k\right)g_i^{k+1} + \left(\frac{1}{\Delta\eta^2} + \frac{f_i}{2\Delta\eta}\right)g_{i+1}^{k+1} = -\frac{2m}{m+1}\left(g_i^k\right)^2,$$

and the solution procedure closely follows that described for the Blasius problem. An initial guess for g_i is inserted and the corresponding f_i is found from trapezoidal-rule integration. The equation is solved as many times as required to reach a prescribed level of convergence.

> The Program **FS** provides the student experience in running such a solution method. The program determines the solution to the Falkner–Skan equation (Eq. 8.90) for any allowed value of β.

Numerical solutions for several values of the pressure-gradient parameter, β, are shown in Fig. 8.21. For negative values of β, an inflection point exists in the parallel velocity profile (u versus η), a point that was deduced previously using a qualitatitive argument. For the separation value (about β = –0.199), the inflection point is farthest away from the wall. Also, for the higher positive values of the pressure gradient, the boundary layer is thinner, as we might expect. The vertical-velocity profile (v versus η) also is shown in Fig. 8.21. Notice that the result is scaled to the parameter U_1 and the square root of the Re number based on the distance x from the leading edge of the surface. Also note the remarkable feature that for favorable pressure gradients (β > 0.1, approximately), the normal velocity component is *less* than zero; that is, the flow is toward the surface and the boundary layer is thinner. It was demonstrated that boundary-layer thickness decreases for values of β greater than 1. Because increased β corresponds to an increasingly favorable pressure gradient, we see that the effect is to make the boundary-layer thickness increasingly smaller.

Table 8.2 lists the values of displacement thickness, momentum thickness, wall-shear stress, and skin-friction coefficient as a function of β.

Viscous Flow on Wedges

The pressure gradients that allow the Falkner–Skan equation to be an ordinary differential equation, given by:

$$U_1 = cx^m \quad \text{or} \quad p_1 = p_0 - \frac{1}{2}\rho U_1^2 = p_0 - \frac{1}{2}\rho(cx^m)^2 \tag{8.92}$$

correspond to the potential flow over wedges, as shown in Fig. 8.22.

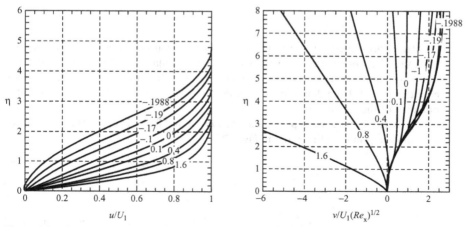

Figure 8.21. Solution of the Falkner–Skan equation for various values of the pressure-gradient parameter, β.

Table 8.2. *Properties of the numerical solutions for the Falkner–Skan equation*

β	A(β)	B(β)	f″(β,0)	η₈(β)
-0.1988383	3.497798	0.868110	0.000000	5.0
-0.198800	2.332770	0.585389	0.005260	5.0
-0.190000	2.006747	0.576523	0.085702	4.7
-0.170000	1.778859	0.559664	0.162116	4.4
-0.140000	1.595897	0.538560	0.239737	4.2
-0.100000	1.442694	0.515044	0.319270	4.0
-0.050000	1.312361	0.490464	0.400323	3.9
0.000000	1.216779	0.469600	0.469600	3.7
0.100000	1.080320	0.435458	0.587035	3.5
0.200000	0.984159	0.408231	0.686708	3.4
0.300000	0.910995	0.385737	0.774755	3.2
0.400000	0.852636	0.366693	0.854422	3.1
0.500000	0.804550	0.350272	0.927681	3.0
0.600000	0.763974	0.335910	0.995837	2.9
0.800000	0.698683	0.311849	1.120269	2.8
1.000000	0.647904	0.292346	1.232589	2.6
1.200000	0.606902	0.276114	1.335723	2.5
1.400000	0.572872	0.262324	1.431587	2.4
1.600000	0.572872	0.262324	1.431587	2.3

$$\delta(x) = \eta_\delta(\beta)\sqrt{\frac{2}{1+m}\frac{vx}{U_1}}, \delta^*(x) = A(\beta)\sqrt{\frac{2}{1+m}\frac{vx}{U_1}}, \theta(x) = B(\beta)\sqrt{\frac{2}{1+m}\frac{vx}{U_1}}$$

$$\tau_w = f''(\beta,0)\mu U_1\sqrt{\frac{1+m}{2}\frac{U_1}{vx}}, C_f = 2f'(\beta,0)\sqrt{\frac{1+m}{2}\frac{v}{U_1 x}}$$

There are several interesting special cases for various choices of the β value. For example, if $m = 0$ and $\beta = 0$, the Falkner–Skan equation collapses to the Blasius flat-plate equation. The factors of 2 that appear to be different between the two equations result from definitions of some of the parameters but do not change the outcome of the calculation.

A useful special case is produced when $m = 1$ and $\beta = 1$. This yields the flow near a stagnation point, as illustrated in Fig. 8.23.

$\theta = \dfrac{2m}{m+1}\pi$ Figure 8.22. Wedge flow for the Falkner–Skan equation.

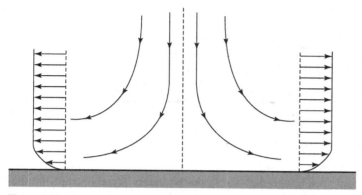

Figure 8.23. Flow near a stagnation point using the Falkner–Skan solution.

The external flow is expressed as:

$$U_1 = u_1 x,$$

where for the case of the stagnation-point flow over a circular cylinder of radius R, the constant, U_1, is defined by:

$$u_1 \equiv 2 \frac{U_\infty}{R}.$$

This is an important result because it can be helpful in starting numerical solutions in certain CFD applications, because the boundary-layer equations lose their validity near the stagnation point.

Momentum Integral Method for Arbitrary Pressure Gradients

A powerful approximate method obtained by applying the momentum theorem to a boundary-layer flow is discussed next. This method of solution does not yield the boundary-layer-velocity profile directly, as did the Blasius and Falkner–Skan solutions. Rather, a velocity-profile function is assumed that contains a dimensionless parameter incorporating an arbitrary pressure gradient. Substituting this assumed profile into the momentum integral produces expressions for $\delta(x)$, $\delta^*(x)$, $\theta(x)$, and $C_f(x)$ in the presence of a specified but arbitrary pressure gradient, expressed in the form of the external flow, $U_e(x)$. The resulting velocity profiles are no longer similar because the mathematical requirements that led to that special condition are no longer satisfied. Numerical methods for solving the momentum-integral equation are described after the equation is derived.

Momentum-Integral Equation

The momentum-integral relationship (attributed to von Kàrmàn) may be derived by either combining the continuity and momentum-differential equations for the

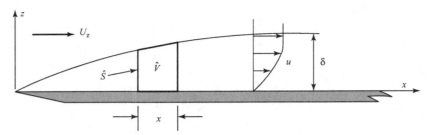

Figure 8.24. Control volume in boundary layer.

boundary layer and then integrating across the boundary layer, or applying the momentum theorem to a fixed control volume within the boundary layer. The latter approach is used here because it represents another application of the Conservation of Momentum Principle discussed in Chapter 3. Steady flow is assumed, and the resulting equation is applicable to both laminar and turbulent boundary layers.

Consider a boundary layer growing on a flat plate in the presence of a streamwise pressure gradient, $U_e(x)$, as illustrated in Fig. 8.24.

A fixed-control volume, $\hat{V}$, is chosen that extends from the wall to the edge of the boundary layer and is Δx in length. It is assumed that the boundary-layer approximations hold so that $w/u \ll 1$ (i.e., the boundary-layer flow is normal to the vertical sides of $\hat{V}$) and $\partial p/\partial z = 0$. Also, the shear stress along the top surface of $\hat{V}$ is negligible compared to the pressure stress there, and the normal viscous stresses on the vertical sides of $\hat{V}$ are ignored compared to the pressure stress. Then, from Eq. 3.1, the momentum theorem in the x-direction applied to $\hat{V}$ may be written as follows:

$$\iint_{\hat{S}} \rho u(v \cdot n) d\hat{S} = -\iint_{\hat{S}} p(n \cdot i) d\hat{S} + \iint_{\hat{S}} (\tau \cdot i) d\hat{S} \qquad (8.93)$$

$$[1] \qquad\qquad [2] \qquad\qquad [3]$$

Each term is detailed in turn, as follows:

[1] The net-momentum-flux expression consists of the following three terms, as shown in Fig. 8.25a:

(a) Inflow through the left face, $\int_0^\delta \rho u(-u) dz$.

(b) Outflow through the right face, $\int_0^\delta \rho u(+u) dz + \dfrac{d}{dx}\left[\int_0^\delta \rho u(u) dz\right]\Delta x$.

(c) Momentum entering the slant upper surface of $\hat{V}$, namely, $(-U_e \dot{m})$.
Recall that the edge of the boundary layer is not a streamline. The mass flux $\dot{m}$ entering through the upper surface is the difference between the mass flux out through the right face of $\hat{V}$ and the mass flux in through the left face of $\hat{V}$, or:

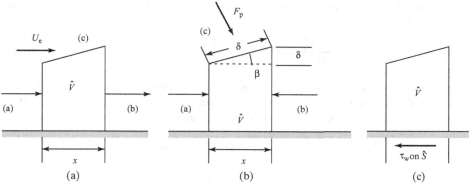

Figure 8.25. Control-volume details.

$$-U_e \dot{m} = -U_e \left\{ \left[\int_0^\delta \rho u\, dz + \frac{d}{dx} \left(\int_0^\delta \rho u\, dz \right) \Delta x \right] - \left[\int_0^\delta \rho u\, dz \right] \right\}.$$

[2] The net pressure force on the control surface consists of the following three terms, as shown in Fig. 8.25b:

(a) The pressure force on the left face, $(p\delta)$.

(b) The pressure force on the right face, $-\left[p\delta + \dfrac{d}{dx}(p\delta)\Delta x \right]$.

(c) The horizontal component of F_p; namely, $F_H = F_p \sin\beta = F_p(\Delta\delta/\Delta s)$, where F_p is evaluated at the midpoint of the interval Δx as:

$$F_p = (\text{normal pressure})(\text{area}) = \left[p + \frac{dp}{dx}\left(\frac{\Delta x}{2} \right) \right](s)(1),$$

where the area is $(s)(1)$ because the flow is two-dimensional. Thus,

$$F_H = \left[p + \frac{dp}{dx}\left(\frac{\Delta x}{2} \right) \right](\Delta\delta).$$

[3] The net shear stress acting on the surface of the control volume is shown in Fig. 8.25c. The shear stress on the upper surface is negligible, whereas the shear stress on the two vertical faces of the control volume is at right angles to the x-direction. The only shear-stress term of interest is the shear force acting on the lower surface of the control volume. The fluid pulls the plate to the right (i.e., drag); hence, the force on the control surface due to the plate is directed to the left. Thus, the shear force is $-\tau_w(x)(1)$.

Now, we substitute all of these seven terms into Eq. 8.93. Then, we divide through by (Δx) and take the limit as $\Delta x \to \delta x$, which, in [2](c) makes $\Delta\delta/\Delta x \to d\delta/dx$.

Expanding, simplifying, and neglecting one higher-order term in [2](c) results in the following equation:

$$\frac{d}{dx}\left[\int_0^\delta \rho u^2 dz\right] - U_e \frac{d}{dx}\left[\int_0^\delta \rho u dz\right] = -\delta \frac{dp}{dx} - \tau_w. \tag{8.94}$$

This is the Kàrmàn momentum integral for steady flow. The student should fill in the details of the derivation of Eq. 8.94. The upper limit in this equation is δ rather than ∞ because an assumed velocity profile can be expressed in terms of z/δ so that at $z = \delta, u = U_e$.

Eq. 8.72 can be written in an equivalent form, as follows:

$$\frac{d}{dx}\left[\int_0^\delta \rho u^2 dz\right] - \left\{\frac{d}{dx}\left[U_e \int_0^\delta \rho u dz\right] + \frac{dU_e}{x}\left[\int_0^\delta \rho u dz\right]\right\} = \delta \rho U_e \frac{dU_e}{dx} - \tau_w. \tag{8.95}$$

Collecting and multiplying through by a minus sign, Eq. 8.95 becomes:

$$\frac{d}{dx}\left[\int_0^\delta \rho u(U_e - u)dz\right] + \frac{dU_e}{x}\left[\int_0^\delta \rho(U_e - u)dz\right] = \tau_w,$$

or:

$$\frac{d}{dx}\left[\rho U_e^2 \theta\right] + \frac{dU_e}{dx}\left[\rho U_e \delta^*\right] = \tau_w. \tag{8.96}$$

Finally, expanding the first term of Eq. 8.74 by the chain rule and then dividing by U^2_e:

$$\frac{d\theta}{dx} + (2 + H)\frac{\theta}{U_e}\frac{dU_e}{dx} = \frac{C_f}{2}, \tag{8.97}$$

where $H = \delta^*/\theta$, the shape factor defined previously. This is the most commonly used form of the Kàrmàn momentum integral for steady, incompressible, boundary-layer flow.

Solving the Momentum-Integral Equation

A powerful method for solving the boundary-layer equations involves representing the velocity profile (i.e., the parallel part of the velocity in terms of the coordinate normal to the undisturbed flow direction) by an approximate analytical expression and integrating the equations in the direction normal to the flow. Integrating the equations removes the normal independent variable, leaving an ordinary differential equation in the streamwise direction, as demonstrated in the previous subsection.

The expression for the velocity profile is represented by:

$$\frac{u}{V_\infty} = f\left(\frac{z}{\delta}\right) = f(\eta), \tag{8.98}$$

where f is any function of the dimensionless normal coordinate that satisfies the necessary boundary conditions. For example, it is likely that f must go to zero at the surface to satisfy the no-slip boundary condition.

We now rewrite Eq. 8.96 to put it in the form most convenient for calculations. The wall shear stress and boundary-layer thickness are related to the momentum thickness by:

$$\tau_w = \mu \frac{\partial u}{\partial z}\bigg|_{z=0} = \frac{\mu U_e}{\delta} f'(0), \tag{8.99}$$

where the primes denote derivatives relative to the dimensionless, normal boundary-layer variable ($\eta \equiv z/\delta$). It also is useful to define a second shape factor based on the boundary-layer thickness, δ:

$$H' = \frac{\delta}{\theta} \tag{8.100}$$

Hence, the momentum-integral relationship, Eq. 8.96, can be written as:

$$\frac{H\theta}{2} \frac{dU_e^2}{dx} + \frac{d}{dx}(U_e^2 \theta) = \frac{\nu U_e}{H'\theta} f'(0). \tag{8.101}$$

The second term on the left side can be expanded and recombined with the first term to give:

$$\frac{d\theta^2}{dx} + (2+H)\frac{1}{U_e^2}\frac{dU_e^2}{dx}\theta^2 = \frac{2\nu U_e}{H'} f'(0), \tag{8.102}$$

where the equation was multiplied through by 2θ.

For simple problems such as flow over a flat plate and flows with favorable or only weakly unfavorable pressure gradients, the shape factor, H, can be approximated as a constant. In such cases, Eq. 8.101 has an integrating factor:

$$(U_e^2)^{2+H}.$$

The momentum integral then becomes:

$$\frac{d}{dx}(U_e^{2+H}\theta^2) = \frac{2\nu f'(0)}{H'} U_e^{1+H}. \tag{8.103}$$

Integration yields:

$$\theta^2 = \frac{2\nu f'(0)}{H'U_e^{2+H}} \int_0^x U_e^{1+H} dx. \tag{8.104}$$

This simple expression can be evaluated for the momentum thickness, and the shape factor, H, then can be used to determine the displacement thickness.

Flat-Plate Solution with a Polynomial Velocity Profile

For an example, we consider the flow over a flat plate and assume that the velocity profile is given by a third-order polynomial in η:

$$\frac{u}{V_\infty} = a\eta^3 + b\eta^2 + c\eta + d. \tag{8.105}$$

The constants in the equation can be evaluated using the boundary conditions that:

$$
\begin{cases}
\dfrac{u}{V_\infty} = 0 & \text{at} \quad \eta = 0 \\[2mm]
\dfrac{u}{V_\infty} = 1 & \text{at} \quad \eta = 1.
\end{cases}
\tag{8.106}
$$

These lead to the requirements that $d = 0$ and $a + b + c = 1$. For a flate plate, the external pressure gradient is zero, so this provides no additional information. To complete the evaluation of the coefficients, we require that derivatives of the velocity profile at the freestream condition go to zero. This is consistent with an asymptotic approach to the freestream. Thus, we write:

$$
\begin{cases}
\dfrac{d}{d\eta} \dfrac{u}{V_\infty} = 0 & \text{at} \quad \eta = 0 \\[2mm]
\dfrac{d^2}{d\eta^2} \dfrac{u}{V_\infty} = 1 & \text{at} \quad \eta = 1,
\end{cases}
\tag{8.107}
$$

leading to the requirement that $a = 1$, $b = -3$, and $c = 3$. Therefore, a physically consistent polynomial for the flat-plate boundary-layer profile is:

$$
\frac{u}{V_\infty} = \eta^3 - 3\eta^2 + 3\eta.
$$

The displacement thickness and momentum thicknesses divided by the boundary-layer thickness now can be found from their definitions:

$$
\frac{\delta^*}{\delta} = \int_0^1 \left(1 - \frac{u}{V_\infty}\right) d\eta = \int_0^1 (1 - \eta^3 + 3\eta^2 - 3\eta)\, d\eta = 0.25
$$

$$
\frac{\theta}{\delta} = \int_0^1 \frac{u}{V_\infty}\left(1 - \frac{u}{V_\infty}\right) d\eta = \int_0^1 (\eta^3 - 3\eta^2 + 3\eta)(1 - \eta^3 - 3\eta^2 + 3\eta)\, d\eta = 0.10714.
$$

Hence, the two shape factors become:

$$
\begin{cases}
H = \dfrac{\delta^*}{\theta} = \dfrac{\delta^*/\delta}{\theta/\delta} = 2.333 \\[3mm]
H' = \dfrac{\delta}{\theta} = 9.334.
\end{cases}
$$

The skin-friction coefficient also is given by:

$$
C_f = \frac{\tau_w}{\frac{1}{2}\rho V_\infty^2} = \frac{\mu \left.\dfrac{\partial u}{\partial z}\right|_0}{\frac{1}{2}\rho V_\infty^2} = \frac{\mu V_\infty f'(0)}{\frac{1}{2}\rho V_\infty^2 \delta} = \frac{6\nu}{V_\infty \delta}.
$$

To complete the calculation, the momentum-integral relationship, Eq. 8.103, is used to find the momentum thickness, θ. The momentum-integral relationship, with a zero pressure gradient, becomes:

$$\frac{\theta^2}{\nu} = \frac{2f'(0)x}{9.334V_\infty} = 0.6428\frac{x}{V_\infty},$$

or:

$$\theta = \frac{\theta^2}{\nu} = \frac{2f'(0)x}{9.334V_\infty} = 0.8017\sqrt{\frac{\nu x}{V_\infty}}$$

and, finally:

$$\begin{cases} \delta^* = 1.8704\sqrt{\dfrac{\nu x}{V_\infty}} \\[2mm] \delta = 7.4816\sqrt{\dfrac{\nu x}{V_\infty}} \\[2mm] C_f = \dfrac{0.802}{\sqrt{R_{ex}}}. \end{cases}$$

These compare fairly well with the exact Blasius result, as summarized in Section 8.8. The discrepancies arise from the somewhat crude representation we used for the boundary-layer-velocity profile and the assumption-of-constant-shape factor.

For flows with pressure gradients, Pohlhausen (Pohlhausen, 1921) proposed the fourth-order polynomial:

$$\frac{u}{V_\infty} = 2\eta - 2\eta^3 + \eta^4 + \frac{\Lambda}{6}\left(\eta(1-\eta)^3\right),$$

where Λ is a pressure-gradient parameter defined as:

$$\Lambda = \frac{\delta^2}{\nu}\frac{dU_e}{dx}.$$

The pressure-gradient parameter arises from applying the differential-momentum equation at the wall so that:

$$\frac{dp}{dx} - \nu\frac{\partial^2 u}{\partial z^2} = 0.$$

The method is described fully in Pohlhausen's original paper and also in Schlichting, 2003. This polynomial provides reasonably accurate results for favorable pressure gradients but yields disappointing accuracy for adverse pressure gradients.

Thwaites Method

Unfortunately, the shape factors are not usually constant as assumed in the previous subsection, and the single-parameter approach to computing the boundary layer for more severe pressure gradients cannot be accurately accomplished this way. Another approach is to use empirical correlations for some of the integrated boundary-layer quantities derived from experimental observations. We discuss one of the more popular of these in this subsection and leave a more thorough treatment for textbooks devoted more exclusively to the topic of boundary-layer theory.

The momentum-integral relationship can be written in terms of both the shape factor, H, and a shear factor, S. These are defined as follows:

$$H = \frac{\delta^*}{\theta} \qquad S = \frac{\tau_w \theta}{\mu U_e}. \tag{8.108}$$

When placed into the momentum-integral relationship we obtain:

$$\frac{d}{dx}\left(\frac{\theta^2}{\nu}\right) + \frac{2}{U_e}\left[(2+H)\frac{dU_e}{dx}\left(\frac{\theta^2}{\nu}\right) - S\right] = 0. \tag{8.109}$$

Unfortunately, the previous approach assumed that the shear and shape factors were constant, which eliminates many potentially interesting flows from consideration. Instead, it is found that the factors H and S very nearly depend only on the quantity:

$$\lambda = \frac{\theta^2}{\nu}\frac{dU_e}{dx}. \tag{8.110}$$

This parameter has a maximum value of 3.55, which corresponds to the point separation. Separated flows and the problem of turbulent transition are discussed later in this chapter. Using these expressions, the momentum-integral relationship reduces to:

$$U_e\frac{d}{dx}\left(\frac{\lambda}{dU_e/dx}\right) \approx 2S - 2\lambda(H+2) = F(\lambda). \tag{8.111}$$

By analyzing a large volume of experimental and analytical results for laminar boundary layers in terms of these parameters, Thwaites proposed a simple linear relationship for $F(\lambda)$ given by:

$$F(\lambda) = 0.45 - 6.0\lambda. \tag{8.112}$$

This allows an immediate integration of the integral-momentum relationship as follows:

$$\frac{\theta^2(x)}{\nu} = \frac{0.45}{U_e^6(x)}\int_{x_0}^{x} U_e(x')dx' + \frac{\theta^2(x_0)}{\nu}\frac{U_e^6(x_0)}{U_e^6(x)}. \tag{8.113}$$

Note that once θ^2/ν is found, the value of λ can be computed. The shear and shape factors then can be computed from a table of H and S values as a function of λ. Analytic curve fits of these data are shown in (White, 1974), as follows:

$$\begin{cases} H(\lambda) = 2.0 + 4.14z - 83.5z^2 + 854z^3 + 3{,}337z^4 + 4576z^5 \\ S(\lambda) = (\lambda + 0.09)^{0.62}, \end{cases} \tag{8.114}$$

where the variable z in the H equation is defined as:

$$z = 0.25 - \lambda.$$

These then can be used to determine the displacement thickness and the wall shear stress. A few examples demonstrate this simple but powerful method.

Constant-Speed Freestream over a Flat Plate

Here, $U_e(x) = U_0$, a constant, and the flow begins at the leading edge of the flat plate. In this case, the Thwaites integral becomes:

$$\frac{\theta^2}{\nu} = \frac{0.45x}{U_0},$$

which compares favorably with the exact (i.e., Blasius) value of $0.441x/U_0$. We also note that here, $\lambda = 0$ because the derivative of U_e is zero. Then, we find that:

$$H = 2.55 \text{ and } S = 0.225.$$

These compare well with the exact values of $H = 2.59$ and $S = 0.220$. The skin friction on the plate is given by:

$$C_f = \frac{\tau_w}{\frac{1}{2}\rho U_e^2} = \frac{2\nu S}{U_e \theta},$$

which now becomes:

$$C_f = \frac{2\nu(0.09)^{0.62}}{U_0\sqrt{\frac{0.45\nu x}{U_0}}} = \frac{0.671}{\sqrt{\frac{U_0 x}{\nu}}} = \frac{0.671}{\sqrt{Re_x}}.$$

This is within 1 percent of the exact result of $0.664/\sqrt{Re_x}$.

Linearly Retarded Flow of Howarth

In this example case, the external flow is linearly decelerating on a flat plate, as represented by:

$$U_e(x) = U_0(1 - x/L).$$

Then, the Thwaites equation is integrated easily to obtain:

$$\frac{\theta^2(x)}{\nu} = \frac{0.45}{U_0^6(1-x/L)^6}\int_{x_0}^{x} U_0^5(1-x/L)^5 \, dx = 0.075\frac{\nu L}{U_0}\left[\left(1-\frac{x}{L}\right)^{-6} - 1\right].$$

Then, λ is given by:

$$\lambda(x) = \frac{\theta^2}{\nu}\frac{dU_e}{dx} = -0.075\left[\left(1-\frac{x}{L}\right)^{-6} - 1\right].$$

From this, we can find the variation of the wall shear stress and displacement thickness along the plate from the previous correlation expressions. For example, the skin friction is given by:

$$C_f = \frac{\tau_w}{\frac{1}{2}\rho U^2} = \frac{\mu S(\lambda)}{\frac{1}{2}\rho U \theta} = \frac{2\nu(\lambda+0.09)^{0.62}}{U_0(1-x/L)\theta}.$$

In addition, separation is predicted at $\lambda = -0.09$, which corresponds to:

$$\lambda_{sep} = -0.09 = -0.075 \left[\left(1 - \frac{x_{sep}}{L} \right)^{-6} - 1 \right],$$

leading to:

$$\frac{x_{sep}}{L} = 0.123,$$

which is within 3 percent of the exact result of 0.120.

Flow over a Circular Cylinder

Here, the potential flow is given by:

$$\frac{U_e}{U_0} = 2 \sin \phi.$$

This is transformed into the distance along the cylinder, $\phi = x/R$, where R is the cylinder radius and x is the distance along the cylinder. The expression then is expanded in a Taylor series to give:

$$\frac{U_e}{U_0} = 2 \left(\frac{x}{R} \right) - 0.333 \left(\frac{x}{R} \right)^3 + 0.0167 \left(\frac{x}{R} \right)^5 - \dots$$

The presence of the boundary layer on such a bluff body, however, significantly alters the potential flow and the polynomial:

$$\frac{U_e}{U_0} = 1.814 \left(\frac{x}{R} \right) - 0.271 \left(\frac{x}{R} \right)^3 - 0.0471 \left(\frac{x}{R} \right)^5 + \dots,$$

which is a curve fit of the data found by Hiemenz in 1911. It represents a closer fit (White, 1974). Using this second polynomial results in a significantly more accurate solution. Separation is predicted at $\phi = 78.5°$, which is close to the experimentally observed value of $80.5°$. The first polynomial produces a separation point of $\phi = 104.5°$.

Direct Numerical Solution of the Boundary-Layer Equations

In solving the boundary-layer equations, there are numerical alternatives to the integral methods using fully coupled, finite-difference methods. Integral methods tend to be fast, but they have limited range of applicability because they rely on the velocity profiles being specified—sometimes by guesswork—and being of the same geometric family at each axial location, which may not always be the case. Hence, they lack generality. Direct solutions of the governing equations by finite-difference methods are significantly slower due to the coupled nature of the equations. However, it is possible to take advantage of the rapid convergence of a novel, uncoupled procedure to give a significantly faster boundary-layer solution, which is then coupled to the inviscid-flow model.

The boundary-layer equations used in the finite difference scheme are two-dimensional, unsteady, and incompressible. The z-momentum equation simply reduces to:

$$\frac{\partial p}{\partial z} = 0. \tag{8.115}$$

The remaining equations (i.e., continuity and x-momentum) are as follows:

$$\begin{cases} \dfrac{\partial u}{\partial x} + \dfrac{\partial w}{\partial z} = 0 \\[2mm] u \dfrac{\partial u}{\partial x} + w \dfrac{\partial u}{\partial z} - \dfrac{1}{Re} \dfrac{\partial^2 u}{\partial z^2} = U_e \dfrac{dU_e}{dx}. \end{cases} \tag{8.116}$$

Solution Algorithm

The momentum and continuity equations are differenced using an *implicit method* similar to that used for the Falkner–Skan equation. Because the coefficients are non-linear, the equations must be solved iteratively. Variable spacing for z can be used in this formulation. The derivatives in the momentum equation become:

$$u \frac{\partial u}{\partial x} = \bar{u}^n \frac{u_{ij}^{n+1} - u_{i-1j}}{x_i - x_{i-1}}$$

$$w \frac{\partial u}{\partial y} = \bar{w}^n \frac{u_{ij+1}^{n+1} - u_{ij-1}^{n+1}}{z_{ij+1} - z_{ij-1}}. \tag{8.117}$$

Thus,

$$\frac{1}{Re} \frac{\partial^2 u}{\partial z^2} = \frac{2}{Re(z_{ij+1} - z_{ij-1})} \left[\frac{u_{ij+1}^{n+1} - u_{ij}^{n+1}}{z_{ij+1} - z_{ij}} - \frac{u_{ij}^{n+1} - u_{ij-1}^{n+1}}{z_{ij} - z_{ij-1}} \right] \tag{8.118}$$

and:

$$U \frac{dU}{dx} = \tfrac{1}{2}(U_i + U_{i-1}) \frac{U_i - U_{i-1}}{x_i - x_{i-1}}, \tag{8.119}$$

where:

$$\bar{u}^n = \tfrac{1}{2}(u_{i-1j} + u_{ij}^n).$$

The superscripts refer to the iteration level. The equation is rearranged to form a tridiagonal system for the unknown values of u at the ith station:

$$Cu_{ij-1} + Bu_{ij} + Au_{ij+1} = D,$$

where:

$$
\begin{cases}
C = \dfrac{\bar{w}^n}{z_{ij+1} - z_{ij-1}} - \dfrac{2}{\mathrm{Re}} \dfrac{1}{(z_{ij+1} - z_{ij-1})(z_{ij} - z_{ij-1})} \\[3mm]
B = \dfrac{\bar{u}^n}{x_i - x_{i-1}} - \dfrac{2}{\mathrm{Re}(z_{ij+1} - z_{ij-1})} \left[\dfrac{1}{(z_{ij} - z_{ij-1})} + \dfrac{1}{(z_{ij+1} - z_{ij})} \right] \\[3mm]
A = \dfrac{\bar{w}^n}{z_{ij+1} - z_{ij-1}} - \dfrac{2}{\mathrm{Re}} \dfrac{1}{(z_{ij+1} - z_{ij-1})(z_{ij+1} - z_{ij})} \\[3mm]
D = \dfrac{\bar{u}^n u_{i-1j}}{x_i - x_{i-1}} + \tfrac{1}{2}(U_i + U_{i-1}) \dfrac{U_i - U_{i-1}}{x_i - x_{i-1}}.
\end{cases}
$$

The continuity equation is solved for a new value of w by a trapezoidal-rule integration:

$$
w_{ij} = w_{ij-1} - \frac{(z_{ij} - z_{ij-1})}{2} \left[\frac{U_{ij} - U_{i-1j}}{x_i - x_{i-1}} + \frac{U_{ij-1} - U_{i-1j-1}}{x_i - x_{i-1}} \right].
$$

The equations can be solved in an uncoupled manner, as described by Gad-El-Hak, 1989. The procedure is as follows:

1. The Initial values of u and w are obtained from the i-1 station.
2. The nonlinear coefficients are evaluated.
3. The momentum equation then is solved for u_{ij}.
4. The continuity equation then is used to obtain a new value of w_{ij}.
5. The convergence of u_{ij} is checked.
6. If not converged, then the procedure is repeated from Step 2.

The method described is fast compared with other finite-difference calculations, even though iterations are used at each station. The iterative updating of the nonlinear coefficients, instead of simply using the value at i-1, improves accuracy and allows coarser spacing to be used. The computer code **BL** uses this solution procedure and is set up to demonstrate a number of flows for various freestream pressure distributions, as well as to accept pressure distributions from data files.

8.9 Free-Shear Layers, Wakes, and Jets

Space limitations do not permit a detailed discussion of these topics. The interested student is referred to the book by Schlicting, 2003. The following discussion describes the physical phenomena of interest and introduces terminology. Only two-dimensional flows of this type are presented.

Free-Shear Layers

A *free-shear layer* is a viscous flow bounded on one side by a uniform flow and on the other side by a fluid at rest. Consider, for example, a viscous flow coming off

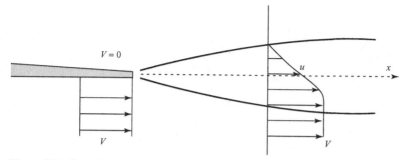

Figure 8.26. Free-shear layer.

the lower surface of a plate with zero-velocity flow above (Fig. 8.26). For simplicity, assume that the boundary layer on the plate surface is negligibly thin.

The shearing action tends to pull along some of the fluid at rest and retard some of the fluid coming from the plate surface; the result is a velocity profile as shown. The nonuniform portion of this profile becomes wider with increasing x. A two-dimensional free jet (i.e., slot jet) has a velocity profile (as in Fig. 8.26) on either side of the jet as it emerges into still air. A short distance downstream of the exit, the two profiles merge to form a rounded-jet profile near the center (as in the case of a jet).

Wakes

As in the discussion of the drag of an airfoil in Chapter 3, a wake forms behind a body because of the boundary layer leaving the body (as illustrated) or because of separation on the surface of the body (wake flow downstream of a cylinder). Behind the body trailing edge shown in Fig. 8.27, the two boundary layers merge. Then, the wake width increases with increasing downstream distance and the mean velocity in the wake decreases.

Free Jet

A *free jet* (here, a free-slot jet) is one that exhausts into a fluid at rest. Initially, the free jet at the exit has a uniform "core" with a free shear layer on either side. As the jet length increases, viscous-shear effects at the sides retard the flow at the center of

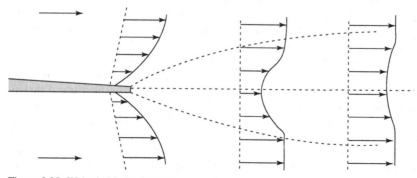

Figure 8.27. Wake behind a flat plate.

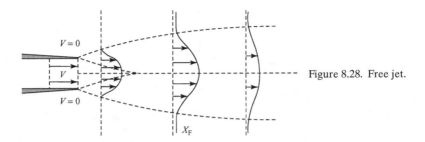

Figure 8.28. Free jet.

the jet and the jet has a rounded profile (e.g., station X_F). When the constant-velocity jet core disappears, the jet is said to be fully developed. With increasing distance downstream, the jet spreads and the maximum velocity decreases; both effects are caused by friction. Ultimately, the jet is dissipated.

All of these flows are unstable and quickly tend to become turbulent shear flows. It is difficult to generate them in the laboratory as laminar flows.

8.10 Transition to Turbulence

In most practical aerodynamic applications, the boundary-layer flow is predominantly turbulent. We consider laminar flows in detail to appreciate the role of viscosity in a fluid flow, as well as to establish definitions and ideas in a simple-flow case. Also, for a distance downstream of an airfoil or wing leading edge, the boundary layer is laminar. However, the laminar boundary layer soon goes through a complicated series of changes, after which a turbulent boundary layer emerges. A turbulent-boundary-layer flow exhibits unsteady, chaotic motion that has a random character and that contains large-scale eddies. The series of changes within the boundary layer that convert a laminar to a turbulent boundary layer usually occur over a finite distance, called a *transition region*. Because of the major differences in laminar and turbulent skin friction (and heat transfer), it is desirable to be able to predict (or estimate) both the start and the end of the transition region. Much experimental and analytical/numerical effort has been expended in this regard. We limit our discussion to an overview of the transition process from a physical perspective, with estimates of the streamwise extent of the transition region.

Recall that a laminar boundary layer in a steady flow consists of flow in smooth, steady laminas. However, the laminar shear layer is subject to small disturbances that occur in any flow (e.g., surface vibration, noise, freestream turbulence, surface imperfections, or roughness). These external disturbances usually are infinitesimal and random. They act as forcing functions that trigger two-dimensional wave-like disturbances within the laminar boundary layer. These waves, which travel downstream in the boundary layer, are called *Tollmein–Shlichting waves* (T–S waves), named after two researchers who conducted basic studies on the stability of these disturbance waves in the early 1930s.

Below a certain minimum Re number, these T–S waves damp out. This number is called the *critical Reynolds number*, which usually is based on a boundary-layer-thickness dimension, such as δ^*, rather than a streamwise dimension, x. For a flat-plate boundary layer, Re_{crit}, δ^* typically has a value of about 700, corresponding to a value of $Re_{cr,x}$ of about 165,000. Values of Re numbers used in this discussion, of necessity, must be prefaced by "about" or "approximately" because they depend so strongly on the environment (e.g., freestream turbulence level) of a particular boundary layer. This is not the transition point; it simply represents the lowest value of the Re number at which small disturbances in the boundary layer can be amplified. For the remainder of the discussion, we think of holding the flow velocity and properties constant and increasing the Re number by increasing the downstream distance, x, from the plate leading edge. At a value above the critical Re number, the disturbance waves within a certain frequency band are found to amplify with downstream distance. Linear-stability theory is used to predict the frequency band and the rate of amplification. A classical series of measurements was made by Schubauer and Skramstad in the 1940s that confirmed this amplification of the T–S waves (See discussion in Schlichting, 2003). Local velocities were measured in a flat-plate boundary layer with a fast-response instrument called a *hot-wire anemometer*; the results are shown in Fig. 8.29. Notice that at about 5 ft. (1.52 m), wave amplification is evident; at 6 ft. (1.83 m), the local flow is becoming nonlinear; and at 8 ft. (2.44 m), it is fully turbulent.

As the T–S waves are amplified, a complicated three-dimensional structure evolves. The waves begin to exhibit spanwise variations, and distorted vortex structures appear. Turbulent spots are formed, which are small regions of intense velocity fluctuations. These spots grow in size and number, finally coalescing into a fully turbulent boundary layer. The process of boundary-layer transition is shown in Fig. 8.30 (White, 1974).

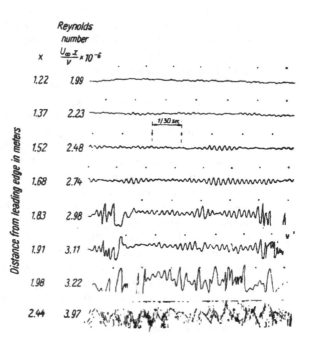

Figure 8.29. Oscillograms showing natural transition from laminar to turbulent flow (See discussion in Schlichting, 2003).

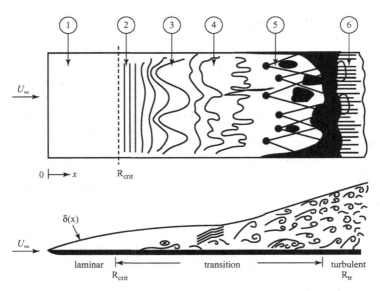

Figure 8.30. Schematic of the transition process on a flat plate (White).

The transition Re number, $Re_{tr,x}$, is the value of the Re number at which the boundary layer becomes fully turbulent. For a smooth flat plate with a low-turbulence level in the external flow, this value can be as large as (about) $Re_{tr,x} = 1 \times 10^7$. This value must be used with caution. For example, the boundary layer on a swept wing becomes turbulent much closer to the leading edge than for a straight wing (or flat plate) due to three-dimensional effects. Looking at Fig. 8.29, note that the caption states that this experimental result, as well as the value for $Re_{tr,x}$ quoted previously, is for "natural" transition (i.e., transition of a boundary layer in the presence of minimal outside disturbances). At times, it is desirable to force transition to a more upstream location (i.e., smaller value of x). For example, consider the wind-tunnel test of a model wing that is conducted to predict the behavior of a much larger full-scale wing at the same velocity. On the full-scale wing, the turbulent boundary layer may be established close to the leading edge; whereas on the model, the same transition Re number (i.e., the same x-dimension) corresponds to a chordwise distance much closer to the trailing edge. Recall from a previous discussion that the location of the boundary-layer separation point is sensitive to whether the boundary layer is laminar or turbulent. Thus, wind-tunnel tests at high angles of attack may be misleading because the boundary-layer separation point can not be located correctly. In such a case, the boundary layer on the model often is "tripped" (i.e., forced transition) by gluing a roughness strip (e.g., sandpaper) to the model wing surface in a spanwise direction a short distance downstream of the leading edge. Such a comparatively large disturbance causes a turbulent boundary layer to be established much closer to the leading edge than in the natural-transition case. Care must be taken that the artificial disturbance introduced is not so large as to lead to a badly distorted (hence, unrealistic) turbulent boundary-layer profile downstream.

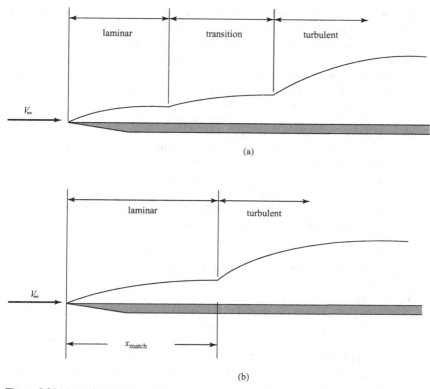

Figure 8.31. Actual and assumed boundary-layer growth.

If we want to calculate simply the behavior of an entire boundary layer from laminar through turbulent, the details in the transition region must be avoided. This normally is accomplished by making an assumption about the behavior of transition. The actual boundary layer, containing a transition region, is like that shown in Fig. 8.31a. For purposes of calculation, it is assumed that the boundary layer is first laminar and then turbulent, as shown in Fig. 8.31b; the change occurs instantly at some Re number value. In similarity analyses, a virtual origin for the turbulent boundary layer may be found by matching δ^* or θ for the two boundary layers at $Re_{match,x}$ (i.e., at x_{match}). In numerical analyses, the downstream matching calculation is switched from laminar-to-turbulent defining equations at x_{match}. The value of $Re_{match,x}$ is somewhat arbitrary and comes from experience. Because it must be larger than Re_{crit}, a typical value of Re_{match} used is several hundred thousand.

The following factors stabilize a laminar boundary layer and thus delay transition in an incompressible flow:

1. *Favorable streamwise pressure gradient.* Laminar-flow airfoils are specially shaped so as to maintain a chordwise favorable pressure gradient over the airfoil for as long as possible. Boundary layers stabilized in this way are said to have natural laminar flow (NLF).

2. *Suction.* Bleeding off the laminar boundary layer from a wing surface through spanwise suction slots or through porous surfaces stabilizes the laminar boundary layer. The suction inhibits boundary-layer growth and also alters the boundary-layer profile, which affects stability characteristics. Designs that use an artificial mechanism such as suction to maintain a laminar boundary layer are said to have laminar flow control (LFC).

3. *Cooling the surface.* Cooling a surface increases the value of $Re_{crit.}$

4. *Making the surface smoother.* Increasing surface roughness has a destabilizing effect on a laminar boundary layer.

5. *Decreasing the turbulence level in the freestream.* This applies to wind-tunnel tests. Turbulence in the atmosphere has a negligible effect on the transition of a boundary layer on a flight vehicle. However, the freestream turbulence level in a wind-tunnel test section has a major influence on transition because it acts as a trigger for boundary-layer disturbance waves. We must be careful when comparing detailed boundary-layer turbulence measurements taken in different wind tunnels.

A good discussion of transition is in White, 1986, and a detailed review of transitional boundary layers is provided in, Lauchle, 1991. A comprehensive review of flow control and control devices is found in Gad-El-Hak, 1989.

The increasing cost of jet fuel has made the subject of transition and LFC increasingly important. Holmes et al. found that for a high-performance business jet operating at $M_{cruise} = 0.7$, NLF on the wings, fuselage, and engine nacelles leads to a drag reduction of 24 percent of total airplane drag (Holmes et al., 1988). Further information on full-scale and wind-tunnel tests on NLF surfaces is found in Holmes et al., 1984.

For design purposes (see Raymer, 1989), a typical conventional aircraft (i.e., no special boundary-layer treatment) may be considered to have laminar flow over perhaps 10 to 20 percent of the wings and tails and virtually no laminar flow over the fuselage. With some NFC incorporated into the design, laminar flow can extend over as much as 50 percent of the wings and tails and about 20 to 35 percent of the fuselage. Thus, any discussion of incompressible boundary layers is incomplete without treating the turbulent boundary layer, which is the topic of the next section.

8.11 Turbulent Flow

A fully developed turbulent boundary layer is a region of large-scale velocity fluctuations (i.e., up to 10 percent fluctuations about a mean value). The flow is chaotic and irregular. Thus, although the external mean flow at the outer edge of the boundary layer is steady, the boundary-layer flow is unsteady. There is no chance—even with modern computers—of being able to predict the turbulent-flow behavior at a fixed point in the boundary layer as a function of time. However, experiments have shown that the turbulent fluctuations in the boundary layer are random, so that a time-averaged approach may be used. Thus, we focus on being able to predict what happens "on the average" at a point within the turbulent boundary layer. As in the laminar-boundary-layer problem, our goal is to be able to predict the growth and velocity profile of an incompressible, turbulent boundary layer. To accomplish this, the first step is to derive the appropriate differential equations and then seek solutions for them.

Time-Averaged Boundary-Layer Equations

We could begin by developing the time-averaged Navier–Stokes equations and then apply the boundary-layer approximations (i.e., $d/x \ll 1$, and so on). However, we choose instead to derive the time-averaged, boundary-layer equations of continuity and momentum directly from the laminar-boundary-layer equations. As before, we use an airfoil coordinate system (x, y, z), where x is the streamwise direction, y is the transverse direction, and z is normal to the surface. Based on experimental measurements, it is assumed that the turbulent flow is characterized by random fluctuations about a mean. Thus, we let:

$$\tilde{U} = U + u'; \quad \tilde{V} = V + v'; \quad \tilde{W} = W + w', \tag{8.120}$$

where:

$$\tilde{U} \Rightarrow \text{instantaneous velocity}$$
$$U \Rightarrow \text{mean velocity}$$
$$u' \Rightarrow \text{turbulent fluctuation,}$$

and so on. Now, if a fast-response instrument were used to measure the instantaneous velocity components at a fixed point in a two-dimensional boundary layer, the output would look like Fig. 8.32, which recognizes that in a two-dimensional flow, the transverse mean velocity, V, is zero (although the fluctuations in the transverse direction are not).

Because the velocity fluctuations are random, a mean-velocity component in any direction may be defined as the integral of the instantaneous velocity component over a time interval $t = T$, where T is a long-enough interval so that the value of the integral is independent of time. Thus,

$$U \equiv \frac{1}{T} \int_0^T \tilde{U} dt \quad \text{and} \quad \bar{u}' \equiv \frac{1}{T} \int_0^T u' dt = 0, \tag{8.121}$$

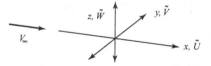

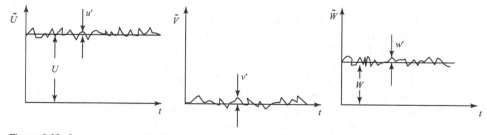

Figure 8.32. Instantaneous velocity components in a two-dimensional, turbulent boundary-layer.

where the bar over the u' denotes a time average. Likewise, $\overline{v'} = 0$ **and** $\overline{w'} = 0$ These equations are stating that, because there is an equal likelihood that u' is positive or negative at any instant, a summing of the value of u' over a long time interval T is zero. However, the time average of the squares of the velocity fluctuations (to appear later in the boundary-layer equations) is not zero; in fact, it is positive. This follows if we imagine that at each instant of time, the value of u' is measured and the value squared. Whether u' is instantaneously positive or negative, the square is positive and the sum of the squared quantities over time is not be zero. Thus,

$$\overline{u'^2} \equiv \frac{1}{T}\int_0^T u'^2 dt > 0 \quad \text{and, likewise} \quad \overline{v'^2} > 0, \overline{w'^2} > 0. \tag{8.122}$$

Be careful when writing (or reading) the time-averaged notation. Thus,

$$\overline{u'^2} \neq \overline{u'}^2.$$

The time average of the cross-products of velocity fluctuations in general is not zero. However, in a two-dimensional boundary layer:

$$\overline{u'w'} \equiv \frac{1}{T}\int_0^T (u'w')dt \neq 0, \tag{8.123}$$

(a fact that is needed later), whereas:

$$\overline{u'v'} = \overline{v'w'} = 0.$$

Before proceeding to time-average the defining boundary-layer equations, the following rules are needed, which follow from the definition of *mean value*. If A is a mean quantity, a and b are fluctuating quantities, and t is any Cartesian or time coordinate, then:

(a) $\quad \overline{a+b} \equiv \frac{1}{T}\int_0^T (a+b)dt = \overline{a} + \overline{b}.$

(b) $\quad \overline{Aa} \equiv \frac{1}{T}\int_0^T (Aa)dt = (A)\frac{1}{T}\int_0^T (a)dt = A\overline{a}.$

(c) $\quad \overline{\dfrac{\partial a}{\partial \xi}} \equiv \frac{1}{T}\int_0^T \left(\frac{\partial a}{\partial \xi}\right)dt = \frac{\partial}{\partial \xi}\left(\frac{1}{T}\int_0^T (a)dt\right) = \frac{\partial \overline{a}}{\partial \xi}.$ $\qquad$ (8.124)

We proceed now to substitute instantaneous velocity components into the laminar boundary-layer equations for continuity and x-momentum, Eq. 8.37, with the steady-flow velocity components (u, w) in Eq. 8.37 replaced by the instantaneous velocity components $(\tilde{U}, \tilde{W})$. Because the flow now is unsteady, a $\partial/\partial t$ term must be added to the steady-flow momentum equation in Eq. 8.37. If the velocity field within the boundary layer is unsteady, then pressure fluctuations must exist as well. Thus, we also replace the steady pressure p by the instantaneous pressure $\tilde{p} \equiv P + p'$, where P is the mean value of the pressure and p' is the pressure fluctuation. As with the velocity fluctuations, $\overline{p'} = 0$. Finally, because the flow is incompressible, $r = $ constant. Then, Eq. 8.37 becomes:

continuity $$\frac{\partial \tilde{U}}{\partial x} + \frac{\partial \tilde{W}}{\partial z} = 0 \qquad (8.125)$$

momentum $$\rho \frac{\partial \tilde{U}}{\partial t} + \rho \tilde{U} \frac{\partial \tilde{U}}{\partial x} + \rho \tilde{W} \frac{\partial \tilde{W}}{\partial z} = -\frac{\partial \tilde{p}}{\partial x} + \frac{\partial}{\partial x}\left[\mu \frac{\partial \tilde{U}}{\partial z}\right]. \qquad (8.126)$$

Substituting for $\tilde{U}$ and $\tilde{W}$ from Eq. 8.113 and replacing p by $(\bar{p} + p')$, Eqs. 8.117 and 8.118 become:

$$\frac{\partial}{\partial x}(U + u') + \frac{\partial}{\partial z}(W + w') = 0 \qquad (8.127)$$

and

$$\rho\left[\frac{\partial}{\partial t}(U + u') + (U + u')\frac{\partial}{\partial x}(U + u') + (W + w')\frac{\partial}{\partial z}(W + w')\right] =$$
$$= -\frac{\partial}{\partial x}(P + p') + \frac{\partial}{\partial z}\left[\mu \frac{\partial}{\partial z}(W + w')\right]. \qquad (8.128)$$

We now time-average these two equations over a suitably long time interval T so that the equations reflect what is happening in the boundary layer on the average. The process begins by taking the average of both sides of each equation; that is, Eqs. 8.118 and 8.120 are rewritten with an overbar above each side. Then, by appealing to Eq. 8.117a, the two sides are expanded and then decomposed into the time average of each term. Thus,

$$\underbrace{\frac{\partial U}{\partial x} + \frac{\partial \overline{u'}}{\partial x}}_{\hat{1}} + \underbrace{\frac{\partial W}{\partial z} + \frac{\partial \overline{w'}}{\partial z}}_{\hat{2}} = 0$$

and

$$\rho\left[\underbrace{\frac{\partial U}{\partial t}}_{3} + \underbrace{\frac{\partial \overline{u'}}{\partial t}}_{4} + \underbrace{U\frac{\partial U}{\partial x} + U\frac{\partial \overline{u'}}{\partial x}}_{5} + \underbrace{\overline{u'\frac{\partial U}{\partial x}} + \frac{\partial(\overline{u'^2}/2)}{\partial x}}_{6} + \underbrace{W\frac{\partial U}{\partial z} + W\frac{\partial \overline{u'}}{\partial z}}_{7} + \underbrace{\overline{w'\frac{\partial U}{\partial z}} + \overline{w'\frac{\partial u'}{\partial z}}}_{8}\right] =$$

$$= \underbrace{-\frac{\overline{\partial P}}{\partial x} - \frac{\overline{\partial p'}}{\partial x}}_{9} + \underbrace{\frac{\partial}{\partial z}\left(\mu \frac{\partial U}{\partial z}\right) + \frac{\partial}{\partial z}\left(\mu \frac{\partial \overline{u'}}{\partial z}\right)}_{10},$$

where $\overline{u'\dfrac{\partial u'}{\partial x}}$ has been written as $\dfrac{\overline{\partial u'^2/2}}{\partial x}$.

All of the numbered terms in the previous equations are zero, for the following reasons:

1. Terms 1, 2: Eqs. 8.114 and 8.117c.
2. Term 3: U is independent of time if steady mean flow is assumed.

3. Term 4: Eqs. 8.114 and 8.117c with ψ playing the role of t.
4. Terms 5, 6, 7, and 8: Eqs. 8.114 and 8.117b, c.
5. Term 9: Eq. 8.114.
6. Term 10: Eqs. 8.114 and 8.117b.

(The student should verify all of these statements.)

The time-averaged boundary-layer equations become:

$$\frac{\partial U}{\partial x} + \frac{\partial W}{\partial z} = 0 \tag{8.129}$$

$$\rho\left[U\frac{\partial U}{\partial x} + W\frac{\partial U}{\partial z}\right] + \rho\left[\overline{\frac{\partial(u'^2/2)}{\partial x}} + \overline{w'\frac{\partial u'}{\partial z}}\right] = -\frac{\partial P}{\partial x} + \frac{\partial}{\partial z}\left[\mu\frac{\partial U}{\partial z}\right]. \tag{8.130}$$

In examining these equations, we note that Eq. 8.10 is the same as the continuity equation for laminar flow but now is written with mean-velocity values as dependent variables. Eq. 8.11 has the same character, but a new term (underlined) appeared that was not present in the laminar-boundary-layer equations. This new term must be investigated further. We begin by writing it in a more convenient form. Because the numbered Terms 1 and 2 are both zero, their sum is zero; namely:

$$\overline{\frac{\partial u'}{\partial x}} + \overline{\frac{\partial w'}{\partial z}} = 0.$$

Now, manipulating the underlined term in Eq. 8.11:

$$\rho\left[\overline{\frac{\partial(u'^2/2)}{\partial x}} + \overline{w'\frac{\partial u'}{\partial z}}\right] = \rho\left[\overline{\frac{\partial(u'^2/2)}{\partial x}} + \overline{\frac{\partial(u'w')}{\partial z}} - \overline{u'\frac{\partial w'}{\partial z}}\right] =$$

$$= \rho\left[\overline{\frac{\partial(u'^2/2)}{\partial x}} + \overline{\frac{\partial(u'w')}{\partial z}} + \overline{u'\frac{\partial u'}{\partial x}}\right] = \frac{\partial}{\partial x}\left(\rho\overline{u'^2}\right) + \frac{\partial}{\partial z}(\rho\overline{u'w'}).$$

Then, Eq. 8.122 may be written as follows:

$$\rho\left[U\frac{\partial U}{\partial x} + W\frac{\partial U}{\partial z}\right] + \rho\left[\frac{\partial}{\partial x}\left(\overline{u'^2}\right) + \frac{\partial}{\partial z}\left(\overline{u'w'}\right)\right] = -\frac{\partial P}{\partial x} + \frac{\partial}{\partial z}\left[\mu\frac{\partial U}{\partial z}\right]. \tag{8.131}$$

Experiments have shown that for boundary-layer flows, where $\partial/\partial x \ll \partial/\partial z$, it is observed that:

$$\frac{\partial}{\partial x}\left(\overline{u'^2}\right) \ll \frac{\partial}{\partial z}\left(\overline{u'w'}\right).$$

Then, Eq. 8.12 may be simplified to:

$$\rho\left[U\frac{\partial U}{\partial x} + W\frac{\partial U}{\partial z}\right] + \frac{\partial}{\partial z}\left(\overline{\rho u'w'}\right) = -\frac{\partial P}{\partial x} + \frac{\partial}{\partial z}\left(\mu\frac{\partial U}{\partial z}\right). \tag{8.132}$$

Recall from Chapter 3 that the differential-momentum equation is derivable from the application of Newton's Second Law to a fixed control volume. Hence, we may interpret Eq. 8.124 as stating that the net time-averaged momentum outflow from the control volume (left side) is equal to the net pressure and viscous forces acting on the control surface (right side). In this equation, the time-averaged momentum flux is that due to mean velocity (first term on the left side) and fluctuating velocities (second term on the left side). Now, say that there is a change in the pressure and viscous forces that leads to an increase in overall momentum flux. Any corresponding increase in the momentum flux due to velocity fluctuations must reduce the effect of the force change on the time-averaged momentum flux due to the mean velocities. Thus, the turbulence term has the same role as a stress; namely, it tends to extract momentum associated with the mean flow. The turbulence-momentum term in Eq. 8.124 then is moved conventionally to the other side of the equation and considered as a stress. It is called an *apparent stress* or a *Reynolds stress*. Be careful: The term $(\rho u'w')$ is physically not a stress; however, mathematically, it is considered to play the role of a stress. The term $\overline{(\rho u'w')}$ is an apparent shear stress. The term (ρu^2) neglected in Eq. 8.12 is an apparent normal stress. As in the molecular-viscosity case studied previously, the apparent normal stress is negligible compared to the apparent shear stress.

Rewriting Eq. 8.121 and moving the apparent shear-stress term in Eq. 8.124 to the right side, the turbulent boundary-layer equations are:

$$\frac{\partial U}{\partial x} + \frac{\partial W}{\partial z} = 0 \tag{8.133}$$

$$\rho\left(U\frac{\partial U}{\partial x} + W\frac{\partial U}{\partial z}\right) = -\frac{\partial P}{\partial x} + \frac{\partial \tau}{\partial z}, \tag{8.134}$$

where:

$$\lambda \equiv \tau_L + \tau_T = \mu\left(\frac{\partial U}{\partial y}\right) - \overline{\rho u'w'}.$$

In turbulent flows, $\tau_T \gg \tau_L$, except near the surface, where τ_T approaches zero.

Inspection of Eqs. 8.125 and 8.126 shows that the solution of these equations is problematic. There now is an extra unknown, $\overline{\rho u'w'}$, with no information on how to find it. Unfortunately, it is not a property of the fluid but rather a property of the turbulence. Hence, we expect this unknown to have different values within the boundary layer (Fig. 8.33) and also for different turbulent shear flows (e.g., for a boundary-layer flow and for a free-shear layer). Specifying the turbulent-shear-stress term is called the *closure problem*, meaning that it must be modeled empirically. This is discussed, when solutions are addressed.

A final comment on the turbulent boundary layer: The mean turbulent kinetic energy per unit volume in the boundary layer is defined as:

$$KE = \frac{1}{2}\rho\left[\left(\tilde{U}^2 + \tilde{V}^2 + \tilde{W}^2\right) - \left(U^2 + V^2 + W^2\right)\right] = \frac{1}{2}\rho\left[\overline{u'^2} + \overline{v'^2} + \overline{w'^2}\right]. \tag{8.135}$$

Large-scale eddies in the boundary layer extract kinetic energy from the freestream. This kinetic energy becomes increasingly more randomly distributed as it is transferred to increasingly smaller eddies. Ultimately, it is dissipated as thermal energy (i.e., heat) at the molecular level.

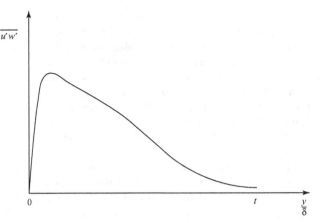

$\overline{u'w'}$

Figure 8.33. Variation of turbulent shear stress through a turbulent boundary layer.

The quality of the freestream flow in a "low-turbulence" wind tunnel often is defined as the magnitude of the freestream turbulent intensity,

$$\text{turbulent intensity } (\%) \equiv \frac{100}{V_\infty} \sqrt{\frac{\overline{u'^2} + \overline{y'^2} + \overline{w'^2}}{3}}. \qquad (8.136)$$

If a wind tunnel is to be used for turbulence research, the turbulent intensity of the freestream flow in the test section should be only a few hundredths of 1 percent.

PROBLEMS

8.1   A steady, incompressible, viscous flow at standard atmospheric conditions has velocity components given by $u = 8y$, $v = 3x^2$ m/s. What is the shear stress at $x = 2$m, $y = 4$m on the positive face of a plane normal to the x-axis? Give the notation and units of the answer.

8.2   A wing of 2-meter chord is flying through sea-level air at a speed of 50 m/sec. A microorganism with a characteristic length of 10 microns ($10 \cdot 10^{-6}$m) is swimming at a speed of 20 microns/sec through water with kinematic viscosity of $1.15 \cdot 10^{-6}$ m²/s. Show that the characteristic Re numbers of these two flows are $6.88 \cdot 10^6$ and $1.74 \cdot 10^{-4}$, respectively.

8.3   From the Navier–Stokes equations developed in this chapter, show that the fully developed flow in a pipe of any cross section is:

$$\frac{\partial^2 u}{\partial y^2} + \frac{\partial^2 u}{\partial z^2} = \frac{1}{\mu} \frac{\partial p}{\partial x}.$$

Solve this equation for the general case of a pressure gradient that decreases linearly with distance along the pipe (in the x-direction).

8.4   Within the boundary layer on a certain surface, the horizontal component of velocity is found to vary as:

$$u = z / \sqrt{x}.$$

Find the expression for the vertical component of velocity, v, by appealing to the continuity equation. Assume the flow to be incompressible. Do not use the Blasius result.

8.5 A two-dimensional, steady flow has velocity components given by $u = 8z$, $w = 3x^2$ m/s. Assume = constant (set equal to unity for convenience) and that the normal stresses in the flow are due to pressure only.

(a) If is assumed to be zero, what are the components of the gradient of the pressure at any point in the flow field:

(b) Suppose that $\mu = 17.8 \times 10^{-6}$ kg/(m s). What is the shear stress at $x = 2$m, $z = 4$m, on the positive face of a plane normal to the x axis.

8.6 Consider two-dimensional viscous flow of air at standard conditions over a flat plate, as shown in the figure. The Blasius solution holds. The wall shear stress at Point A is to be determined experimentally. This is accomplished by making a static-pressure measurement on the plate surface at Point A (PA) and a total-pressure measurement (p_0) at Points B and E, where E is just outside the boundary layer, and then assuming a linear-velocity profile between A and B, very near the wall. The pressure measurements are:

$$[p_{0,B} - p_A] = 0.3233 \text{ psf}$$
$$[p_{0,E} - p_A] = 2.9725 \text{ psf,}$$

where Point B is located at $z_B = 0.020$ in.
Determine the wall shear stress at Point A by using:
(a) the experimental data
(b) the Blasius-theory prediction

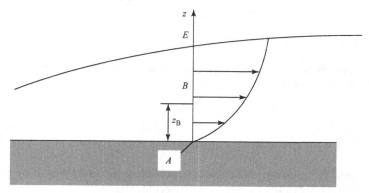

8.7 A flat plate is in a flow that has a viscosity $\mu = 4 \times 10^{-7}$ slug/ft/s. At a Point P, downstream of the flat-plate leading edge, a direct measurement of wall shear stress is made by using a small floating element set flush with the plate surface and attached to a sensitive balance. The measured value of shear stress is $\tau = 2.656 \times 10^{-5}$ lbf/ft^2. Using the Blasius solution, predict the distance, z, above the flat-plate surface where the velocity within the boundary layer at Point P is 91.3 percent of the freestream velocity.

8.8 A flat plate is immersed in a flowing liquid ($\rho = 2 \times 10^{-5}$ slug/ft^3). Assume that the Blasius solution holds. At a certain point 5 ft. downstream of the leading edge, the dynamic pressure at the edge of the boundary layer is 1.78 times its value at a distance of 1.00 in. above the flat plate. A measured velocity profile at this same station was used to infer a value of wall shear stress of 6.5×10^{-5} lbf/ft^2. What is the density of the flowing liquid? Give units of the answer.

8.9 In a certain flat-plate experiment in a gas, the freestream velocity is 20 ft/s and the kinematic viscosity, v, of the flow is $v = 1.6 \times 10^{-4}$ ft^2/s. At a certain x location, the local velocity at a distance $z = 0.20$ in. is 16.9 ft/s. What is the value of the skin-friction coefficient at this particular station? Use the Blasius solution.

8.10 A thin plate is aligned with the flow of a certain liquid. The flow velocity is 7.0 ft/s and the flow density is $\rho = 2.0$ slug/ft^3. The velocity in the boundary layer at a certain downstream location B is measured at 5.25 ft/s at a certain distance, h, above the plate surface. The Blasius solution applied. What is the velocity in the boundary layer at Station B at $z = 2h$?

8.11 A man-powered airplane has a total lift of 200 lbs. and flies at 20 mph. The wing planform is rectangular, with a 3-foot chord and a 60-foot span. This airplane is in steady flight at sea-level standard conditions.

(a) Assuming that the wing of the airplane is a flat plate at zero (small) angle of attack and that Blasius results hold, calculate the wing frictional drag, lbf. Use strip theory; that is, assume that the flow over the wing is two-dimensional flow at every spanwise station. Remember that the wing has two wetted surfaces.

(b) Assuming that the wing has an elliptical-lift distribution (i.e., minimum induced drag), calculate the induced drag, lbf, on this wing. What percentage of the total wing drag is the frictional drag?

8.12 Consider the following polynomial boundary-layer profile:

$$u/U_e = A + B[z/\delta] + C[z/\delta]^2.$$

Evaluate the coefficients of this assumed profile by applying the appropriate boundary conditions.

8.13 A flat plate is in a flow of water with a streamwise pressure gradient. The velocity history at the edge of the flat-plate boundary layer is given by:

$$U_e = 5 + x - x^2,$$

where U_e has units of ft/s and x is in feet. The density of water is 2.0 slug/ft^3.

(a) What is the strength of the streamwise pressure gradient at $x = 2$ft? Give units.

(b) Could this pressure gradient lead to possible separation? Why or why not?

8.14 Consider the following assumed velocity-profile shapes:

1. $u/U_e = z/\delta$
2. $u/U_e = \sin[z/2]$.

(a) Find the expressions for the displacement thickness and skin-friction coefficient for two flat-plate (i.e., zero pressure gradient) boundary layers with profiles 1 and 2. Do this by using the Kàrmàn momentum-integral equation. Solve first for the momentum-thickness expression, then for θ in terms of δ and δ^* in terms of δ.

(b) The required expressions are of the form (constant $/\sqrt{Re_x}$). Compare these with the Blasius solution for a flat plate. Find the percentage difference between δ^* and C_f predicted for these assumed profile shapes compared to the Blasius result.

8.15 A flat plate is immersed in a liquid stream with $U_e = 10$ ft/s, $p = 2.0$ slugs/ft^3, $\mu = 2 \cdot 10^{-2}$ slug/ft/s. Calculate the displacement thickness (δ^*, in.) at $x = 5$ ft. along the plate by using:
(a) the Blasius solution
(b) the Kàrmàn momentum-integral equation, assuming the velocity profile is:

$$u/U_e = (z/\delta)[2 - (z/\delta)].$$

(c) Compare the two results, with the Blasius solution assumed to be the standard.

8.16 Consider a steady flow along a flat plate that is parallel to the flow. At a certain streamwise station, the boundary-layer velocity profile may be taken as:

$$u = (4 \cdot 10^4)z - (2 \cdot 10^6)z^2$$

and $dU_e/dx = 0$ because the body is a flat plate. Find the expression for the rate of change of momentum thickness (d/dx) at this same station. Note: This expression is not a general result.

8.17 Imagine that a flat plate is warped so that there is a streamwise velocity gradient impressed on the edge of the boundary layer. Suppose that at some point along this warped plate, the velocity profile may be described as:

$$u = (4 \cdot 10^4)z - (2 \cdot 10^6)z^2.$$

Find the value of dU_e/dx necessary to make $d/dx = 0$ at that same point. Take $v = 1.5 \times 10^{-4}$ ft^2/s and give units of the answer.

8.18 A curved surface is in a flow of glycerine, $p = 2.0$ slug/ft^3, $\mu = 0.02$ slug/ft/s. At a certain station on this surface, the boundary-layer thickness $\delta = 6.0$ in., the local velocity at the edge of the boundary layer is $U_e = 10$ ft/s, and the rate of change of the momentum thickness is $d/dx = 0.004$ (i.e., dimensionless). Use an assumed linear velocity profile, $u/U_e = z/\delta$, to determine the value of the local streamwise pressure gradient, dp/dx, at this station. What are the units of the answer?

8.19 Calculate the friction drag per unit span on the rectangular wing (chord 3 m) of a light aircraft flying at 90 km/h at an altitude of 3 km and assuming two different cases:
(a) The boundary layer over the entire chord length is laminar; i.e.,

$$C_d = (2)(\overline{C}_f),$$

where the factor 2 appears because there are two wetted surfaces. Assume the wing to be a flat plate.
(b) The boundary layer over the entire chord length is turbulent. Assuming an empirical "one-seventh" power law for the velocity distribution in the turbulent boundary layer:

$$u/U_e = \left[\frac{z}{\delta}\right]^{1/7},$$

it may be shown (see Schlichting and Gersten, 2003) that the corresponding drag coefficient is given by:

$$C_d = 2C_f = 2(0.074)/\left[\text{Re}_c\right]^{0.2},$$

where Re_c is the Re number based on chord length. Again, there are two wetted surfaces.

(c) What percentage drag reduction per unit span [(turbulent-laminar)/(turbulent)] is theoretically possible if the LFC could be established over the entire chord length of this wing and the wing boundary layer initially was completely turbulent?

8.20 A man-powered airplane has a total lift of 200 lbs. in steady, level flight at a speed of 20 mph at standard sea-level conditions. The wing planform is rectangular, with a 5-foot chord and an 80-foot span. There is no boundary layer separation on the wing.

(a) Assuming the wing is a two-dimensional flat plate, calculate the frictional drag of the wing. Be sure to use the correct wetted area.

(b) Calculate the induced drag on the wing, assuming an elliptical-lift distribution.

(c) What percentage of the wing total drag is due to viscous effects?

8.21 The measured local wall shear stress along the center chord line of a thin rectangular wing in low-speed flow is found to be well represented by:

$$\tau_w = \frac{0.06}{x^{1/3}},$$

where x is the chordwise distance from the wing leading edge. Find the total frictional drag of the wing if the chord is 8 ft. and the span is 90 ft. Assume two-dimensional flow across the entire surface.

8.22 A thin plate is aligned with a flow of a certain liquid that has a density of 2.0 slug/ft^3. The liquid is flowing with a freestream velocity of 7 ft/s near atmospheric pressure (15 psia). Assume that the flat-plate laminar-boundary-layer theory applies in this case.

(a) At a certain station where the Re number based on the distance downstream from the leading edge of the plate is 10,000, the boundary-layer velocity profile is measured and integrated to yield a value of the displacement thickness of 1.63 inches. Predict the dynamic pressure (psi) at this same station at a height $z = 3.22$ in. above the plate surface.

(b) The local velocity is measured at 5.25 ft/s at a certain distance, h, above the plate surface at a second streamwise station. Predict the velocity (in ft/s) at this new postion at a distance of $2h$ above the plate surface.

8.23 Consider the flow over a flat plate in a water tunnel. The plate is held at zero angle of attack and the tunnel freestream velocity is 7 ft/s at a static pressure of 14 psia. Assume that the properties of the water are as follows:

$$\rho = 2.0 \text{ slug/ft}^3, \mu = 2 \times 10^{-5} \text{ slug/ft s.}$$

(a) If the slope of the boundary-layer displacement thickness is $d\delta^*/ds = 0.0005$ at the trailing edge of the plate, how long is the plate (in feet)?

(b) At a distance of 33 inches downstream of the leading edge of this plate, what would you predict the stagnation pressure to be at a height of 0.09 inch above the plate surface (in psia)?

(c) If we were to define the boundary-layer thickness, δ, as the height above the plate surface where the velocity is 75 percent of the freestream velocity, what is the boundary-layer thickness, according to this (nonstandard) definition, at a location 2 feet downstream of the plate leading edge?

8.24 Velocity profiles in laminar boundary layers sometimes are approximated by simple algebraic expressions. Some examples are as follows:

$$\frac{u}{U_e} = \frac{z}{\delta} \quad \text{(linear);} \qquad \frac{u}{U_e} = \sin\left(\frac{\pi}{2}\frac{z}{\delta}\right) \quad \text{(sinusoidal);}$$

$$\frac{u}{U_e} = \frac{3}{2}\left(\frac{z}{\delta}\right) - \frac{1}{2}\left(\frac{z}{\delta}\right)^3 \quad \text{(cubic);} \qquad \frac{u}{U_e} = 2\left(\frac{z}{\delta}\right) - \left(\frac{z}{\delta}\right)^2 \quad \text{(parabolic).}$$

Compare the shapes of these velocity profiles by plotting z/δ versus u/U_e. Also compare to the Blasius velocity profile.

8.25 The velocity profile in a turbulent boundary layer sometimes is approximated by the *power-law formula,* as follows:

$$\frac{u}{U_e} = \left(\frac{z}{\delta}\right)^{1/7}.$$

Compare the shape of this profile with the parabolic laminar boundary profile defined in Problem 8.8 and also the Blasius profile. Comment on the differences in the profiles. Give special attention to the slope of the profile near the surface. What does this imply in terms of the shear stress at the surface and, therefore, the skin-friction drag for the turbulent profile as compared to the laminar profiles?

REFERENCES AND SUGGESTED READING

Falkner, V. M., and Skan, S. W., "Some Approximate Solutions to the Boundary-Layer Equations," Philosophical Magazine, Vol. 12, (pp. 865–96), 1931.

Gad-El-Hak, M., "Flow Control," Applied Mechanics Reviews, Vol. 42, no. 10, (p. 251), October 1989.

Hiemenz, K., "Die Grenzschicht an einem in den gleichformigen Flussigkeitsstrom eingetauchten geraden Kreiszylinder," Dingler's Polytechnical Journal, Vol. 326, (pp. 321–26), 1911.

Holmes, B. J., Croom, C. C., Hastings, E. C., Obara, C. J., and Van Dam, C. P., "Flight Research on Natural Laminar Flow." NASA Tech. Paper No. 88-14950, 1988.

Holmes, B. J., Obara, C. J., and Yip, L. P., "Natural Laminar Flow Experiments on Modern Airplane Surfaces," (NASA Tech. Paper No. 84-2660), 1984.

Howarth, L., "On the Solution of the Laminar Boundary Layer Equations," Proc. Royal Society London Ser. A, Vol. 164, (pp. 547–579), 1938.

Lauchle, G. C., "Hydroacoustics of Transitional Boundary Layer Flow," Applied Mechanics Reviews, Vol. 44, No. 12, pt. 1, (pp. 517–31), December 1991.

Liepmann, H. W., and Dhawan, S., "Direct Measurement of Local Skin Friction in Low-Speed and High-Speed Flow," Proc. First U.S. Nat. Congress of Appl. Mech., 1951.

Nikuradse, J., "Laminare Reibungsschichten an der Längsangeströmten Platte," Monograph, Zentrale f. wiss. Berichtswesen, Berlin, 1942.

Pohlhausen, E., "Der Warmeaustausch zwischen festern Korpen und Flussigkeiten mit kleiner Reibung und kleiner Warmeleitung," Z. Angew Math Mech., Vol. 1, (pp. 115–121), 1921.

Raymer, D. P., Aircraft Design, A Conceptual Approach, AIAA Education Series, American Institute of Aeronautics and Astronautics, Washington, DC 1989.

Roach, R. L., Nelson, C., Sakowski, B. A., Darling, D. D., and van de Wall, A. G., "A Fast, Uncoupled, Compressible, Two-Dimensional, Unsteady Boundary Layer Algorithm with Separation for Engine Inlets," AIAA 92-3082, 1992.

Schlichting, E. H. and Gersten, E. H., *Boundary Layer Theory*, 8th ed., Springer-Verlag, Berlin, 2003.

Schubauer, G. B. and Skramstad, H. K., "Laminar Boundary Layer Oscillations and Stability of Laminar Flows," Journal of the Aeronautical Sciences, Vol. 14, (pp. 69–78), 1947.

Thwaites, B., "Approximate Calculation of the Laminar Boundary Layer," Aeronautical Quarterly, Vol. 1, (pp. 245–280), 1949.

White, F. M., *Fluid Mechanics*, 2nd ed., McGraw-Hill Book Company, New York, (pp. 269–270), 1986.

White, F. M., *Viscous Fluid Flow*, McGraw Hill Book Company, New York, 1974.

9 Incompressible Aerodynamics: Summary

It is now becoming clear that it is also mistaken to assume that computers could produce optimum designs in an empirical manner: it cannot be carried out in practice.

D. Küchemann
"The Aerodynamic Design of Aircraft"
Pergamon Press, 1978

9.1 Introduction

The preceding eight chapters take wholesale advantage of the assumption that the flow field for low-speed flight is incompressible. This allows considerable simplification in the formulation of the governing equations and in the solution of key aerodynamic problems. However, results of the calculations are limited in an important way that is emphasized in this summary chapter. What we attempt to do here is:

1. Summarize the most important elements of the first eight chapters.
2. Demonstrate how the results are incorporated in actual vehicle design.
3. Define the limits of application of the results.

Modeling of Airflows

What is accomplished to this point is the application of basic fluid mechanics in constructing detailed models for the airflow over aerodynamic surfaces (e.g., wings and bodies) at speeds low enough that compressibility effects do not seriously affect the results. These models are intended to provide accurate estimates of the aerodynamic forces and moments needed in solving the basic problem of aerodynamics as it was defined in Chapter 1. Although there is much discussion centered on the application of modern computational tools, for the most part, we rely on simplified mathematical representations. We try to emphasize the role of valid, simplifying assumptions in arriving at useful representations for the airflow. As Küchemann described the process in his famous book on the aerodynamic design of aircraft (Küchemann, 1978), "... the most drastic simplifying assumptions must be made before we can even think about the flow of gases and arrive at equations which are amenable to treatment. Our whole science lives on highly idealized concepts and ingenious abstractions and approximations." First-class examples of this approach are demonstrated in this book, including Prandtl's elegant models describing the creation of lift by an airfoil, three-dimensional wing theory, and boundary-layer flows. These provide the backbone of the subject of aerodynamics.

The Main Assumption

We now emphasize a key assumption made in deriving all of these important results: The flow field is incompressible; density variations are ignored. This asumption may not seem important until we realize that it leads to strict limitations in the domain of applicability of the results herein, especially in terms of speed range of an aerodynamic vehicle. We put these limitations in perspective by examining the speeds of various common flight vehicles with the object of determining which classes of vehicles are governed by the analyses carried out to this point.

We show in previous chapters that aerodynamic performance often can be described in terms of the cost in drag associated with the production of lift. Lift is the main output of the aerodynamic system and it is used directly in countering gravitational forces, thereby rendering atmospheric flight possible. We demonstrate in Chapter 1 that a useful figure-of-merit for aerodynamic performance is the L/D ratio.

Consider the effect of speed on the achievable L/D ratio by an optimally designed flight vehicle. Figure 9.1 is a typical presentation of such information. The horizontal axis covers speeds from the lowest possible (corresponding perhaps to a man-powered aircraft) to the highest velocities for atmospheric flight vehicles that may nearly reach the speed required for earth orbit. The Mach number (i.e., dimensionless speed) is used to reflect the velocity on the horizontal axis because it directly reflects the importance of compressibility.

Several useful observations should be apparent on inspection of Fig. 9.1. The first is that there is a definite optimal aerodynamic configuration associated with each speed range. We classify those in the manner suggested by Küchemann. It may be disappointing to see that all of our labors are applicable only in the lower part of the subsonic speed range (i.e., to "classical" airplane configurations) and appropriate only in the speed range up to approximately 400 mph (640 km/hr). We refer to this as the speed range for the classical aircraft-design configuration because, indeed, it is the earliest and most common type and still absorbs a major effort by aircraft designers.

For emphasis, it is crucial to understand that the aerodynamic models developed thus far apply only to low speeds (i.e., up to about 600 km/hr)—that is, to flight Mach numbers less than, say, 0.6. The aerodynamic performance of a classical airplane shape falls off drastically at higher speeds, and it may not even be capable of stable flight due to shifting of the center of pressure as compressibility effects become important. It is clear, then, that there is much to accomplish if we are to understand the design process for high-speed aircraft. Fortunately, as demonstrated for the low-speed range, there exists a multitude of powerful simplifying assumptions and "ingenious abstractions and approximations" of the type we already use up to this point for incompressible flows.

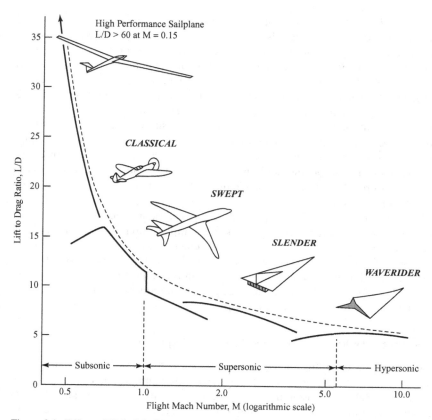

Figure 9.1. Effect of flight Mach number on aircraft performance.

It remains to explain in more detail how we can use what we have learned in the prediction of performance and in the practical analysis and design of efficient, classical
flight vehicles. Many of these applications of aerodynamics qualify for textbooks
in their own right; therefore, only an introduction is possible here. The subjects
of vehicle performance, stability and control, and so on, most often are treated in
separate courses. Students will find in their study of those subjects that a thorough
understanding of the aerodynamic principles examined herein is indispensable. Students may find it useful as we start this discussion to review Chapter 1 to examine
the photographs and descriptions of classical subsonic aircraft.

9.2 Prediction of Lift and Drag on a Flight Vehicle

We treat aerodynamic forces as though they could be separate entities acting on
various components, such as wings and bodies. For example, we show that the drag
on a wing consists of several parts, including profile drag contributed by the pressure
forces over the surfaces; skin-friction drag due to viscous interactions on the surfaces; and induced drag, which largely is due to three-dimensional effects involving

vortices produced near the wing tips in the process of lift generation. It is implied that when these effects act in unison, our estimates can be combined in a simple, linearly additive way to determine the characteristics of the complete aerodynamic assembly. However, it is not clear that these models are dependable when the wing is attached to a fuselage or when nacelles containing engines, fuel, or external stores are attached to the wing. Is there some type of mutual influence or interference among the components? What happens if the surfaces are degraded by the presence of rivet heads or lap joints in the skin? What happens if the tail surfaces lie in the wake of the wing? What changes in the flow-field results when a propulsion system is operating? What happens if turbulent eddies from a separated flow in one component enter the otherwise laminar flow on another? There is a seemingly endless list of such questions: Some were adequately addressed in the past in a practical manner; some await satisfactory resolution.

We now must ask questions regarding the application of basic aerodynamic analyses of the type presented in this textbook to a complete aircraft or watercraft (i.e., a boat or submarine) or, perhaps, even ground vehicles. Such questions immediately invoke specialized answers; generalities are difficult to make. However, we attempt to set the stage for the student's further studies in this important area. A short but useful bibliography is included to aid in the process of reducing the basics to a form that can be applied in solving real problems.

Experimental Verification of Lift and Drag on a Flight Vehicle

It should be clear from this discussion that predicting the behavior of the flow field and the resulting forces and moments for a complete flight vehicle using analytical or computational models is an involved process. The time may not be far off when this can be accomplished by fully three-dimensional computations on supercomputers. However, even if this were possible, there still would be the need to verify the results by actual testing of the vehicle. Consider that in such testing, many environments must be represented. That is, to fully map out the aerodynamic performance, it is necessary to cover a range of speeds, fluid densities, and possibly even atmospheric temperatures that are compatible with design specifications and intended use of the vehicle.

Several methods are used to verify predictions made by application of analytical and computational models. Those most often used are as follows:

- Wind-tunnel testing using scale models or, when possible, the full-sized vehicle
- Full-scale (or scale-model) flight testing

Testing of models or even the full-scale vehicle in a wind tunnel (i.e., for flight or land vehicles) or a water channel or basin (i.e., for watercraft) is a routine part of the design procedure. The tests provide direct verification of the predictions but usually also introduce an additional set of unknowns. For example, was the test carried out with full static and dynamic similarity to the full-scale prototype? That is, in addition to a precise representation of the geometry, are the Re numbers and Mach-number ranges appropriate? Other similarity parameters such as the Froude number also must be matched in some cases (e.g., testing of boats).

To see how important such questions may be, we examine the sensitivity of the airfoil measurements in Chapter 5 to the test Re numbers. It is clear that this type of testing may provide confidence in our predictions, but it does not completely verify them.

The second method—testing with a full-scale prototype of the vehicle in the intended operational environment—is often considered the only truly reliable verification method. However, many uncertainties still must be addressed in interpreting test results. For example, in flight testing, we must ask questions such as: Were the test instruments properly calibrated? Were atmospheric disturbances such as turbulence present that would influence the test data? Was there a patina of squashed insects layering the leading edges? And so on, *ad infinitum*.

Drag and Lift Estimation on an Actual Airplane

We consider what must be done to predict the aerodynamic performance of a flight vehicle that operates in the subsonic speed range (e.g., a World War II fighter). The Messerschmitt Bf-109G, shown in Fig. 9.2, was subjected to a long period of aerodynamic and powerplant refinement during its operational history. What we review briefly here is an elegant analysis by the famous aerodynamicist, Dr. Sighard Hoerner, who was involved in the refinement and testing of the Bf-109 in Germany in the 1940s (Hoerner, 1993).

To achieve a realistic assessment of the drag, it is clear that we must include the following:

- skin friction and profile drag on the fuselage
- skin friction and profile drag on the wing and tail surfaces
- induced drag
- drag due to surface imperfections such as bolt heads, rivet heads, sheet-metal blisters housing cannon components, aileron gaps, gaps around landing-gear wells, Pitot tube mounting, and sheet-metal lap joints and edges.
- drag of engine components such as air scoop, exhaust stacks, oil cooler, wing radiators, and ventilation openings
- interference drag due to tail wheel, antenna mast, canopy, and gun installation

Figure 9.2. Messerschmitt Bf-109G.

Table 9.1. *Messerschmitt Bf-109G aerodynamic configuration*

Total wing area	$S = 172 \text{ ft}^2$
Wing span	$b = 32$ ft
Aspect ratio	$AR = 6.1$
Overall length	$\ell = 29$ ft
Gross weight	$W = 6{,}700$ lbf
Wing loading	$W/S = 39 \text{ lbf/ft}^2$
Maximum cruising speed	$V_{max} = 610$ km/hr
Maximum power (at 22,000 ft)	$P = 1{,}200$ hp

The various contributions usually are separated into two groups: those directly due to the production of lift (i.e., the induced drag) and those not associated with the production of lift (i.e., the parasite drag). As we analyze the Bf-109G, we carefully distinguish between, these two types of drag. This is not always an easy task, because clearly, the tail surfaces and even the fuselage may contribute to the generation of lift.

Only the high-speed configuration with the landing gear and flaps retracted is considered here. To fully evaluate the performance and other qualities of the aircraft, other configurations covering the entire speed range must be analyzed similarly. For example, the configuration needed for landing or takeoff, with the flaps and landing gear deployed, involves an additional set of drag elements and related assumptions. We begin by examining the drag characteristics of the Bf-109G. Table 9.1 describes the aerodynamic configuration of this aircraft for high-speed flight at operational altitude.

Equivalent Drag Area

Hoerner use the *drag area* as a convenient means to represent the relative drag of the various airplane components. These are summed at the end of the analysis to determine overall drag. It is necessary to understand that this sum must include corrections to account for the interference drag created by the influence of one element on another. An important source of such drag is the interference between the fuselage and the wing. The smooth fairing, or fillet, at the juncture of the fuselage and wing of the Bf-109 was intended to reduce this drag contribution to a practical minimum. *Drag area* is defined as:

$$(\text{drag area})_i = \frac{D_i}{q} = (C_d S)_i, \qquad (9.1)$$

where q is the dynamic pressure, D_i is the drag of the component i, and S is an appropriate reference area. Notice that depending on the component, different reference areas may be used. When working with aerodynamic surfaces such as the wing and empennage, the planform area almost always is used. For other parts, the projected frontal area often is used. Thus, use of this drag area combines the area and the related drag coefficient in a way that prevents errors in applying the results in a summation to determine the complete vehicle drag.

In defining the overall drag, the usual convention is to use the projected wing area, S, as the reference. S is the area found by extending the wing planform to the fuselage centerline. Hoerner stated that the maximum speed of the 1944 Bf-109G was 610 km/hr (380 mph) at an altitude of about 22,000 feet. This corresponds to an effective dynamic pressure of:

$$q = \frac{1}{2}\rho V^2 = 184\frac{1bf}{ft^2}.$$

The thrust of the Daimler Benz DB605 (i.e., 12-cylinder, inverted "V" engine) plus the jet thrust from the ejector exhaust stacks was about 1,140 lbf. Therefore, because the drag is equal to the thrust in unaccelerated level flight, the overall drag area of the Bf-109G was:

$$(\text{drag area})_{total} = \sum_{i}(C_d S)_i = 6.2\ ft^2.$$

Because the projected wing area ($S = 172\ ft^2$ for the Bf-109G) usually is used in describing the overall aerodynamic performance, the resultant *overall* drag coefficient is $C_D = 0.036$ at the stated flight condition. We follow Hoerner in compiling a breakdown of all drag sources that contribute to this total. It is important to follow the analysis, because many of the concepts developed in previous chapters are reviewed and an understanding of the capabilities as well as the limitations of the analysis are demonstrated. Will careful application of all that has been learned give a total drag area in fair agreement with the one from experimental flight data? We shall now proceed to find out.

1. **Drag Due to Lift**

   The methods used in Chapter 6 are applicable here. The lift coefficient for level, unaccelerated flight follows directly from the information in Table 9.1 because the lift must equal the weight. We find:

   $$C_L = \frac{W}{qS} = 0.21.$$

   Then, the induced-drag coefficient is found from:

   $$C_{Di} = \frac{C_L^2}{\pi e AR} = \frac{(0.21)^2}{\pi(0.98)5.8} = 0.0025,$$

   where the effective AR was reduced by Hoerner from 6.1 to 5.8 to account for planform losses related to the wing-tips shape. Thus, the drag area representing the induced drag is:

   $$(\text{drag area})_{induced} = C_{Di}S = 0.42\ ft^2.$$

   Adjustment of the AR represents an unusual method for handling the effect of the planform shape. The departure from the ideal (i.e., elliptic) lift distribution usually is accounted for entirely by the efficiency factor, e. Hoerner used a value of 0.98, which appears too large—although it may have been the value for which the designers were striving. If the actual AR is used in the formula and the efficiency factor is adjusted (assuming that the final calculated induced-drag coefficient is correct), then we find that e should be 0.93. This seems to be in reasonable agreement with values of span efficiency based on the work of Glauert and others for straight-tapered wings with rounded tips.

2. **Wing Contribution to Parasite Drag**
Parasite drag on the wing is the sum of the profile drag, skin friction, drag due to flow separation in various gaps (i.g., around the landing-gear fairings), and surface imperfections (e.g., protruding rivet and bolt heads). The drag areas for each is summarized in Table 9.2. In estimating the skin-friction drag, Hoerner noted that the sheet-metal gaps behind the leading-edge slats (which opened automatically at low speed to improve stalling characteristics) render the boundary-layer flow fully turbulent. He also indicated that the camouflage paint used for these aircraft exhibited a roughness height on the order of 1 mil. With this in mind, he then estimated that the skin-friction drag coefficient is $C_f = 0.0035$, which is considerably larger than an estimate (0.0028) assuming a smooth surface with a Re number of 1.1×10^7. The projected surface area of the part of the wing outside of the fuselage is adjusted by a factor of 1.28 to account for the influence of the wing-section thickness. These considerations lead to the skin-friction drag area shown in the table. Most of the table entries are based on drag data from wind-tunnel tests simulating each type of drag source or surface imperfection. Hoerner's book is a useful compendium of such data, with practical techniques for application in drag estimation.

3. **Fuselage Drag**
In estimating the skin friction drag, the boundary layer is assumed to be every where turbulent due to the propeller slipstream. Again, the rough camouflage-paint surface is taken into account. This yields a skin-friction coefficient of $C_f = 0.0025$. Account also must be taken of various appendages, as listed in Table 9.3.

The fuselage wetted area is 250 ft^2, so the basic drag area for the fuselage is 0.625 ft^2. This must be adjusted to account for rivet and bolt heads and the dynamic-pressure increase along the fuselage sides (Hoerner multiplied the result by 1.07 to account for this). Therefore, the adjusted drag area is 1.75 ft^2, as shown in Table 9.3.

A various "appendages" must be accounted for in the drag estimate. Table 9.3 shows the estimates for the canopy, tail wheel, and antenna components. The canopy has numerous edges around the window panes so that the drag of the canopy is almost double that of the same shape without these irregularities. Again, the estimates are based on wind-tunnel studies of many different shapes. A correction must be added to account for the increase in dynamic pressure because the fuselage flow field is within the slipstream of the propeller. Hoerner estimated that there is a 10 percent increase in q, as indicated in Table 9.3.

4. **Engine Installation Drag**
Some of these effects are due to fuselage appendages and usually are included with the fuselage drag; others are additions to the wing drag. Hoerner chose to include them in a detailed estimate of the drag caused by the various air scoops for the radiators and oil coolers, as well as exhaust stacks needed for the engine installation, as tabulated in Table 9.4. The high-drag contribution from the two

Table 9.2. *Contributions to Bf-109G wing pvarasite drag (Hoerner, 1993.)*

Drag Source	Drag Area, ft^2	
Wing skin friction (turbulent, rough surface)	1.350	
Surface imperfections common to both upper and lower surfaces:		
Aileron gaps	0.018	
Aileron hinges	0.030	
Aileron balance weights	0.027	
Side gaps on slots	0.090	
Side gaps at ailerons and flaps	0.030	
Pitot-static tube	0.010	
Position lights at wing tips	0.002	
Upper-wing side blisters	0.020	
Holes around landing gear	0.140	
Subtotal	0.367	
Upper-wing surface imperfections:	0.011 (1.16)*	0.013
Lower-wing surface imperfections:		
Lateral sheet-metal edges (29 ft)	0.013	
Lateral surface gaps (36 ft)	0.038	
Longitudinal sheet edges (50 ft)	0.001	
Bolt heads (500)	0.004	
Flush rivet heads (3,500)	0.002	
Sheet-metal blisters (e.g., over cannons.)	0.007	
Subtotal	0.075(1.42)*	0.107
Additional skin friction due to surface imperfections:	0.090	
Total Wing-Parasite Drag Area:	1.927	

* Factors in parentheses are corrections for dynamic pressure on upper and lower-wing surfaces above the freestream reference value.

wing-mounted radiators, according to Hoerner, is due to "poor aerodynamic design and considerable internal leakage." Much work was accomplished more recently by NACA (and later NASA), as well as the aircraft industry, in reducing drag of air intakes, motivated mainly by the advent of jet propulsion. Compressibility effects are important in the design process for high-speed flight.

Table 9.3. *Contributions to Bf-109G fuselage-parasite drag (Hoerner, 1993)*

Drag Source	Drag Area, ft^2
Fuselage skin friction (based on wetted area)	0.750
Fuselage appendages:	
Canopy	0.120
Tail wheel	0.290
Antenna mast	0.030
Antenna wires, etc.	0.030
Direction finder	0.050
Gun installation	0.030
Other irregularities	0.080
Subtotal	0.630
Wing/fuselage interference	0.220
Total	1.600

Correction for increase in dynamic pressure due to propeller slipstream (q is 10 percent higher on fuselage and wing fillets); therefore, multiply 1.6 by 1.1=1.75

Total Fuselage-Parasite Drag Area:	1.750

Table 9.4. *Contributions to Bf-109G engine-installation drag*

Drag Source	Drag Area, ft^2
Engine appendages:	
Fuselage appendages:	
Carburetor air scoop	0.067
Intake momentum drag	0.080
Exhaust stacks	0.056
Oil cooler	0.168
Ventilation openings	0.100
Wing radiators	0.660
Subtotal	1.131

Correction for increase in dynamic pressure due to propeller slipstream (q is 10 percent higher on engine parts); therefore, multiply $1.131 \times 1.1 = 1.24$

Total Engine-Installation Drag Area:	1.240

5. Empennage (Tail Surfaces) Drag Area

The horizontal (i.e., outside the fuselage) and vertical tail-surface areas are 25 and 11 ft^2, respectively. Again assuming the rough camouflage paint, the skin-friction drag coefficient is estimated to be $C_f = 0.004$. Correcting for the thickness effect, the gaps at the control-surface hinge lines, and the interference drag at the junctions between the fuselage and the tail surfaces, the total parasite drag area for the empennage is:

$$(D/q)_{\substack{\text{empennage} \\ \text{parasite}}} = 0.360 \text{ ft}^2.$$

Because the horizontal tail produces lift at some trim conditions (usually negative lift in high-speed flight), a correction for induced drag is necessary. It is small because the lift coefficient needed to accomplish the required downward trim force is small. An estimate is:

$$(D/q)_{\substack{\text{empennage} \\ \text{induced}}} = 0.010 \text{ ft}^2.$$

Therefore, the total empennage-drag area is approximately 0.370 ft^2.

Summary of BF-109G Drag Calculation; Comparison to Measured Drag

If the various estimated drag-area terms are summed, the total contributions to the overall airplane drag are as fo.llows:

$$\left\{ \begin{array}{l} \text{overall parasite} - \text{drag area} = (D/q)_{\text{parasite}} = 5.277 \\ \text{overall induced} - \text{drag area} = (D/q)_{\text{induced}} = 0.430 \\ \\ \text{Total airplane} - \text{drag area} = (D/q)_{\text{total}} = \overline{5.707} \end{array} \right.$$

The results are shown graphically in Figure 9.3. The total estimated drag area based on the assumption of incompressible flow is somewhat lower than the value of 6.2 deduced from actual level-flight performance data. Why is there a discrepancy? It appears that every possible drag contribution was taken into account. Hoerner explained this in terms of compressibility effects. Notice that at 22,000 ft., the speed of sound is 295 m/s (try using the standard atmosphere program to verify this), so that the flight Mach number under these conditions is about:

$$M_\infty = 0.58,$$

which indicates that conditions clearly are pushing the upper bound of the subsonic approximation. Hoerner estimated by means of the Prandtl-Glauert compressibility correction (Liepmann and Roshko, 2001) that there should be about a 10 percent drag increase because of compressibility, which accounts for the discrepancy between the calculated and measured drag results. The results show that the aerodynamic performance for a complete airplane with all of its imperfections realistically can be predicted if care is taken with all of the details. A student who wants to learn this process should examine Hoerner's book even though it may seem dated; it is

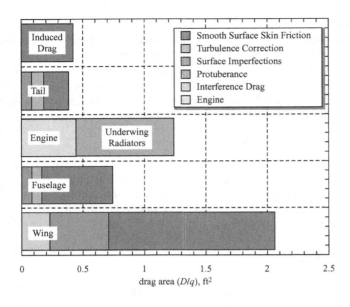

Figure 9.3. Drag summary for Bf-109G.

difficult to find a better summary of the practical approach to drag estimation. If this is supplemented with modern information covering the high-speed (i.e., high-subsonic, transonic, supersonic, and hypersonic) flight regimes, then a powerful performance-prediction tool can be devised. Such tools are an important design resource in the aerospace industry. Each company and government laboratory has preferred methods for carrying out analyses such as we reviewed herein.

Aerodynamic Efficiency

Hoerner defined the aerodynamic efficiency as the following ratio:

$$\eta_{aero} = \text{aerodynamic efficiency} = \frac{D_{useful}}{D_{total}} \tag{9.2}$$

where the useful drag is the part that is either useful (as in the induced drag) or necessary (or unavoidable). For the Bf-109G, the result is $\eta_{aero} = 0.4$, or 40 percent. This indicates that "greater than half the total drag of this airplane could theoretically be avoided by extremely clean design and faultless construction of the skin and details." However, the amount of labor and time (and, therefore, expense) to accomplish this would be significant. What Hoerner described is typical of the concentration on drag reduction used in the design and construction of world-class, high-performance sailplanes. Such efforts result in high costs due to the required labor-intensive procedures. For example, a competitive racing-class (i.e., 15 meter span with camber-changing flaps) sailplane can cost more than $150,000.

Hoerner demonstrated that if the Bf-109 were to be redesigned such that the ratio of "useful" to total drag were to be increased to unity, the top speed would increase from 610 to 800 km/hr! The benefit of such a speed increase in terms of the mission of this fighter is obvious. However, we always must ask the question: How much does it cost to reduce the drag? Clearly, in the case of the Bf-109G, the answer was apparently "too much." The aircraft performance was increased over its service life mainly by increases in engine horsepower, although minor improvements resulted from drag reduction. For example, the tail struts, squared wing tips, and large undercowl radiator found in earlier versions of the Messerschmitt 109 were modified to lower the drag to the levels demonstrated herein for the G version.

9.3 Aircraft-Performance Calculations

In previous chapters, we introduce other ways to describe the aerodynamic efficiency. One that is useful is the "lift-to-drag ratio," or L/D. For the Bf-109G, the L/D ratio for the high-speed cruise condition we analyze is:

$$L/D = \frac{C_L}{C_D} = \frac{0.21}{0.036} = 5.83,$$

which is based on the lift coefficient for cruise and the total drag coefficient (i.e., the total drag area divided by the wing area). This is a fairly good value, indicating that the Bf-109 is a relatively clean aircraft; however, it would not make a good glider (at least, not at 610 km/hr) because from a 1-km altitude, it would be on the ground in less than 6 km with the engine off. Why is this so? Is this L/D ratio a good indication of the *overall* aerodynamic performance of the aircraft? Is there a speed at which the L/D ratio is higher? These questions deserve careful consideration, and we do this in the remainder of this chapter. We seek the means to represent the overall performance over the entire speed range, along with useful information such as the maximum rate of climb, ceiling (i.e., maximum altitude), stalling speed, power required as a function of flight speed, maximum range, maximum endurance, speeds to fly for maximum climb rate, best range maximum endurance, and so on. These matters usually are considered as part of the subject of airplane performance. However, it is useful to complete this book with a short introduction, because it summarizes what has been accomplished to this point and emphasizes key results.

Extension to Other Flight Conditions: Drag Polar

Because the detailed analysis refers to only one flight condition—namely, high-speed level flight—it is useful to find a way to extend the collected data to estimate the airplane performance at other speeds. For the moment, we assume the following:

1. The parasite-drag coefficient stays effectively constant over a wide range of speeds.
2. The compressibility effect can be neglected.
3. The flaps, slats, cowl flaps, and landing gear remain undeployed.
4. There is no wind or atmospheric turbulence.

Clearly, these assumptions result in an approximate analysis because compressibility was shown to account for a substantial drag rise at the top-speed condition. Also, parasite drag is likely to depend on the angle of attack of the airplane, which must increase as the speed decreases so that the necessary lift is produced. However, a reasonable estimate of the Bf-109G performance over its operational speed range can be made in the manner suggested. Using the wing area and other information already assembled, we can summarize the aero-dynamic behavior of this example airplane in the following manner, recalling that we separated the total drag into two categories: drag due to lift (induced drag) and parasite drag. Then, if we assume minor influence of speed and aircraft attitude on the parasite drag, the value for the Bf-109G is:

$$C_{Do} = \text{overall parasite-drag coefficient} = \frac{5.227}{172} = 0.0307.$$

The induced-drag coefficient:

$$C_{Di} = \frac{C_L^2}{\pi e AR}$$

depends on flight speed because the lift coefficient must be adjusted to provide the necessary lift as dynamic pressure changes. The total drag is expressed by:

$$C_D = C_{Do} + C_{Di} = C_{Do} + \frac{C_L^2}{\pi e AR}, \qquad (9.3)$$

which often is referred to as the drag polar for the airplane. If it is assumed that the airplane is in level, equilibrium (i.e., unaccelerated) flight with the drag exactly balanced by the thrust produced by the propulsion system and the weight balanced by the lift, then we can write:

$$\begin{cases} T = D = qSC_L & (9.4) \\ W = L = qSC, & (9.5) \end{cases}$$

where $q = \frac{1}{2}\rho V^2$ and S is the projected wing area, including the part enclosed by the fuselage. Equations 9.3–9.5 now can be evaluated using the Bf-109G data to determine useful information about its performance.

Thrust Required for Level Flight

If Eq. 9.4 is divided by Eq. 9.5, we find that the ratio of thrust required to total airplane weight is equal to the ratio of the drag coefficient to the lift coefficient, or:

$$T = T_R = \text{thrust required} = \frac{W}{L/D}, \qquad (9.6)$$

which illustrates again the importance of the L/D ratio in performance characteristics. That is, to minimize the amount of thrust that must be used in level flight, the L/D ratio must be as large as possible. It is useful to determine how the thrust required varies with flight speed; Eq. 9.3 provides the needed information. First, we calculate the required lift coefficient to provide the lift to balance the weight as a function of flight speed:

$$C_L = \frac{W}{qS} = \frac{2W}{\rho V^2 S}. \qquad (9.7)$$

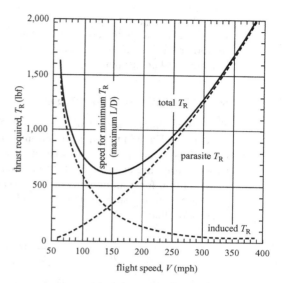

Figure 9.4. Thrust required for Bf-109G.

Substituting this result into Eq. 9.3 gives the drag coefficient as a function of flight speed; namely:

$$C_D = C_{Do} + \frac{1}{\pi e AR}\left(\frac{2W}{\rho V^2 S}\right)^2. \tag{9.8}$$

Notice that the induced-drag coefficient decreases rapidly (as the inverse fourth power of the speed) as V increases. This indicates that induced drag is most important at lower speeds, whereas the parasite drag dominates at high speed. The thrust required as a function of velocity is found readily by means of Eq. 9.4. We find:

$$T_R = qSC_{Do} + \frac{W^2}{\pi e AR qS}. \tag{9.9}$$

This is plotted in Fig. 9.4 for the Messerschmitt Bf-109G. All values shown are for standard sea-level atmospheric conditions. (The changes that result from higher altitudes are checked in an exercise at the end of the chapter.) Both the induced-and parasite-drag contributions are plotted separately and compared to the total thrust required.

Stalling Speed

The curves shown in Fig. 9.4 are extended into a low speed-range that is not actually accessible to the clean Bf-109G. The airplane probably would not fly in its clean configuration at a speed of less than about 90 mph. This stalling speed can be estimated by using the largest possible lift coefficient that reasonably could be expected without deflected flaps or slats. When we solve Eq. 9.5 for the flight speed for a given maximum lift coefficient, the result is:

$$V_{Stall} = \sqrt{\frac{2W}{\rho S C_{L\,max}}}. \tag{9.10}$$

If a value of $C_{L\max} = 1.4$ is used in this result, the minimum (i.e., stalling) flight speed at sea level in the clean aerodynamic condition is about 104 mph. The need for flaps and other high-lift devices is clear because this is an uncomfortably high speed at which to land an airplane.

Speed for Minimum Thrust Required

Notice that there is a special speed at which the thrust required is a minimum. It is clear (see Eq. 9.6) that this condition corresponds to the speed at which the L/D ratio is maximum. It is curious that this occurs at the point where the induced and parasite drags are exactly equal. The reason for this becomes clear if we determine the condition for a minimum value by taking the derivative of the thrust required with respect to the speed (or corresponding dynamic pressure) and setting it equal to zero. We find:

$$\frac{dT_R}{dq} = SC_{Do} - \frac{W^2}{\pi e ARq^2 S} = SC_{Do} - SC_{Di} = 0, \qquad (9.11)$$

which shows that for minimum thrust required we must have:

$$C_{Do} = C_{Di},$$

as shown graphically in Fig. 9.4. What is important from a performance standpoint is that the speed we must fly to reduce the thrust (and drag) to a minimum corresponds to the condition of Eq. 9.11. Solving for the value of dynamic pressure, q, that corresponds to the minimum, we find:

$$q_{\min T_R} = \frac{W/S}{\sqrt{\pi e ARC_{Do}}}$$

and the associated flight velocity is:

$$V_{\min T_R} = V(L/D)_{\max} = \sqrt{\frac{W/S}{\rho\sqrt{\pi e ARC_{Do}}}}. \qquad (9.12)$$

This flight-speed information is important in the effective operation of an airplane because flying at maximum L/D ratio (i.e., minimum drag or thrust required) is required to achieve the maximum range or minimum fuel consumption, as demonstrated in a subsequent subsection. For the Bf-109G at sea level, Eq. 9.12 yields a value of about 210 ft/sec (143 mph), as indicated in Fig. 9.4. Notice that this speed changes with altitude, as indicated by the appearance of density in Eq. 9.12.

We now can determine the maximum L/D ratio for the Messerschmitt Bf-109G. This is easy because it is clear that the drag coefficient is equal to twice the parasite-drag coefficient at the best L/D speed. We take the ratio of the lift and drag coefficients and insert the value for the best speed from Eq. 9.12; the result is:

$$\left(\frac{L}{D}\right)_{\max} = \frac{1}{2}\sqrt{\frac{\pi e AR}{C_{Do}}}, \qquad (9.13)$$

which yields a value of about 12 for the Bf-109G:

$$\left(\frac{L}{D}\right)_{max} = 12.1 \quad \text{for Bf} - 109\text{G}.$$

If a pilot suffers an engine failure, he or she should glide at 143 mph to achieve the best chance to find a suitable field for a successful off-field landing. Flying slower may increase the time in the air but results in less distance covered in search of a good landing field.

Maximum Flight Speed for a Jet Aircraft

In rating the performance of propulsion systems, it is the usual practice to indicate the power available for reciprocating engines and the thrust available for jet engines. Thus, for a jet-propelled aircraft, the maximum flight speed is estimated easily by superposing a plot of thrust available on the thrust-required curve, as calculated in Eq. 9.9. The intersection of the two curves indicates graphically the point at which the available thrust is exactly equal to that required for level flight. The corresponding speed represents the maximum flight speed.

Equivalently, we equate the expression for the thrust available (usually a function of speed) to the thrust available and then solve for the corresponding speed. In carrying out this calculation, it is necessary to use a model of the propulsion-system performance as a function of flight speed. A manufacturer of a system usually provides these data to an airframe designer.

To estimate the maximum flight speed for a reciprocating-engine propulsion system, it is most useful to work with power instead of thrust. We accomplish this in the next subsection and illustrate results for the Bf-109G.

Power Required for Level Flight

Recalling that power is the rate at which work is done and that the work done by an engine is the thrust force times the distance traveled, then the power required for unaccelerated, level flight is:

$$P_R = \text{power required} = T_R V = \frac{W}{L/D}V, \tag{9.14}$$

where V is flight speed. Using Eq. 9.9, we find:

$$P_R = \left(\frac{1}{2}\rho V^3\right) S C_{Do} + \frac{W^2}{\pi e \text{AR}\left(\frac{1}{2}\rho V\right)S}$$

$$= \frac{1}{2}\rho V^3 S \left(C_{Do} + \frac{1}{\pi e \text{AR}}\left(\frac{W}{\frac{1}{2}\rho V^2 S}\right)^2\right). \tag{9.15}$$

This expression is plotted in Fig. 9.5 for the example airplane at two altitudes (i.e., sea level and 22,000 ft) to illustrate the effect of atmospheric density on the results.

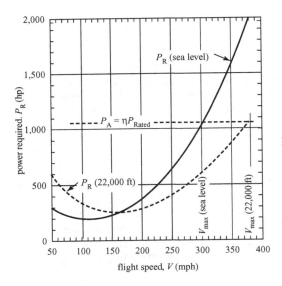

Figure 9.5. Power required for Bf-109G.

Notice that at low speed, the power required is higher for the high-altitude case, whereas for high speeds, the opposite is true. This is due to of the parasite-drag penalty that dominates in the high-speed range. The effect of the induced drag becomes much smaller at high speeds, as indicated by Eq., 9.15. Therefore, the higher density for the low-altitude case results in more power required at a given flight speed.

Maximum Flight Speed for a Propeller-Driven Aircraft

To find the maximum speed for the propeller-driven Messerschmitt fighter, we follow a procedure similar to that described previously for jet-propelled vehicles. To do this, it is necessary to determine the actual power available for producing thrust. This power can be estimated by reducing the rated power (sometimes called the *shaft brake horsepower*) by a factor that considers the aerodynamic losses of the propeller. The power available can be written as:

$$P_A = \text{power available} = \eta P_{\text{Rated}}, \qquad (9.16)$$

where η is the propeller efficiency. A typical value for a good propeller design is on the order of $\eta = 0.88$, which was used in the estimated power available shown in Fig. 9.5. There usually is a correction to account for the effects of altitude on engine performance, but this is not indicated in the Figure. The usually significant P_A variation with speed also is not represented properly. To graphically find the maximum speed, we locate the point where the power available is equal to the power required. For the case shown, the maximum speed at altitude is about 379 mph (610 km/hr), which is in good agreement with the values shown in Table 9.1. Due to the higher power required at the lower altitude, the maximum speed is reduced, so that at sea level, the Bf-109G is limited to a top speed of about 300 mph (482 km/hr). These

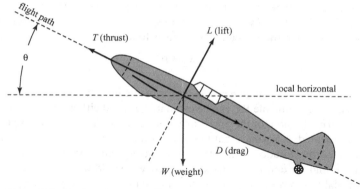

Figure 9.6. Airplane in steady climb.

numbers are in good agreement with published data for this airplane. In working with the power equations, recall that:

$$1 \text{ horsepower} = 550 \frac{\text{ft lbf}}{\text{sec}} = 746 \text{ watts.}$$

Calculation of the Rate of Climb

An important performance parameter is the rate at which an aircraft can increase altitude and the time needed to achieve a particular altitude. Estimates for these quantities are determined readily by using tools that we already assembled. We consider Fig. 9.6, which show an airplane in climbing configuration. The angle θ between the velocity vector and the horizontal is the *climb angle*, which obviously has an important role in the calculation. The force balance in steady-climbing attitude indicates that we must have:

$$\begin{cases} T = D + W \sin \theta & (9.17) \\ L = W \cos \theta, & (9.18) \end{cases}$$

where lift and drag still are given by the usual formulas. Notice that the thrust force now must carry part of the weight as well as balance the drag.

What is required is the vertical speed, or *rate of climb*, (RC). This is clearly given as the component of flight speed in the vertical direction:

$$RC = V \sin \theta. \tag{9.19}$$

Equation 9.17 can be used to determine RC by solving for and then multiplying through by the flight speed. The result is:

$$RC = \frac{TV - DV}{W}. \tag{9.20}$$

Recalling that force times speed is the power (i.e., rate of doing work), a useful physical interpretation of Eq. 9.20 as well as a practical method for calculating the RC, is obtained. The products of thrust and drag with velocity are the power available and the power required, respectively. Their difference often is called *excess power*—that is, the power available to climb or to execute other maneuvers. If the excess power is zero, then an airplane is flying at maximum speed in unaccelerated, level attitude. Therefore, we see that the RC can be written as follows:

$$RC = \frac{\text{excess power}}{W}. \tag{9.21}$$

A pilot controls RC by changing excess power by means of the engine throttle.

The maximum available RC can be estimated easily if either numerical data or a mathematical expression for the power available is known. In Fig. 9.5, a simple model for P_A is illustrated. It assumes that the power available is not affected significantly by flight speed or altitude. In fact, there are usually rather important sensitivities to the speed and air density. In actual practice, we would have the detailed information available. However, we can use the simple model to illustrate the procedure. Referring to Fig. 9.5, we see that the maximum excess power corresponds to the minimum point in the power-required curve because the power available is assumed to be insensitive to speed. Therefore, it is necessary to estimate this low point and to determine the corresponding flight speed. Comparison of Figs. 9.4 and 9.5 shows that the minima in the thrust required and power-required curves do not occur at the same flight speed. The necessary information is found by setting the derivative of the power required with respect to speed equal to zero to locate the extremum. The value of speed corresponding to minimum power required is found by solving:

$$\frac{dP_R}{dV} = \left(\frac{3}{2}\rho V^2\right) S C_{Do} - \frac{W^2}{\pi e AR\left(\frac{1}{2}\rho V^2\right) S} = 0 \tag{9.22}$$

for this special value of V. The result is:

$$V_{\min P_R} = \sqrt{\frac{2W/S}{\rho\sqrt{3\pi e AR C_{Do}}}} = \frac{1}{3^{1/4}} V_{\min T_R}, \tag{9.23}$$

which shows that the speed to fly for maximum RC is about 24 percent less than the speed to fly for minimum power required (i.e., maximum L/D). Thus, for the Bf-109G, the best climbing speed at sea level is approximately 163 ft/sec (111 mph). The speed for best RC increases with altitude approximately as the inverse square root of the ratio of density to the sea-level density, as indicated in Eq. 9.23 and verified in Fig. 9.5. Because the power available decreases similarly with altitude, it clear that the RC diminishes with altitude.

Inserting the best climb speed into Eq. 9.15 yields all of the necessary information for determining RC and climb angle, as summarized in Table 9.5. The results shown agree closely with published data for this aircraft.

Table 9.5. *Climb performance of the Bf-109G*

Altitude (ft)	Speed, V (mph)	Excess Power* (horsepower)	Climb Rate, RC (ft/min)	Climb Angle, θ (degrees)
sea level	111	863.8	4,255	23.5
22,000	154	791.8	3,900	16.1

* Assumes that constant power available = $P_A = 0.88 \cdot 1,200 = 1,056$ horsepower.

Maximum Flight Altitude, Absolute and Service Ceilings

The RC information can be used to estimate the flight altitude that can be reached by an airplane. Again, we use the Messerschmitt fighter as an example. Clearly, the RC decreases with altitude so that the maximum achievable altitude or *absolute ceiling*, corresponds to the condition at which no further RC is available—that is, the altitude for which there remains no excess power. Such a flight condition is of little use because there is no margin left for maneuvering. For example, if an aircraft is required to execute a simple turn under this flight condition, it would descend because of insufficient thrust for level flight. It would stall if the stall speed happens to be at or near the speed for maximum RC at this altitude. Part of the lift then would be required to produce the turn, and no power would be available to adjust the speed to increase the vertical component of lift needed to balance the weight. Thus, we define the *service ceiling*, which represents a practical maximum altitude with some excess power remaining for minimal maneuvering capability. The service ceiling is usually defined as that altitude at which the maximum RC drops to 100 ft/min.

If we again neglect the effect of altitude on power available, we can estimate this altitude for an aircraft like the Bf-109G. First, we solve for the power required for the stated RC by using Eq. 9.20:

$$RC = 100 = \frac{TV - DV}{W} = \frac{550(0.88 \cdot 1,200 - P_R)}{6,700}.$$

This indicates that for this flight condition, the power required is:

$$P_R = 162.2 \text{ horsepower.}$$

Notice the handling of the units in this calculation; it is necessary to convert from horsepower to ft-lbf/sec. The simplest procedure is to vary the density until the required condition is met. Directly solving equations for the density is difficult algebraically. To make a realistic calculation, it is necessary to account for the degradation of power available as the altitude increases. Because we do not have the Daimler-Benz power-available curves, we can only make educated guesses. The simplest possibility is that the power available drops off in direct proportion to the ratio of the density to the sea-level density. A more common assumption is that it drops off as the square root of this ratio, as does the power required. Both cases are displayed in Fig. 9.7, which is a plot of maximum RC versus density. The service ceilings for the cases are 34,500 ft. for the linear dependence on density and 46,500 for the square-root density dependence of the power available. The published service ceiling for the Bf-109G is 37,900 ft., so a reasonable estimate was achieved without knowledge of the exact power-available curves for this aircraft.

Maximum Range of a Subsonic Airplane

We complete this brief set of applications of basic aerodynamics by estimating the range of a propeller-driven aircraft. The key here is to introduce the rate at which fuel is used by the propulsion system and to use it in determining the rate at which

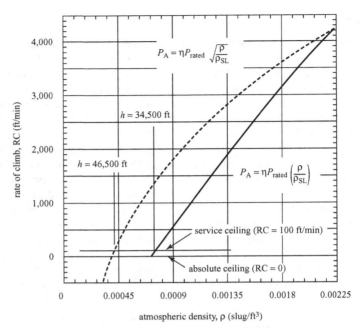

Figure 9.7. Determination of service ceiling for Bf-109G.

the weight of an airplane changes with distance. We define the *specific fuel consumption, c,* as:

$$c \equiv \frac{\dfrac{dw_f}{dt}}{P_{\text{rated}}} = \frac{\text{weight of fuel consumed per unit time}}{\text{rated engine power output}}, \tag{9.24}$$

where the power, P_{rated}, represents the brake-horsepower rating of the engine. From this, we see that the weight of an airplane changes at the same rate as fuel is consumed so that the weight change per unit time is:

$$\frac{dW}{dt} = -c\, P_{\text{rated}}, \tag{9.25}$$

where $W(t)$ represents the aircraft weight. The negative sign is inserted to emphasize that W decreases as fuel is consumed. Working with the differential form of Eq. 9.25, we see that the length of time required for a weight change, dW, is:

$$dt = -\frac{dW}{c P_{\text{rated}}} \tag{9.26}$$

and, assuming constant flight speed, the corresponding distance traveled is:

$$ds = V dt = -\frac{V dW}{c P_{\text{rated}}}. \tag{9.27}$$

To find the range, we integrate this expression over the weight change from the initial (full tanks) to the final (empty tanks) condition. Thus, the range, $\mathfrak{R}$, is found from:

$$\mathfrak{R} = \int_0^R ds = -\int_{W_i}^{W_f} \frac{V dW}{cP_{rated}}. \tag{9.28}$$

To simplify the problem, we neglect changes in speed and rate of fuel consumption in initial takeoff and climb to cruising altitude, the effects of wind and atmospheric disturbances, and changes in power settings at the end of the flight. We note that in level, unaccelerated flight the power required and power available are equal and that:

$$P_A = \eta P_{rated} = P_R = DV. \tag{9.29}$$

Then, the range equation becomes:

$$\mathfrak{R} = \int_{W_f}^{W_i} \frac{\eta V dW}{cDV} = \int_{W_f}^{W_i} \frac{\eta dW}{cD} = \int_{W_f}^{W_i} \frac{\eta}{c} \frac{L}{D} \frac{dW}{W}. \tag{9.30}$$

The last series of changes involves multiplying the top and bottom of the integrand by the weight and then replacing the weight in the numerator by the lift because they are equal in level, unaccelerated flight. If it is reasonably assumed that the parameters c, η, and L/D remain constant during flight, then Eq. 9.30 can be integrated to give the final formula for range:

$$\mathfrak{R} = \frac{\eta}{c} \frac{L}{D} \ln\left[\frac{W_i}{W_f}\right], \tag{9.31}$$

the famous Breguet range equation mentioned in Chapter 1. This was introduced previously to demonstrate the importance of the aerodynamic efficiency in the operation of an airplane. To achieve great range, it is necessary to start with a large mass ratio, as represented by the logarithmic term. This ratio depends on how much fuel can be carried and how efficiently the airplane structure was designed so that its weight is as low as possible. The need for an efficient propeller is illustrated by the direct dependence on η. Finally, there is no question that the L/D ratio must be as large as possible. These ideas were used by Charles Lindbergh in his 1927 flight from New York to Paris in the Ryan *Spirit of St. Louis,* using the best technology of the time. Material limitations precluded his use of high AR wings, as used in the more recent *Voyager* around-the-world flight to obtain the best possible L/D ratio. The necessary range for the *Spirit* was achieved mainly by accommodating the necessary amount of fuel in a relatively light airframe.

To illustrate the use of the Breguet equation, we use it to estimate the range of the Ryan *Spirit of St. Louis*. This aircraft took off from New York in the spring of 1927 carrying 450 gallons of fuel (weighing 2,750 lbf). The initial gross weight was 5,250 lbf. The best L/D ratio was about L/D = 9.8 (considerably less than the more modern Bf-109G design). The specific fuel consumption of the 220-hp Wright Whirlwind (J-5C) engine at 200 hp was:

$$c = 0.53 \text{ lbf/hr/hp},$$

so that assuming a propeller efficiency of $\eta = 0.82$ and a cruising flight at a throttle setting producing 200 hp, the range is given by:

$$\mathfrak{R} = \frac{\eta}{c}\frac{L}{D}\ln\left(\frac{W_i}{W_f}\right) = \frac{0.82}{0.53/(3,600\cdot550)}(9.8)\ln\left(\frac{5,250}{5,250-2,750}\right)$$

$$= 2.23 \cdot 10^7 \text{ft} = 4,220 \text{ miles},$$

which is comparable to published values for this aircraft. Examine this calculation to see how the units are adjusted in the specific fuel consumption to give the range in feet. The result is divided by 5,280 ft/mile to determine the range in statute miles.

9.4 Extension to High-Speed Flight

Although the material covered in this book is a necessary prerequisite for understanding basic aerodynamics, it is important to realize that the principles set forth may not be directly applicable to problems involving high-subsonic, supersonic, and hypersonic flight. Nevertheless, the analyses and concepts introduced are an important foundation on which to build high-speed aerodynamics. In particular, we find that more attention to the thermodynamic aspects of fluid mechanics is required. The density now becomes an important variable and considerably more use of the energy equation is demanded. That is, compressibility effects play a major role in the aerodynamics and performance of flight vehicles at high subsonic, supersonic, and hypersonic speeds.

PROBLEMS

9.1   Test your understanding of the boundary-layer methods presented in Chapter 8 by verifying Hoerner's estimate of the turbulent viscous drag on the Bf-109G wing surfaces. Estimate the skin-friction coefficient using a Re number of $R_e = 1.1 \cdot 10^7$.

9.2.   Suppose that Germany developed laminar-flow airfoils in time for its application in the Messerschmitt 109 program (in fact, the Germans did not know of laminar-flow airfoils until they captured a North American P-51 fighter that employed a NACA six-digit series laminar airfoil). Assuming that the Bf-109G is modified to use an airfoil on which the forward 30 percent of the wing surface maintains laminar flow, estimate the new parasite-drag coefficient for the airplane using the data presented in this chapter. Base your calculations on an average wing chord of 5.38 ft.

9.3.   Using the modified parasite-drag coefficient in Problem 9.2 for the hypothetical laminar-flow Bf-109G, find the effect on maximum speed at sea level and at 22,000 ft.

9.4.   Using the modified parasite-drag coefficient in Problem 9.2, estimate the increase in RC at sea level for the modified Bf-109G.

9.5.   Use the modified parasite-drag coefficient in Problem 9.2 to estimate the effect on service altitude.

9.6. Find the maximum L/D ratio for the Bf-109G using the modified laminar-wing parasite-drag coefficient in Problem 9.2. At what speed must the airplane fly to achieve this L/D ratio?

9.7. Assume that the Bf-109G carries a total fuel load of 150 gallons of aviation gasoline (with a specific weight of 5.64 lbf/gal). The gross weight (including the fuel weight) of the aircraft is 6,700 lbf. Estimate the range with the original aerodynamic configuration as determined in this chapter and find the range improvement if the laminar airfoil of Problem 9.2 is used.

REFERENCES AND SUGGESTED READING

Glauert, H., *Elements of Aerofoil and Airscrew Theory*, Cambridge University Press, Cambridge, 1926/1947.

Hoerner, S. F., *Fluid-Dynamic Drag*, Hoerner Fluid Dynamics (publisher), Bakersfield, California, June 1993.

Küchemann, D., *The Aerodynamic Design of Aircraft*, Pergamon Press, Oxford, 1978.

Liepmann, H. W., and Roshko, A., *Elements of Gasdynamics*, (pp. 253–256), Dover Publications, Inc., Mineola, New York, 2001.

McClamroch, N. H. *Steady Aircraft Flight and Performance*, Princeton University Press, 2011.

Ojha S. K., *Flight Performance of Aircraft*, AIAA Education Series, American Institute of Aeronautics and Astronautics, Washington, DC, 1995.

Index